S0-AWV-408

Principles of Environmental Toxicology

Second Edition

Sigmund F. Zakrzewski
State University of New York at Buffalo

ACS Monograph 190

American Chemical Society, Washington, DC

Library of Congress Cataloging-in-Publication Data

Zakrzewski, Sigmund F., 1919–

Principles of environmental toxicology / Sigmund F. Zakrzewski — 2nd ed.

p. cm.—(ACS monograph : 190)

Includes bibliographical references and index.

ISBN 0–8412–3380–2 (cloth)

1. Environmental toxicology.

I. Title. II. Series.

RA1226.Z35 1997
615.9′02—dc21 96–52362
CIP

The paper used in this publication meets the minimum requirements of American National Standard for Information Sciences—Permanence of Paper for Printed Library Materials, ANSI Z39.48-1984.

PRINTED IN THE UNITED STATES OF AMERICA

About the Author

SIGMUND F. ZAKRZEWSKI, a soldier in the Polish army, was captured when Germany invaded Poland in 1939 and spent the war years as a prisoner of war in Germany. He later studied at the University of Hamburg, where he earned an M.S. degree in organic chemistry. Studies in biochemistry at the Western Reserve University (Cleveland, Ohio) and Yale University provided material for his Ph.D. dissertation at the University of Hamburg. This degree was followed by a postdoctoral fellowship at Yale University's Department of Pharmacology.

Dr. Zakrzewski has served as senior cancer research scientist (Department of Experimental Therapeutics) and principal cancer research scientist (Department of Clinical Pharmacology) at the Roswell Park Memorial Institute in Buffalo, New York. In 1976 and 1977, he spent six months as an airplane pilot for missionaries in Papua New Guinea. This experience gave him some insight into the lives and problems of people in developing countries. Following an EPA-sponsored course in toxicology at Massachusetts Institute of Technology, he began teaching environmental toxicology at the Roswell Park Graduate Division, State University of New York at Buffalo. Although he retired from the research position in 1987, he still offers the toxicology course.

He has published 101 scientific papers, chapters, and abstracts.

Foreword

ACS MONOGRAPH SERIES was started by arrangement with the interallied Conference of Pure and Applied Chemistry, which met in London and Brussels in July 1919, when the American Chemical Society undertook the production and publication of Scientific and Technological Monographs on chemical subjects. At the same time it was agreed that the National Research Council, in cooperation with the American Chemical Society and the American Physical Society, should undertake the production and publication of Critical Tables of Chemical and Physical Constants. The American Chemical Society and the National Research Council mutually agreed to care for these two fields of chemical progress.

The Council of the American Chemical Society, acting through its Committee on National Policy, appointed editors and associates to select authors of competent authority in their respective fields and to consider critically the manuscripts submitted. The first Monograph appeared in 1921. Since 1944, the Scientific and Technological Monographs have been combined in the series.

These Monographs are intended to serve two principal purposes: first, to make available to chemists a thorough treatment of a selected area in a form usable by persons working in more or less unrelated fields so that they may correlate their own work with a larger area of physical science; and second, to stimulate further research in the specific field treated. To implement this purpose, the authors of Monographs give extended references to the literature.

Contents

Preface to the First Edition

Toxicology is traditionally defined as the study of the harmful effects of drugs, chemicals, and chemical mixtures on living organisms. Within the past two decades the environmental branch of toxicology has assumed a wider meaning. The survival of individuals and the human race alike is the ultimate goal of this area of study. However, the survival of humanity depends on the survival of other species (plants and animals alike); on the availability of clean water, air, and soil; and on the availability of energy. Moreover, although preservation of our local and regional environment is vital to our survival, global problems such as the increasing CO_2 content in the atmosphere and depletion of stratospheric ozone are also critical.

Use of poisons is as old as the human race. For centuries, primitive people applied toxic plant extracts to poison their arrows for hunting and warfare. In our civilization, poisons have been studied and used for political, financial, or marital advantages. Doull and Bruce covered this subject in more detail in the introductory chapter of *Cassarett and Doull's Toxicology* (*1*).

The credit for elevating toxicology to a true science goes to a Spanish physician. Mattieu Joseph Benaventura Orfila (1787–1853), who first described the correlation between the persistence of chemicals in the body and their physiological effect. He also developed analytical testing methods to detect the presence of toxins in the body and devised certain antidotal therapies.

Contemporary toxicology has evolved into a study with three branches:

- Clinical toxicology is concerned with the effect of drugs on human patients.
- Forensic toxicology is concerned with the detection, for judicial purposes, of the unlawful use of toxic agents.
- Environmental toxicology is concerned with the effects of toxins, whether purposely applied (such as pesticides) or derived from industrial processes, on health and the environment.

Environmental toxicology is a multidisciplinary science involving many widely diverse areas of study such as

- chemistry, the characterization of toxins;
- pharmacology, the mode of entry and distribution of toxins in the body;
- biochemistry, the metabolism and interaction of toxins with cell components;
- physiology, the effect of toxins on body organs;
- biology, the effect of toxins on the environment;
- genetics, the effect toxins can have on the reproductive system and on future generations by altering genetic codes;
- epidemiology, the effect on the population as a whole of chronic exposure to small quantities of suspected agents;
- law, regulation of the use or release into the environment of toxic substances; and
- economics, evaluation of the environmental cost vs. benefit of economic development and the determination of trade-offs among economy, health, and the environment.

About the Book

The following chapters were prepared as a text for a one-semester introductory course in environmental toxicology. This course is intended mainly for students of chemistry or of other scientific disciplines who have some background in chemistry and for industrial chemists and chemical engineers who wish to learn how chemicals interact with living organisms and how deterioration of the environment affects our lives.

The first four chapters provide a background in basic toxicological principles such as entry, mode of action, and metabolism of xenobiotics. (*Xeno* is a Greek word for "alien" or "strange"; thus, *xenobiotics* means a foreign, biologically active substance.) Chapter 5 presents principles of chemical carcinogenesis. The remainder of the text introduces the student to specific environmental problems.

A one-semester course imposes certain limitations on the depth and amount of coverage when such a great variety of subjects is involved. Despite these limitations, this text will give students an overall view of environmental toxicology and of the environmental problems facing this planet.

Acknowledgments

It gives me great pleasure to acknowledge with gratitude the help of my professional colleague Dr. Debora L. Kramer and my daughter Nina (Dr. Kristina M. Harff) in critically reviewing this manuscript.

Reference

1. Doull, J.; Bruce, M. C. In *Cassarett and Doull's Toxicology,* 3rd ed.; Klaassen, C. D.; Amdur, M. O.; Doull, J., Eds.; MacMillan: New York, 1986; Chapter 1, p 3.

SIGMUND F. ZAKRZEWSKI
Department of Pharmacology
Roswell Park Graduate Division
State University of New York at Buffalo
Buffalo, NY 14263

Preface to the Second Edition

This edition of *Principles of Environmental Toxicology* is essentially patterned on the first edition, but many changes have been introduced. While the first edition was in circulation, several reviews of the book appeared in different journals. Although the reviews were basically favorable, certain shortcomings and omissions were pointed out. I am indebted to the reviewers, and I took their comments into consideration while preparing this edition. Thus, for instance, a section on indoor air pollution was added in Chapter 8, and the scope of the book was widened beyond direct concern with human toxicity. A section on wetlands and estuaries, including a description of the plight of the Chesapeake Bay, was added in Chapter 10. A new chapter, "Population, Environment, and Women's Issues", was added. The original Chapter 6, on air pollution, was split into two chapters: Chapter 8, on the problems of urban and industrial air pollution, and Chapter 9, on the despoilment of the earth's atmosphere, meaning stratospheric ozone depletion and global warming. This change allowed expansion of the scope of both areas.

At the suggestion of the reviewer of this manuscript, the sequence of the chapters was changed somewhat, and all but one of the appendices were moved into the appropriate chapters. The chapter on occupational toxicology (Chapter 10 in the first edition) was moved after Chapter 6 ("Risk Assessment").

Other changes involved updating the information contained in the first edition. Two world events have taken place since the press time of the first edition: the United Nations Conference on Environment and Development, in Rio de Janeiro, and the United Nations Conference on Population and Development, in Cairo. Brief descriptions of the proceedings and accomplishments of these conferences are included in Chapters 1 and 13, respectively. Another major event was the publication by the Environmental Protection Agency (in a preliminary report), as well as by independent scientists, of new findings on the toxicity and environmental impact of chlorinated hydrocarbons. This newest information was also added. Because of the discovery that polychlorinated biphenyls and dioxins affect the human immune system at low doses—below the doses that produce a carcinogenic effect—the basic functioning of the immune system was included in Chapter 7. Another change was the inclusion in Chapter 14 of a section that briefly describes some important environmental acts and international treaties protecting marine life.

Despite these changes, this book is primarily a toxicology, and not an ecology, text. Thus, certain important areas of interest to environmentalists have been omitted. To remedy these shortcomings, a list of subjects for student research and seminars has been included, as in the previous edition. The book was originally prepared as a text for a one-semester introductory three-credit course in environmental toxicology. However, with the expansion of the scope of the topics in the second edition, it may be necessary to upgrade the course to four or five credits to thoroughly cover the book's content.

Certain sections of this new edition were taken, with some modifications and with permission, from my book *People, Health and Environment* (*1*).

Acknowledgments

I am indebted to the reviewers of my manuscript for their constructive criticism and useful suggestions, which helped to improve this book, and also to Jane M. Ehrke for her review and correction of the section on the basic functioning of the immune system.

Reference

1. Zakrzewski, S. F. *People, Health, and Environment;* SFZ Publishing: Amherst, NY, 1994.

SIGMUND F. ZAKRZEWSKI
260 Lakewood Parkway
Amherst, NY 14226

1 Environment: Past and Present

Historical Perspective

Concern for the environment is not an entirely new phenomenon. In isolated instances, environmental and wildlife protection laws have been enacted in the past. Similarly, astute early physicians and scientists occasionally recognized occupationally related health problems within the general population.

Protective Legislation

As early as 500 B.C., a law was passed in Athens requiring refuse disposal in a designated location outside the city walls. Ancient Rome had laws prohibiting disposal of trash into the river Tiber. In 17th century Sweden, legislation was passed forbidding "slash and burn" land clearing; those who broke the law were banished to the New World. Although no laws protecting workers from occupational hazards were enacted until much later, the first observation that occupational exposure could create health hazards was made in 1775 by a London physician, Percival Pott. He observed among London chimney sweeps an unusually high rate of scrotal cancer that he associated (and rightly so) with exposure to soot.

Colonial authorities in Newport, Rhode Island, recognizing a danger of game depletion, established the first closed season on deer hunting as early as 1639. Other communities became aware of the same problem; by the time of the American Revolution, 12 colonies had legislated some kind of wildlife protection. Following the example of Massachusetts, which established a game agency in 1865, every state had game and fish protection laws before the end of the 19th century (*1*). In 1885, to protect the population from waterborne diseases such as cholera and typhoid fever, New York State enacted the Water Supply Source Protection Rules and Regulations Program.

These instances of environmental concern were sporadic. It was not until some time after World War II that concern for the environment and for the effects of industrial development on human health became widespread.

The Industrial Revolution

The industrial development of the late 18th century, which continued throughout the 19th and into the 20th century, converted the Western agricultural societies into industrialized societies. For the first time in human history, pervasive hunger ceased to be a problem. The living standard of the masses improved, and wealth was somewhat better distributed. Throughout the 19th century, the use of steam power and coal as fuel became widespread for manufacturing and transportation. Smoke-spewing factory stacks became a symbol of prosperity. The successful technological development led people to believe that their ability to use resources (which were considered to be inexhaustible) and master nature was unlimited.

As early as 1899, T. C. Chamberlin observed that atmospheric carbon dioxide was increasing because of coal combustion, and in 1903, S. A. Arrhenius made the same observation. They suggested that excessive carbon dioxide in the atmosphere may have an effect on the earth's climate (*2*).

At the end of the 19th century, with the development of the internal combustion engine, the automobile entered the scene. Early automobiles were expensive and were considered a luxury and a plaything of the wealthy. It was not until the Ford Model T was introduced in 1908 that the automobile turned from a luxury into an everyday necessity; this blessing of humanity later became a nightmare of many modern cities. With the popularization of the automobile, the emphasis changed from coal to oil as fuel. Although oil is cleaner-burning than coal, large-scale oil exploitation, processing, and combustion began unnoticeably to take their toll on the environment.

In 1922 a technological breakthrough occurred that left a toxic legacy of lead until today: the introduction of leaded gasoline. This breakthrough was hailed as a great achievement because it allowed an increase, in an inexpensive way, in the compression of the engine, thus yielding more power without the necessity of increasing the size and the weight of the engine.

In the early 1930s, another development took place that haunts us to this day and probably will for another hundred years: the invention of chlorofluorocarbons (CFCs). These compounds, popularly known as freons, are chemically stable, nonflammable, and nontoxic. They proved to be ideal substances to replace toxic ammonia as refrigeration and air-conditioning fluids. They also found many industrial applications. Yet their use is now ending because they keep destroying the earth's protective ozone layer. May these two examples of failed technology be a warning to those who have an unshaken faith that technology alone can solve all our environmental problems.

Good Life Through Chemistry

During and immediately after World War II, chemical industries began to develop rapidly. "Good life through chemistry" was the slogan of those days. Chemical fertilizers, insecticides, and herbicides came into widespread use. These substances, together with the development of new high-yield grains (specifically, rice and wheat), revolutionized world agriculture in the 1960s in what came to be called the green revolution. Thus many developing countries, especially in Asia, became self-sufficient in food production; some even became food exporters.

Between 1950 and 1985, grain production more than doubled; after 1965, nearly half of the increase was contributed by developing countries (*3*). Between 1950 and 1973, the world economy expanded by an average of 5% per year, which resulted in rising income in all countries (*4*). This economic expansion was paralleled by generally improved health throughout the world. For instance, in India and China, the incidence of malaria, which had plagued the population for generations, decreased between 1976 and 1983 as a result of the control of mosquitoes with pesticides.

The progress was possible, at least in part, thanks to an enormous input of energy; however, the yield of grain per unit of energy was constantly decreasing, eventually reaching a constant value (Figure 1.1). This record indicates that a future increase in the world grain supply may be achieved only by increasing the acreage of land under cultivation or by currently unattainable bioengineering of new high-yield crops. The implications of this conclusion will become obvious in the course of further discussion.

Warning Signs

Life appeared to be better for everyone. Then the negative aspects of this progress, manifested by general deterioration of air and water quality, began to surface. Three cases of widespread fatalities due to urban smog were reported (Meuse Valley, Belgium, in 1930; Donora, Pennsylvania, in 1948; and London, England, in 1952). In each of these cases, temperature inversion (the settling of a layer of warm air on top of colder air) contributed to the air pollution by keeping the pollutants near the ground. The number of fatalities was 65, 20, and 4000 for Meuse Valley, Donora, and London, respectively. These events brought worldwide attention to the danger from the emission of toxic substances (sulfur dioxide, nitrogen oxides, etc.) as byproducts of fossil-fuel combustion, especially coal combustion. It became obvious that neither water nor air is a bottomless sink allowing indefinite disposal of toxins.

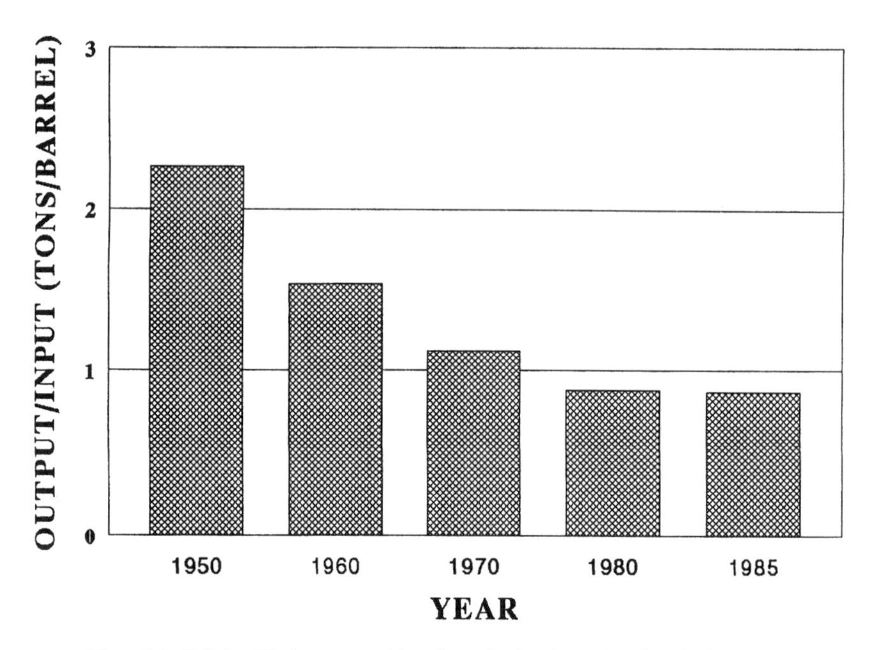

Figure 1.1. Relationship between world grain production (output) and agricultural energy input, 1950–1985. (Adapted from reference 5.)

Thus the use of toxic chemicals, whether applied purposefully or generated as byproducts of industrial processes, had to be restricted. It was also realized that normal human activities threatened the environment. For example, runoff from fields being fertilized with phosphates or nitrogen-containing chemicals caused eutrophication of streams and lakes. Runoff from cattle feedlots had a similar effect. Irrigation of poorly drained fields in a hot climate led to salinization of land, making it irreversibly lost to agriculture.

In 1962, *Silent Spring* (*6*) appeared, written by the then little known biologist Rachel Carson. The gist of this book is summarized on its front flap in these words:

> For as long as man has dwelt on this planet, spring has been a season of rebirth, and the singing of birds. Now in some parts of America spring is strangely silent, for many of the birds are dead—incidental victims of our reckless attempt to control our environment by the use of chemicals that poison not only insects against which they are directed but the birds in the air, the fish in the rivers, the earth which supplies our food, and, inevitably (to what degree is still unknown), man himself.

This controversial book woke the public to the dangers of contaminating the environment with chemical poisons.

Environment and the Economy

Environment is frequently sacrificed for the sake of the economy in our society. This policy is shortsighted because destruction of the environment undermines future economic resources. For example, the Midwestern agricultural loss caused by ozone pollution is estimated to be about $5 billion annually (*7*). Thus, the real tradeoff is not between economy and environment, but between economic prosperity now and in the future. A balance between economic development and protection of resources has to be found. W. U. Chandler's treatise "Designing Sustainable Economics" presents a detailed discussion of this subject (*8*).

The formation of the Club of Rome, an informal international gathering of 30 individuals from a variety of professions, such as scientists, educators, economists, humanists, industrialists, and civil servants, in April 1968 in Accadèmia de Lincei in Rome, marked the beginning of the new era of a holistic approach to environmental problems. The meeting was convened at the urging of Aurelio Pecci, an industrial manager and economist. Recognizing the complexity of interrelated problems afflicting modern societies, such as poverty, overpopulation, and environmental degradation, the meeting discussed the present and future predicament of humanity. The culmination of several deliberations of the club was a decision to initiate a research project on the future of humanity. This research led to the publication in 1972 of a book titled *The Limits of Growth* (*9*). In essence, this book was a computer modeling of the future of humanity, taking into consideration population growth, industrial capital, food production, resource consumption, and pollution. It concluded that "if present trends of population and economic growth continue unchanged, . . . the most probable result will be a sudden and uncontrollable decline in both population and industrial capacity." It also offered hope, suggesting that "it is possible to alter these growth trends and to establish a condition of ecological and economic stability that is sustainable far into the future."

The environmental concern inspired by grassroots movements and by the Club of Rome continued through the 1970s and permeated President Jimmy Carter's political establishment. In the late 1970s, the Carter administration commissioned the preparation of an economic and scientific report that would be a guideline for a future national environmental policy. This report, published in 1980 under the title *Global 2000* (*10*), warns that, unless corrective measures are implemented soon, the world will be facing overpopulation, energy and food shortages, and a general decline in the standard of living.

The warnings of *Global 2000* were not heeded because a different politico-economic philosophy surfaced during the 1980s. This change was reflected in *The Resourceful Earth: A Response to Global 2000* (*11*), a scientific and economic report prepared in 1984 for the Reagan administration. This report contends that long-term economic and population trends "strongly suggest a progressive improvement and enrichment of the earth's natural resource base, and of mankind's lot on earth." In general, this report does not consider environmental deterioration a serious problem and does not anticipate that unchecked population growth will eventually outstrip agricultural production. Nor does it foresee that overuse of land and development of industry may lead to ecological changes.

Although present world grain production, which is sufficient to feed 6 billion people (provided that no grain is used as animal fodder[1]), greatly exceeds that needed for the current population, an increase in food production much above the present level would necessitate the cultivation of more land and further deforestation. In turn, this activity would lead to increased soil erosion, desertification, and, possibly, climatic changes.

In May 1985, a British research team reported that the level of atmospheric ozone over Antarctica had declined sharply. This discovery of an ozone hole in the earth's protective shield created concern in the scientific community. The resultant increase in ultraviolet radiation reaching the earth's surface may increase the incidence of skin cancer, retard crop growth, and affect the food chain of marine species.

Roger Revelle and Hans Suess (*2*) published a paper in 1957 calling attention to the fact that atmospheric carbon dioxide was increasing because of fossil fuel combustion. The paper stated: "The increase is at present small but may become significant during future decades if industrial combustion continues to rise exponentially." For three decades this warning was largely ignored, until a disquieting paper appeared in a July 1986 issue of *Nature* (*12*). The authors suggested that the forecasted climatic changes arising from increasing carbon dioxide levels in the atmosphere were being realized. This *greenhouse effect* and its consequences will be discussed in a later chapter. For now, it suffices to say that adjustment to the new climatic conditions, though gradual, will be costly.

Present State of the World

Environmental problems have assumed dimensions of a global magnitude. What happens in a remote corner of the world concerns all of us, the best example being the nuclear plant accident in Chernobyl (Chapter 12). The burning of tropical forests in Brazil will affect not only the climate in Brazil, but our climate as well. Overpopulation in developing countries may affect our climate, economy, and political stability.

[1]It takes 2 kg of grain to produce 1 kg of poultry meat, 4 kg to produce 1 kg of pork, and 7 kg to produce 1 kg of beef.

Population Growth

In *State of the World 1987,* Brown and Postel wrote, "Sometime in mid-1986, world population reached 5 billion. Yet no celebrations were held in recognition of this demographic milestone. Indeed, many who reflected on it were left with a profound sense of unease about mounting pressure on the earth's forests, soil, and other natural systems" (*3*).

Among other things, the increased population means an increased demand for energy. The absolute number of people is less significant than the rate of population increase. In 1950 there were 2.5 billion people; this number doubled in only 36 years. Population growth has slowed in the last two decades from 2% to 1.68% annually, and it is expected to slow even further in the next decade. However, at the present growth rate the population would double again in the next 42 years. This translates to 11 billion people in the year 2039.[2] Unfortunately, the fastest growth occurs in the economically depressed developing countries, where the average annual growth rate is 2.5% (doubling time, 27.6 years).

In 1981 the United Nations (U.N.) published estimates of expected population growth. The low scenario estimates that the population will stabilize in the year 2050, after reaching 8 billion people. In contrast, the high scenario predicts stabilization around 2125 with 14.2 billion people (*13*). Newer data set the number at 11.5 billion by the year 2150 (Figure 1.2). The number of people the earth can support is difficult to estimate because population growth affects the environment and the availability of resources, which in turn alter the earth's carrying capacity.

Regardless of whether population-control policies are successful, eventually the world population will stabilize. How stabilization will be achieved is another matter. The demographic-transition theory offered by demographer Frank Notestein (*4*) classifies all societies into one of three stages. Stage 1 characterizes primitive societies, in which both birth and death rates are high; consequently, there is little population growth. In stage 2, thanks to improved public health and hygiene, the death rate diminishes while the birth rate remains unchanged; consequently, there is rapid population growth. In stage 3, because of a high employment rate among women and the desire to maintain a high standard of living, there is a tendency to limit family size; consequently, both the birth rate and the death rate decline, and little or no growth occurs.

The industrialized world is now in stage 3 (average growth rate of 0.6%). The developing countries are in stage 2. If nothing is done to arrest this explosive growth, there is danger that the population of the developing world may stabilize by reverting to stage 1, as is now evident in Ethiopia, Somalia, and Sudan.[3]

[2]The formula for calculating doubling time is: doubling time (years) = $\ln 2 \times 100$/percent annual growth. Because $\ln 2 = 0.69$, doubling time = 69/percent annual growth.

[3]In these cases, the political situation is also a factor.

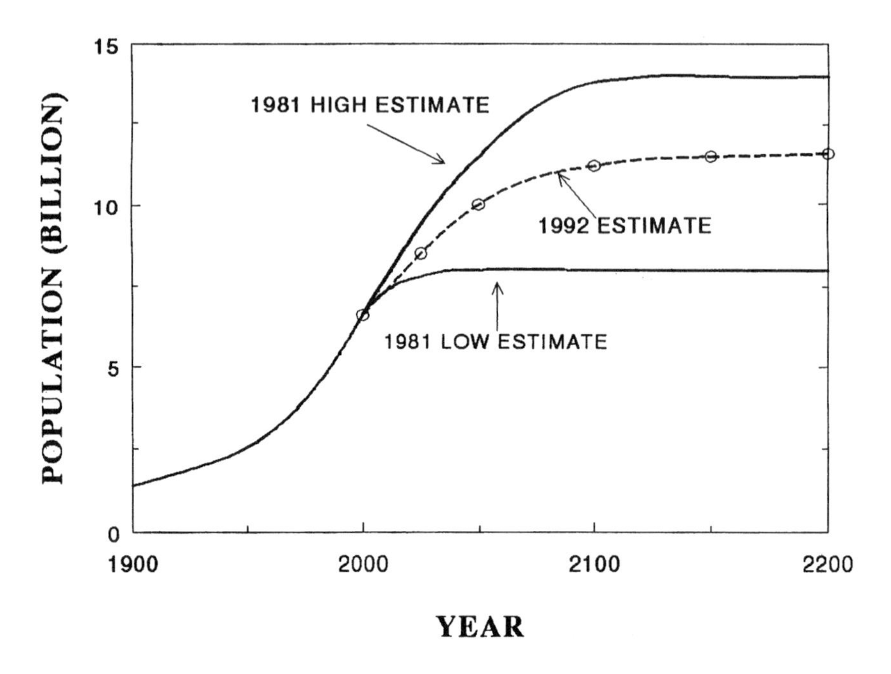

Figure 1.2. The United Nations estimates of expected population growth. (Based on data in references 13 and 14.)

Widespread hunger, high infant mortality, and social and political unrest may result.

Deforestation

Deforestation is a direct consequence of the developing world population explosion. Forests are cut down for land clearing, firewood, and logging. Satellite data show that between 1973 and 1981, India lost 16% of its forest cover (*5*). Removal of forests has serious environmental consequences, such as increased rainfall runoff and accelerated soil erosion. Some of the land is irreversibly lost to agriculture and reforestation, as desertification occurs. The catastrophic floods that occurred in Bangladesh in 1988 were, in part, the consequence of extensive deforestation.

The loss of forests is not only a developing world problem. Although the causes of forest destruction in industrialized countries are different from those of the developing world, the result is the same. As of 1986, 52% of the forests in West Germany were damaged, presumably by acid rain and air pollution. More frightening is the rapidity with which this deterioration occurred; in 1983 the reported damage was 34% (*15*). Forest damage is not restricted to Germany. It has been reported in Scandinavia, the former Czechoslovakia, and the eastern United States.

With the disappearance of forests, the global carbon dioxide balance becomes disturbed. This shift may result in warming of the earth's surface and changes in precipitation patterns. Another consequence of deforestation is a decline in biodiversity, as species disappear. During the 1986 National Forum on Biodiversity in Washington, D.C., scientists warned of the possibility of a mass extinction of species. This development may be compared with the catastrophe that wiped out the dinosaurs and many other species millions of years ago. Whereas then the extinction was due to natural causes, this time it will be due to human handiwork.

Use of Resources

In industrialized countries, population pressure is not the greatest problem. Rather, an insatiable demand for more manufactured goods and energy, as well as the need for economic expansion to provide full employment, stresses the environment. Because of these factors, even a modest increase in the population of industrialized countries increases the demand for energy and other resources to a much greater extent than it would in countries with a low standard of living.

The population of North America, which represents about 5% of the global population, consumes 35% of world resources. The United States alone contributes 21% to the global atmospheric pollution with greenhouse gases (*16*). The growth of urban centers (which is also a problem for the developing countries) causes hydrological changes. Manufacturing, transportation, and energy production cause air and water pollution, with all their ecological consequences. High consumption of goods leads to the growing problem of household, manufacturing, and toxic waste disposal, which presents a threat to groundwater. In cases of sea dumping, this threat is extended to marine life.

Energy Sources

Last, but not least, there is the problem of energy. The supply of energy is vital not only to transportation and to modern conveniences, but to food production as well. The exact amount of world fossil fuel reserves is difficult to estimate because some as yet untapped sources may be discovered. According to Brown and Postel, by 1986 nearly half of the discovered oil had already been consumed. As an estimate, the present proven energy reserves, assuming 1986 production rates, are (*17*)

- oil, 40 years;
- natural gas, 60 years; and
- coal, 390 years.

Of course, how long these reserves actually last will depend on conservation measures and the efficiency of energy use. In addition, both energy production and use have an effect on the environment.

Nuclear energy produces neither carbon dioxide nor acid rain. Still, there is serious concern about the possibility of radioactive contamination of the environment resulting from the operation of nuclear reactors, storage of spent fuel, and nuclear accidents.

The United Nations Conference on Environment and Development

From June 3 to 14, 1992, representatives of 154 nations gathered under the auspices of the United Nations, in Rio de Janeiro, to coin a blueprint for the future sustainable development of the world. This blueprint was called Agenda 21. The conference, referred to as Earth Summit, amassed not only governmental representatives but also representatives of the global scientific community, environmentalists, and many nongovernmental organizations involved in U.N. activities.

The executive director of the U.N. Environment Programme, Mastafa K. Tolba, outlined in his opening speech the problems facing the world: the deterioration of environment, especially in developing countries, the loss of species, climate change, the danger of rapidly growing population, and the steadily increasing imbalance in income and wealth between the industrialized and developing countries. Other keynote speakers emphasized the danger of environmental neglect. Gro Harlem Brundtland, Prime Minister of Norway, expressed her concern this way: "We may temporarily immunize ourselves emotionally to the images of starvation, drought, floods, and people suffocating under the load of wastes we are piling on a nature so bountiful, but there is a time bomb ticking. We cannot betray future generations. They will judge us harshly if we fail at this crucial moment" (*18*). Similarly, the U.N. Secretary-General, Boutros Boutros-Ghali, stated, "We are looking at a time frame that extends far beyond the span of our individual lives. . . . We can waste the planet's resources for a few decades more. . . . We must realize that one day the storm will break on the heads of future generations. For them it will be too late" (*18*).

Despite this lofty rhetoric, the results of the conference were mixed at best, and some parts of the conference were disappointing. Before the Summit, the conference Secretary-General, Maurice Strong, emphasized that the conference "will define the state of political will to save our planet and to make it . . . a secure and hospitable home for present and future generations" (*19*). Unfortunately, the results indicated that perhaps the "political will" was not as strong as expected and narrow national or regional self-interest still prevailed.

On the positive side was the recommendation that the 47th General Assembly establish a high-level U.N. Commission on Sustainable Development. The role of the Commission will be to oversee that the promises made at Rio de Janeiro are kept. Although the Commission lacks enforcement power, it may exert its influence by shining the spotlight on countries that renege on their promises. The other positive outcome was that all 154 nations signed the convention on climate change, and 153, all but the United States, signed the convention on biodiversity. (The biodiversity treaty was eventually signed by President Clinton.) On the negative side, it has to be noted that, because of the obstructive attitude of the United States, the treaty on climate change was watered down, and no definite targets and timetables for stabilizing carbon dioxide emissions were set. As it was finally passed, the treaty set only nonbinding commitments for the industrialized nations to limit their greenhouse-gas emissions. Because of the status of the United States as the indisputable world power, the withdrawal of this nation from signing the biodiversity convention also weakened this treaty.

Another drawback was the statement on forest protection, which was watered down by the attitude of the developing countries. They felt that the industrialized nations destroyed their own forests, and keep destroying what is left from the original growth, yet they preach the need for forest preservation to the developing, impoverished nations. Kamal Nath, Indian Minister of the Environment, put it this way: "If our forests did not sustain fuel needs, I shudder to think what our oil requirement would be. . . . We do not talk of the globalization of oil so we do not talk of globalization of forests" (*18*).

Perhaps the greatest failure of the Earth Summit was that the issue of population and its relation to poverty was not on the agenda at all.

At the conclusion of the conference, Agenda 21 was written to address all the issues that had been discussed. Agenda 21 is a blueprint for international cooperation for sustainable development. It is addressed to governments as well as to civic organizations and to the population at large. The principal aims of the Agenda are (*18*)

1. To ensure that world development proceeds in a sustainable manner, that is, that future generations are taken into consideration in policy making. This goal should be attained by a system of incentives and penalties to motivate economic behavior.
2. To promote a coordinated international effort to eliminate poverty throughout the world; to secure decent shelters, a clean water supply, hygienic facilities, energy, and transportation for all people.
3. To minimize both industrial and municipal waste.
4. To promote efficient and sustainable use of resources, such as energy, land, and water.
5. To promote sustainable use of the atmosphere, the oceans, and marine organisms.
6. To promote better management of chemicals and chemical waste.

The big problem that arose at the conference was financial support for the developing countries for implementation of the Agenda's postulates. Maurice Strong estimated the financial need for implementation at $125 billion annually (the current level of assistance from the industrialized world is $55 billion). This amount could be raised if the industrialized nations contributed, on the average, 0.7% of their gross national product. So far only Norway, Sweden, Denmark, and Netherlands have complied with this requirement. No deadline was set for other countries to achieve this goal. The management of the funds was entrusted to the Global Environmental Facility (which operates under the auspices of the World Bank), regional banks, and certain U.N. agencies. Bilateral aid was not excluded.

It remains to be seen whether the implementation of Agenda 21 will succeed. In spite of its imperfections and failures, the Earth Summit will go down in history as a valiant attempt to avert a global, ecological, and economic disaster.

Antienvironmental Movements in the United States

In contrast to the spirit of the World Summit, an antienvironmental sentiment is brewing in certain circles in the United States. In the last few years, several hundred antienvironmental organizations have sprouted across the nation. They exist under misleading names such as "Citizens for the Environment" or "Oregon Lands Coalition" (*20*). Masquerading as environmental movements, their aim is to weaken the environmental regulatory framework. These organizations are loosely connected and fall under the general designation of "wise use" movement. Their common philosophy is that the earth's resources were meant to be exploited for human gains and profit. This philosophy, however, fails to consider that the resources are not inexhaustible and that they belong to the future as well to the present generations. The wise use movement strategy is a two-pronged attack: One prong is directed toward organizing grassroots support in small Western towns, and the other is engaged in lobbying in Washington, D.C. The immediate aims of the movement are to allow the harvesting of old-growth forests, eliminate or at least reduce the size of many national parks, repeal the Endangered Species Act, and open the Arctic National Wildlife Refuge to oil exploration. Despite its far-fetched and unrealistic objectives, the movement is having some impact on national legislation. Its great success was the inclusion (and the passage), in the new transportation bill, of a provision that designated a part of the proceeds from the gasoline tax to be used for construction of off-road vehicle trails through the wilderness.

Another group, called "People for the West", was formed in 1989 as a lobbying organization aimed specifically at preventing repeal by the Congress of the 1872 Mining Law. This outdated law obliges the federal government to sell federal land for $5 per acre to anyone who discovers mineral deposits. Although the group is heavily funded by mining and oil industries, it is now aiming to broaden its grassroots support and widen its antienvironmental activities.

Whether connected with the wise use movement or not, some well-known syndicated columnists as well as politicians have also taken an antienvironmental stand. The U.N. Conference on Environment and Development in Rio de Janeiro was referred to in the press as a "scientific fraud" (*21*), and environmentalism was called a "green tree with red roots . . . a socialist dream . . . dressed up as compassion for the planet" (*22*).

Such attitudes are frightening, especially when they are so widespread within the educated segment of the society. The message of Agenda 21 still has a long way to go to be generally accepted. Let us hope, however, that the young generation will be more receptive to the message of the Agenda; after all, the young and those unborn are the ones whose fate is at stake.

References

1. Arrandale, T. In *Earth's Threatened Resources;* Gimlin, H., Ed.; Congressional Quarterly: Washington, DC, 1986; pp 21–40.
2. Revelle, R.; Suess, H. E. *Tellus* **1957**, *IX,* 18.
3. World Resources Institute, International Institute for Environment and Development in collaboration with U.N. Environment Programme. *World Resources 1988–89, Food and Agriculture;* Basic Books: New York, 1988; p 51.
4. Brown, L. R. In *State of the World 1987;* Brown, L. R., Ed.; Worldwatch Institute: New York, London, 1987; p 20.
5. Brown, L. R.; Postel, S. In *State of the World 1987;* Brown, L. R., Ed.; Worldwatch Institute: New York, London, 1987; p 3.
6. Carson, R. *Silent Spring;* Houghton Mifflin: Boston, MA, 1962.
7. MacKenzie, J. J.; El-Ashry, M. T. *Ill Winds: Airborne Pollution's Toll on Trees and Crops;* World Resources Institute: Washington, DC, 1988.
8. Chandler, W. U. In *State of the World 1987;* Brown, L. R., Ed.; Worldwatch Institute: New York, London, 1987; p 177.
9. Meadows, D. H.; Meadows, D. L.; Randers, J. *The Limits of Growth;* Universe Books: New York, 1972.
10. *Global 2000, Report to the President;* U.S. Government Printing Office: Washington, DC, 1980.
11. *The Resourceful Earth: A Response to Global 2000;* Simon, J. L.; Kahn, H., Eds.; Basil Blackwell: New York, 1984.
12. Johns, P. D.; Wigley, T. M. L.; Wright, P. B. *Nature (London)* **1986**, *322,* 430.
13. Kuusi, P. *This World of Man;* Pergamon: Oxford, New York, 1985; Chapter 13, p 191 [Figure 1.1].
14. World Resources Institute, International Institute for Environment and Development in collaboration with U.N. Environment Programme. *World Resources 1992–93, Population and Human Development;* Oxford University: New York, 1992; pp 80 and 246.
15. Thompson, R. In *Earth's Threatened Resources;* Gimlin, H., Ed.; Congressional Quarterly: Washington, DC, 1986; p 1.
16. Zurer, P. *Chem. Eng. News* March 5, 1990, p 13.
17. World Resources Institute, International Institute for Environment and Development

in collaboration with U.N. Environment Programme. *World Resources 1992–93, Energy;* Oxford University: New York, 1992; p 143.
18. Hileman, B. *Chem. Eng. News* July 6, 1992, p 7.
19. Hileman, B. *Chem. Eng. News* June 22, 1992, p 4.
20. Ruben, B. *Environmental Action;* Environmental Action: Takoma Park, MD, 1992; p 25.
21. Thomas, C. *Buffalo News* June 6, 1992, p B3.
22. Will, G. *Buffalo News* June 9, 1992, p B3.

2 Review of Pharmacologic Concepts

Dose–Response Relationship

Early scientific knowledge recognized two basic types of substances: beneficial ones (such as foods and medicines), and harmful ones (those that cause sickness or death). The latter were designated as poisons.

Modern science acknowledges that such a strict division is not justified. As early as the 16th century, Paracelsus recognized that "the right dose differentiates a poison and a remedy." Many chemical substances or mixtures exert a whole spectrum of activities, ranging from beneficial to neutral to lethal. Their effect depends not only on the quantity of the substance to which an organism is exposed, but also on the species and size of the organism, its nutritional status, the method of exposure, and several related factors.

Alcohol is a good example. Taken in small quantities, alcohol may be harmless and sometimes even medically recommended. However, an overdose causes intoxication and, in extreme cases, death. Similarly, vitamin A is required for the normal functioning of most higher organisms, yet an overdose of it is highly toxic.

If the biological effect of a chemical is related to its dose, there must be a measurable range between concentrations that produce no effect and those that produce the maximum effect. The observation of an effect, whether beneficial or harmful, is complicated by the fact that apparently homogeneous systems are, in fact, heterogeneous. Even an inbred species will exhibit marked differences among individuals in response to chemicals. An effect produced in one individual will not necessarily be repeated in another one. Therefore, any meaningful estimation of the toxic potency of a compound will involve statistical methods of evaluation.

Determination of Toxicity

To determine the *toxicity* of a compound for a biological system, an observable and well-defined end effect must be identified. Turbidity or acid production, reflecting the growth or growth inhibition of a culture, may be used as an end point in bacterial systems. In some cases, such as in the study of mutagenesis, colony count may be used. Similarly, measures of viable cells, cell protein, or colony count are useful end points in cell cultures. The most readily observable end point with in vivo experiments is the death of an animal, and this is frequently used as a first step in evaluating the toxicity of a chemical. Inhibition of cell growth or death of animals are not the only concerns of toxicology. Many other end points may be chosen, depending on the goal of the experiment. Examples of such choices are inhibition of a specific enzyme, sleeping time, occurrence of tumors, and time to the onset of an effect.

Because the toxicity of a chemical is related to the size of the organism exposed, *dose* must be defined in terms of concentration rather than absolute amount (*1*). (In medical literature and in pharmacokinetics, the total amount administered is frequently referred to as the total dose.) Weight units (milligram, microgram, nanogram, etc.) per milliliter of maintenance medium or molar units (millimolar, micromolar, nanomolar)[1] are used with in vitro systems. In animal experiments doses are expressed in weight or molecular units per kilogram of body weight or per square meter of body surface area.

As an example, a simple experiment is designed to determine the lethality of a chemical in mice. The compound to be tested is administered to several groups of animals, usually 5–10 animals per group, with each successive group receiving a progressively larger dose. The number of dead animals in each group is recorded. Then the percentage of dead animals at each dose minus the percentage that died at the immediately lower dose is plotted against the logarithm of the dose. This plot generates the Gaussian distribution curve, also known as the quantal dose–response curve, which is presented in Figure 2.1. The point at the top of the curve represents the mean of the distribution, or the dose that kills 50% of the animals; it is designated as LD_{50}.[2] The mean minus one standard deviation (SD) corresponds to LD_{16}; LD_{50} minus two SD corresponds to $LD_{2.3}$. The mean plus one SD corresponds to LD_{84}; plus two SD corresponds to $LD_{97.7}$.

This type of plot is not very practical, so the cumulative percentage of dead animals is usually plotted against the logarithm of the dose (Figure 2.2). The use

[1]M always stands for moles per liter and is pronounced as molar. Thus, mM is millimolar, μM is micromolar, and nM is nanomolar.

[2]LD stands for lethal dose. Other terms are also used, depending on the type of experiment. Thus, IC stands for inhibitory concentration and ED for effective dose.

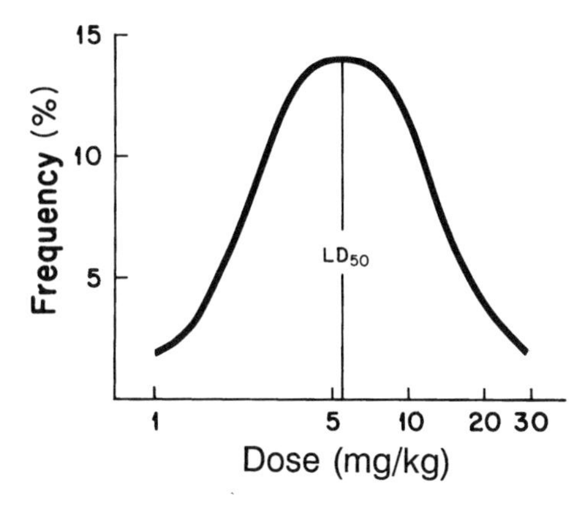

Figure 2.1. Quantal dose–response curve. The frequency represents the percentage of animals that died at each dose.

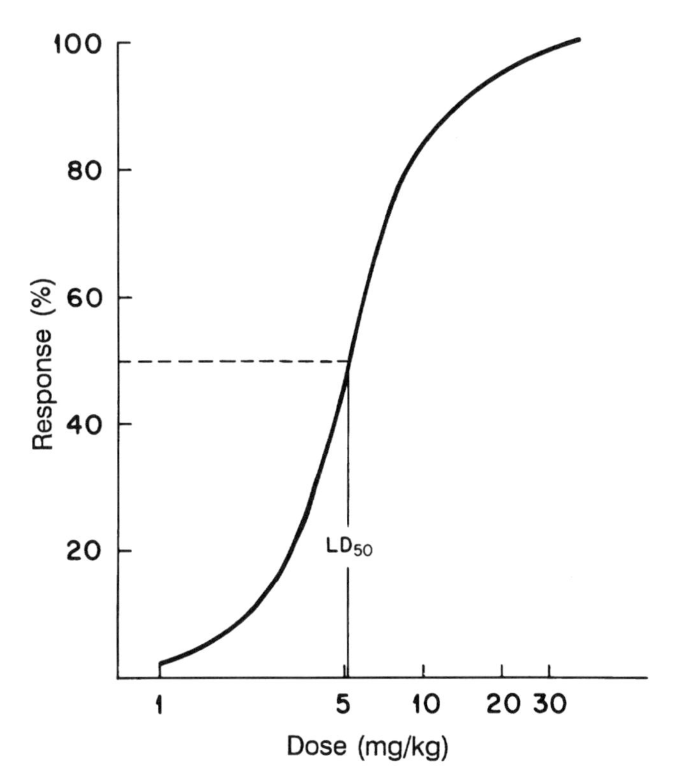

Figure 2.2. Cumulative dose–response curve. The response is the cumulative percentage of animals that died.

of a semilogarithmic plot originated with C. I. Bliss (*1*), who studied the effect of insecticides on insects. He noticed that there were always some dead insects at the minimum dose and always some survivors at the maximum dose. He also observed that doubling the dose always increased the effect by a fixed interval. A mathematical model reflecting these conditions suggested the use of a logarithmic, rather than a linear, dose scale. Because the center portion of the curve is nearly linear, the effect in this segment is proportional to the logarithm of the dose. The two ends of the curve asymptotically approach, but never reach, 0 and 100% effect. Thus, the threshold dose (i.e., the dose below which there is no effect) cannot be determined experimentally. Analysis of the curve in Figure 2.2 reveals that the confidence limits of the data points are greatest in the central segment and lowest at the flat segments of the curve.[3] In these flat segments a small deviation of the observed value from the expected value causes a large error in estimation of the dose. Toxicologists must realize that only those data points that fall along the straight portion of the curve are meaningful.

Probit Transformation

Bliss (*1*) introduced probit transformation (for probability), a different way of plotting the dose–response curve. In this plot, effect is plotted in probit units, LD_{50} being 5; each +SD adds a point to the scale, and each –SD subtracts a point. Table 2.1 shows conversion of percentage effect into probit units. The probit transformation makes the dose–response curve linear (or nearly so), and thus allows its analysis by linear regression ($Y = a + bX$, where b is the slope of the curve) (Figure 2.3).

A graphic method for the determination of LD_{50}, slope, and confidence limits for both parameters (a and b) and for doses other than LD_{50} was described by Lichfield and Wilcoxon (*2*). When this method is used to fit the best line in the probit plot, the data points at both ends of the line should be assigned the least weight.

Table 2.1. Conversion of Percentage into Probit Units

Percent	*Probit*	*Percent*	*Probit*
10	3.72	60	5.25
20	4.16	70	5.52
30	4.48	80	5.84
40	4.75	90	6.28
50	5.00		

[3]Confidence limits are the two points, one on each side of the mean, between which 95% of the data points would fall if the experiment were repeated 100 times. The distance between these points is referred to as the 95% confidence interval. It is equal to the mean $\pm 1.96(SD/\sqrt{n})^2$, where SD is the standard deviation and n is the number of determinations.

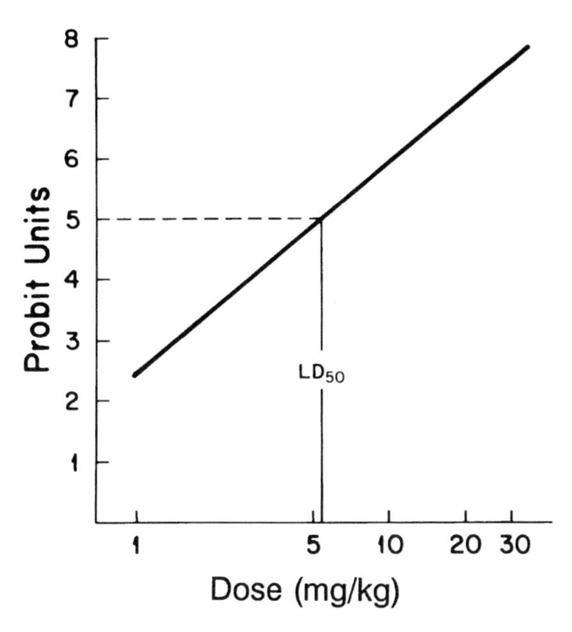

Figure 2.3. Probit transformation of a dose–response curve.

Several computer programs *(3)* are now available for dose–response analysis that can be used with a number of desktop and laptop computers.

Applications of the Dose–Response Curve

The *potency* of a compound, expressed as LD_{50}, is a relative concept and has meaning only for comparison of two or more compounds. Two compounds can easily be compared when their dose–response curves are parallel; the compound with the smaller LD_{50} value is the more potent one. However, two compounds can have a reversed toxicity relationship as LD values vary. Figure 2.4 shows that compound A is more toxic than compound B at the LD_{50} concentrations but less toxic at the LD_{20} concentrations.

The slope of a dose–response curve is also an important factor in determination of the margin of safety. If the slope is steep, a small increase in the dose may produce a significant change in toxicity. Thus the shallower the slope, the greater is the margin of safety. This expression of the margin of safety should not be confused with a concept used in clinical toxicology, where the margin of safety represents a spread between an effective (curative) dose (ED_{50}) and a toxic dose (LD_{50}). The ratio LD_{50}/ED_{50} is referred to as the *therapeutic index*. When the toxicity of a compound is considered, both potency and efficacy are important. Some compounds may have high potency, as expressed by LD_{50}, but low efficacy because their dose–response curve never approaches 100% of the effect.

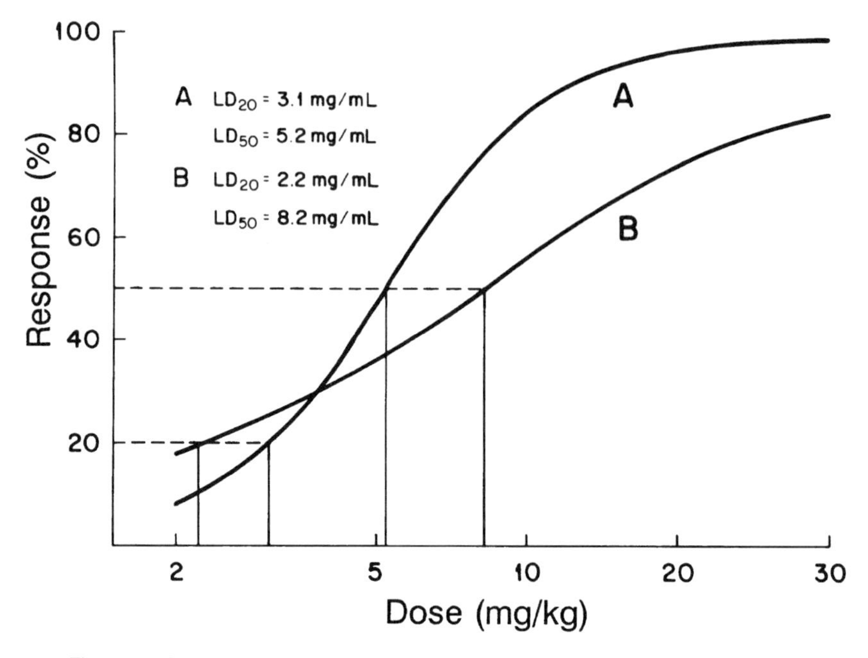

Figure 2.4. Comparison of dose–response curves with different slopes for compounds A and B.

Reversibility of Toxicity

Another aspect to be considered is the reversibility of a toxic effect. In most cases, toxicity induced by a chemical is essentially reversible. Unless damage to the affected organs has progressed too far, so as to threaten the survival of the organism, the individual will recover when the toxin is removed by excretion or inactivated by metabolism. However, in some cases the effect may outlast the presence of the toxin in the tissue. This happens when a toxin irreversibly inactivates an enzyme, and thus deprives the organism of vital functions. In such a case, although no free toxin can be detected in the body, the recovery of the organism will not occur until enough of the affected enzyme has been newly synthesized. A typical example of such an effect is intoxication with organophosphates, which bind essentially irreversibly to acetylcholinesterase.

In some cases, although no irreversible inactivation of an enzyme occurs, the action of a toxin may deprive an organism of a vital substance, and recovery has to await resynthesis of this substance. Such is the case with reserpine, which acts by depleting sympathetic nerve endings of catecholamine; the time required to replenish the reserves of catecholamine is longer than the persistence of reserpine in the tissue.

Compounds that are required in small amounts for the normal functioning of an organism, yet at high concentrations produce toxicity, have a biphasic dose–response relationship, as shown in Figure 2.5. Vitamin A, niacin, selenium, and some heavy metals such as copper and cobalt fall into this category. For such compounds, there is a certain normal range. Concentrations higher than this range cause toxicity and in extreme cases may be lethal. If the concentration is lower than this range, the organism suffers from a deficiency that alters normal functions and again may be lethal.

The Concept of Receptors

Some chemicals, such as strong acids and bases, exert their toxic action in a nonspecific way simply by denaturing protein and dissolving the tissue. Such lesions are referred to as chemical burns. In most cases, however, toxins act by interacting with specific components of the tissue, thus perturbing normal metabolism. Early in the 20th century, Paul Ehrlich *(4)* proposed the concept of specific receptors. He postulated that a chemical, in order to exert biological action, must reach a specific target area and fit into a receptor site.

Many receptors have been identified; in all cases they are proteins. Some of the proteins have enzymatic activity. For instance, dihydrofolate reductase is a receptor for antifolates (Chapter 4), and acetylcholinesterase is a receptor for organophosphates. Some receptors serve as "transport vehicles" across the cellular membranes, such as the receptors for steroid hormones *(5)*. Specific receptors

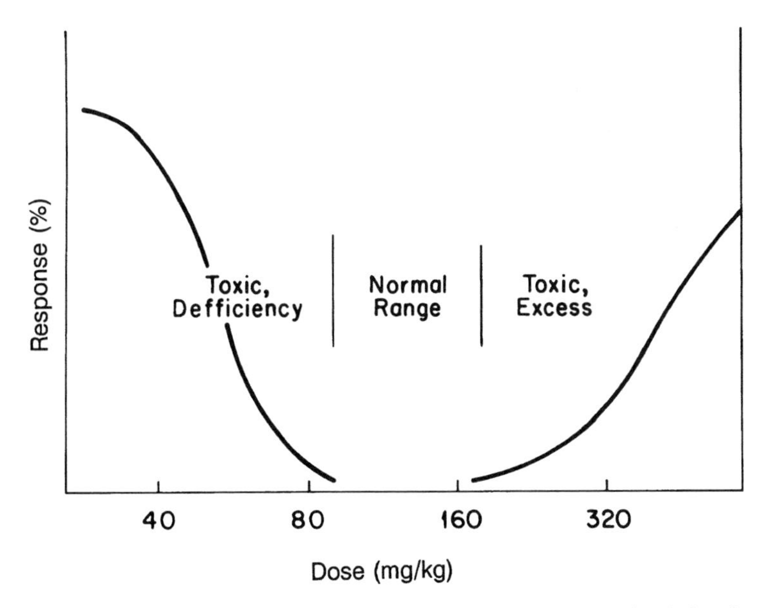

Figure 2.5. Biphasic dose–response curve of compounds required for normal functioning of organisms.

may be confined to certain tissues or may be distributed among all the cells of an organism.

Compounds in circulation are frequently bound, sometimes very tightly, to plasma proteins. Although in many cases this binding is specific for a given chemical, the proteins involved are not considered to be specific receptors. Such interactions simply prevent the compound from reaching target cells and do not result in biological action.

Mode of Entry of Toxins

From the environmental point of view, the three principal routes of entry of xenobiotics into the human body are percutaneous, respiratory, and oral. (The term *xenobiotics* is a general designation of chemical compounds foreign to the organism. It is from the Greek *xeno,* meaning foreign.) In multicellular animals, the extracellular space is filled with interstitial fluid. Thus, regardless of how a compound enters the body (with the exception of intravenous administration), it enters interstitial fluid after penetrating the initial cellular barrier (such as skin, intestinal mucosa, or the lining of the respiratory tract). From the interstitial fluid, the compound penetrates the capillaries and enters the bloodstream, which distributes it throughout the body.

Percutaneous Route

The skin forms a protective barrier that separates the rest of the body from the environment. In the past it was thought that chemicals did not penetrate the skin. In view of more recent research, this view no longer holds. Although penetration of the skin by most substances is slow, this route of entry plays an important role with regard to human and animal exposure to toxic chemicals.

The skin consists of three layers: the outermost protective layer, the *epidermis;* the middle layer, consisting of a highly vascularized connective tissue called the *dermis;* and the innermost layer, consisting of a mixture of adipose and connective tissue, called the *hypodermis.* In addition, the skin contains epidermal appendages (hair follicles, sebaceous glands, and sweat glands and ducts) that penetrate into the dermal layer.

Three possible routes of percutaneous absorption are diffusion through the epidermis into the dermis, entry through sweat ducts, and entry along the hair-follicle orifices. Although the latter routes present relatively easy access to the vascularized dermal layer, it is believed that, because of its large surface area, absorption through the epidermal cells is the major route of entry of toxins.

The main obstacle to percutaneous penetration of water and xenobiotics is the outermost membrane of the epidermis, called the *stratum corneum.* This membrane is made up of several layers of dried, flattened keratinocytes. There is no vascularization and no metabolic activity in the stratum corneum. However, the lower basal layer of epidermis, although not vascularized, has high metabolic activity and is capable of biotransformation of xenobiotics (Chapter 3).

All entry of substances through the stratum corneum occurs by passive diffusion across several cell layers. The locus of entry varies, depending on the chemical properties of a xenobiotic. Polar substances are believed to penetrate cell membranes through the protein filaments; nonpolar ones enter through the lipid matrix (see the section on cellular uptake, later in this chapter). Hydration of the stratum corneum increases its permeability for polar substances. Electrolytes enter mainly in a nonionized form, and thus the pH of the solution applied to the skin affects permeability. Many lipophilic substances, such as carbon tetrachloride and organophosphate insecticides, readily penetrate the stratum corneum. Pretreatment of the skin with solvents, such as dimethyl sulfoxide, methanol, ethanol, hexane, acetone, and, in particular, a mixture of chloroform and methanol *(6)*, increases permeability of the skin. This effect probably results from the removal of lipids from the epidermis, which would alter its structure.

The permeability of skin is not uniform. It varies between species and even within species, depending on the diffusivity and the thickness of the stratum corneum *(7)*. In general, gases penetrate skin more readily than liquids and solutes. Solids do not penetrate as such. However, they may be dissolved into the skin's secretions and subsequently absorbed as solutes.

Percutaneous absorption is a time-dependent process, with passage through the stratum corneum as the rate-limiting reaction. Therefore, duration of exposure to a xenobiotic is critical. It follows that the quick removal of spills is of the utmost importance. The kinetics of percutaneous absorption resembles that of gastrointestinal absorption (Figure 2.6), except that the latter is faster.

Respiratory Route

The *respiratory system* consists of three regions: nasopharyngeal, tracheobronchial, and pulmonary. The *nasopharyngeal canal* is lined by ciliated epithelium through which mucous glands are scattered. The role of this region is to remove large inhaled particles and to increase the humidity and temperature of inhaled air.

The *tracheobronchial region* consists of the trachea, bronchi, and bronchioles. These are branched and successively narrower conduits between the nasopharyngeal and pulmonary regions. They are lined with two types of cells: ciliated epithelium and mucus-secreting goblet cells. The function of these cells is to propel foreign particles from the deep parts of the lungs to the oral cavity, where they can be either expelled with the sputum or swallowed; this function is referred to as the mucociliary escalator. As the tracheobronchial conduits branch, the airways become smaller but the total surface area increases.

The *pulmonary region* consists of respiratory bronchioles (small tubes about 1 mm long and 0.5 mm wide, seeded on one side with alveoli), alveolar ducts (small tubes seeded on all sides with alveoli), and clusters of alveoli (referred to as alveolar sacs).

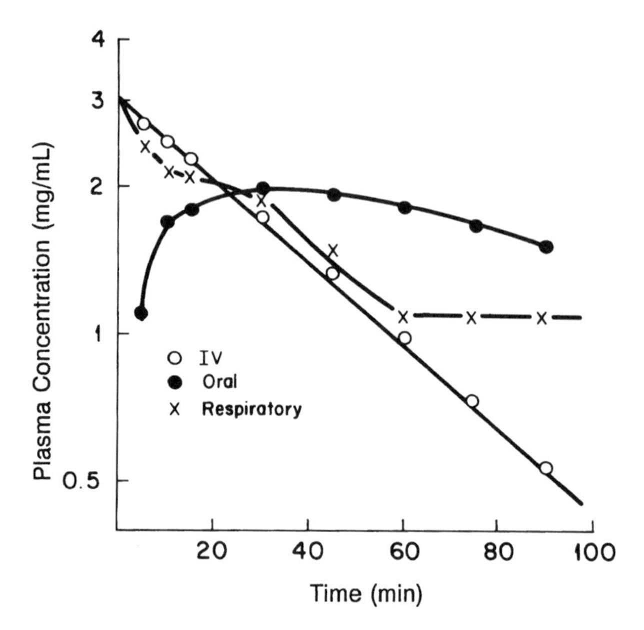

Figure 2.6. Plasma levels of cocaine after intravenous, oral, and respiratory administration (smoking). (Adapted from Chemical and Engineering News, *November 21, 1988. Copyright 1988 American Chemical Society.)*

Alveoli can be described as little bubbles about 150–350 μm in diameter in which the exchange of gases between the environment and the blood takes place. The total alveolar surface area of the human lung is 35 m^2 during expiration and 100 m^2 during deep inhalation. Three types of cells present in the alveolar region deserve to be mentioned: squamous alveolar lining cells (called Type I pneumocytes), surfactant-producing cells (called Type II pneumocytes), and freely floating phagocytic macrophages. Type II pneumocytes, in addition to producing surfactants (required to keep the alveoli inflated), are involved in the repair of injuries. Blood capillaries are in intimate contact with the alveolar lining cells, so that gases as well as solutes can easily diffuse between them.

Inhaled xenobiotics can exert their harmful action either by damaging respiratory tissue or by entering the circulation and causing systemic toxicity. Only the latter situation will be discussed in this chapter.

Readily water-soluble gases are removed, to a certain extent, in the nasopharyngeal and tracheobronchial region. Although this removal protects the lower respiratory system, it does not prevent the entry of these gases into the blood. Poorly water-soluble gases, although somewhat diluted by the humidity of the nasopharyngeal region, reach the alveoli. The amount of a toxin delivered to the lungs (in gaseous form, as liquid aerosols, or as particles) depends on the concentration of the toxin in the air and on the *minute volume* of respiration. The minute volume is a product of *tidal volume* (i.e., normal respiratory volume, about 500 mL) and the number of breaths per minute (about 15).

Gases diffuse readily through alveolar membranes according to Fick's law (*8*):

$$D = c_d \times S/\mathrm{MW}^{1/2} \times A/d \times (P_a - P_b) \quad (2.1)$$

where D is the diffusion rate (g cm^{-2} s^{-1}); c_d is the diffusion coefficient (cm^2 s^{-1}); S is solubility of the gas in blood; MW is molecular weight; A and d are characteristics of the lung (surface area and thickness of the membrane, respectively); and P_a and P_b are partial pressure of the gas in the inspired air and in the blood, respectively. The first two expressions in this equation represent the properties of the gas; the third one represents the properties of the lungs.

Analysis of this equation indicates that as long as P_a is larger than P_b, D is positive and there is uptake of gas by the blood. When $P_a = P_b$, $D = 0$; equilibrium has been established between the gas in the alveoli and in the blood so that no net gas exchange takes place. When P_b is larger than P_a (i.e., the individual was removed from the toxic atmosphere), D becomes negative. In this situation gas diffuses from the blood into the alveoli and is removed by expiration.

Another important factor affecting diffusion rate is the solubility of the gas in blood. When S is large, the diffusion rate is fast and the gas is removed quickly from the alveoli. In this case, the limiting factor in delivery of gas to the blood is the rate of supply of gas to the alveoli. Increasing minute volume (either by deeper respiration or by faster respiration) increases gas delivery. When S is small, the diffusion rate is slow; thus blood flow (i.e., cardiac output) rather than minute volume becomes the rate-limiting factor in toxicity.

Toxins can also reach alveoli as liquid aerosols. If they are lipid-soluble, they readily cross alveolar membranes by passive diffusion.

The toxicity of particulate matter depends on the size of the particles. Particles larger than 5 μm are deposited in the nasopharyngeal region and are either expelled by sneezing or propelled into the oral cavity, where they are swallowed or expelled in the sputum. Particles 2–5 μm in size are deposited in the tracheobronchial region. They are cleared by the mucociliary escalator and eventually end up being expelled in the sputum or swallowed.

Particles 1 μm or smaller are deposited in alveoli. Then the free or phagocytized particles may be carried to the tracheobronchial region, where they are removed from the respiratory system by the mucociliary escalator. Alternately, both free and phagocytized particles may pass through small (0.8–1.0 nm) intercellular spaces between alveolar lining cells and enter the lymphatic system. The latter, however, is a slow and inefficient process.

Particles resulting from combustion frequently carry adsorbed polycyclic aromatic hydrocarbons (PAHs), some of which are carcinogens. These adsorbed hydrocarbons may dissolve in alveolar fluid and enter the circulation as solutes.

Oral Route

The absorption of compounds taken orally begins in the mouth and esophagus. However, in most cases the retention time in this area is so short that no significant absorption takes place.

In the stomach, compounds are mixed with food, acid, gastric enzymes, and bacteria. All of these can alter the toxicity of the chemical, either by influencing absorption or by modifying the compound. It has been demonstrated that there are quantitative differences in toxicity, depending upon whether compounds are administered with food or directly into the empty stomach *(9)*.

Most food absorption takes place in the small intestine. The gastrointestinal tract possesses specialized carrier systems for certain nutrients such as carbohydrates, amino acids, calcium, and sodium. Some xenobiotics use these routes of passage through the cells; others enter through passive diffusion.

Lipid-soluble organic acids and bases are absorbed by passive diffusion only in nonionized form. Equilibrium on both sides of the cell membrane is established only between the nonionized forms, according to the Henderson–Hasselbalch equation:[4]

$$pK_a = \text{pH} + \log(\text{nonionized/ionized}) \text{ for acids} \quad (2.2a)$$

$$pK_a = \text{pH} + \log(\text{ionized/nonionized}) \text{ for bases} \quad (2.2b)$$

Particles several nanometers in diameter can be absorbed from the gastrointestinal tract by pinocytosis and enter the circulation via the lymphatic system. (The lymphatic capillaries are much more permeable to large molecules, such as proteins, than are the blood capillaries.)

A percentage of xenobiotics absorbed in the gastrointestinal cells may be biotransformed before entering the circulatory system; the balance is transported as the parent compound. The absorbed compounds may enter the circulation either via the lymphatic system, which eventually drains into the bloodstream, or via the portal circulation, which carries them to the liver. The proportion of an orally ingested compound that reaches systemic circulation [called *bioavailability* (BA)] can be determined by the following equation:

$$\text{BA} = \text{AUC (oral)} / \text{AUC (iv)} \quad (2.3)$$

where AUC stands for the area under the curve (representing the plot of xenobiotic concentration in plasma versus time) from time 0 to infinity (Figure 2.6).

[4]The pH values of body fluids are as follows: gastric juice, 1.0; contents of the small intestine, 6.5; plasma and interstitial fluid, 7.4; urine, 6.8–7.8.

Translocation of Xenobiotics

To arrive at the receptor site in the target cell, the absorbed xenobiotic must be transported by the blood. The time to the onset of toxicity depends on how quickly plasma levels of the toxic compound may be achieved. Figure 2.6 presents a comparison of cocaine levels in plasma at different times after oral, intravenous, and respiratory administration of the toxin. The similarity between intravenous and respiratory routes is noteworthy. In contrast, the time to reach peak plasma concentration of the toxin is significantly longer after oral administration.

Chemicals enter and exit the circulation at the capillary subdivision of the blood vessels. The capillary walls consist of a single layer of flat epithelial cells, with pores of up to 0.003 μm in diameter between them *(10)*. Water-soluble compounds of up to 60,000 MW enter and exit the bloodstream by filtration through these pores. The velocity of diffusion decreases rapidly with increasing molecular radius.

Two opposing forces determine the flow direction of water and solutes between plasma and interstitial fluid: hydrostatic pressure and osmotic pressure. The difference between these forces on either side of the capillary membrane determines whether solutes enter or exit the capillaries. On the venous end of the blood vessels, the following condition applies:

$$(P_h - P_o)_{plasma} < (P_h - P_o)_{interstitial\ fluid} \qquad (2.4a)$$

where P_h is the hydrostatic pressure and P_o is the osmotic pressure. On the arterial end, the opposite applies:

$$(P_h - P_o)_{plasma} > (P_h - P_o)_{interstitial\ fluid} \qquad (2.4b)$$

Thus, solutes exit the capillaries and enter the interstitial fluid.

Lipophilic compounds diffuse easily through capillary walls. Their diffusion velocity is related to their lipid–water partition coefficient *(10)*.

The entry of a compound into the bloodstream does not necessarily ensure that it will arrive unchanged at its specific receptor. As mentioned before, xenobiotics absorbed from the gastrointestinal tract are carried by the portal vein to the liver. The liver has a very active xenobiotic-metabolizing system in which chemicals may or may not be altered before being released through hepatic veins into the general circulation. Alternatively, they may be excreted into the bile and returned to the gastrointestinal tract. From there they may be excreted, all or in part, or reabsorbed and carried back to the liver. This process is referred to as *enterohepatic circulation.*

Although blood plasma has only a limited metabolic capacity, mostly involving hydrolytic and transaminating enzymes, it may also contribute to the alteration of a chemical. Furthermore, some xenobiotics may be inactivated, at least temporarily, by being bound to plasma proteins.

Cellular Uptake

After leaving the bloodstream at the arterial end of the capillary system, the chemical has to reach the cell to interact with its receptor.

According to the fluid mosaic model (*11*) (Figure 2.7), the *plasma membrane* consists of two layers of lipids with their hydrophobic ends facing each other. Their hydrophilic ends face the aqueous environment of the interstitial fluid on one side and the interior of the cell on the other side. Two types of proteins are embedded into this structure. Peripheral proteins do not penetrate through the membrane and can be removed without disrupting its integrity. Integral proteins extend across the width of the membrane and are probably responsible for the transport of compounds across it.

It is believed that four mechanisms of passage through the cell membrane are possible. Water and small organic and inorganic molecules diffuse through relatively few very small (0.2–0.4 nm) pores in the membrane. Lipid-soluble molecules diffuse easily through the lipid bilayer in the direction of the concentration gradient. Certain molecules are transported across the membrane by specialized enzymatic processes that exhibit saturation kinetics. When this process is energy-independent and the transport occurs in the direction of the concentration gradient, it is called *facilitated diffusion.* If transport occurs against the concentration gradient and therefore requires energy input, it is called *active transport.* The mechanisms of cellular uptake and their characteristics are summarized in Table 2.2.

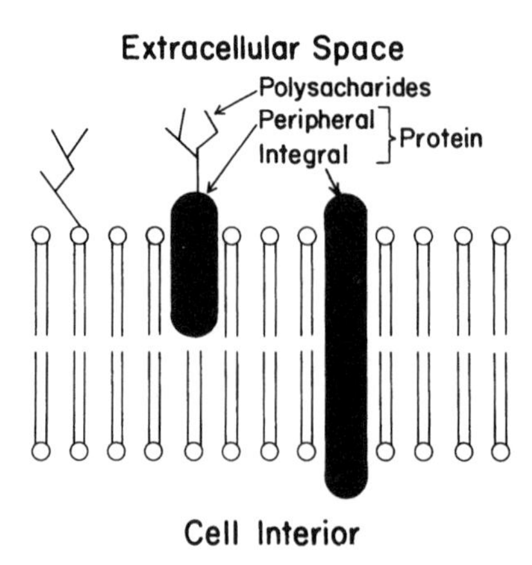

Figure 2.7. Schematic representation of a cell membrane, according to the fluid mosaic model.

Distribution Between Plasma and Tissue (Pharmacokinetics)

At the capillary subdivision, solutes are freely exchangeable between plasma and the interstitial fluid; thus the concentration of a xenobiotic in tissue is proportional to that of the free xenobiotic in plasma. The proportionality factor, a property of the compound, is expressed in terms of an apparent *volume of distribution* (VD). VD expresses what the volume of an animal (in liters) should be if a compound were equally distributed between plasma and tissue. In general, a large VD indicates easy uptake, whereas a small VD indicates poor uptake of a compound by the tissue. However, the true picture is complicated by the binding of a xenobiotic to plasma protein or its deposition in fat.

To determine VD, an animal is injected intravenously with the compound in question. The concentration of the compound in plasma is determined at frequent time intervals, and the logarithms of concentration are plotted versus time. The peak concentration occurs immediately after the injection. Concentration decreases with time through two processes: uptake by tissue, referred to as the α phase, and elimination from plasma, called the β phase. Elimination may include one or more of the following: urinary excretion, fecal excretion, excretion by exhalation, excretion with sweat, or metabolism. When the rate of distribution is of the same order of magnitude as the rate of elimination (but faster, as it usually is), a plot of the logarithm of concentration versus time yields a biphasic curve (Figure 2.8, A). This is referred to as a two-compartment open model *(12)*. The initial part of the plot is a composite curve resulting from two first-order reactions,[5] distribution and elimination, proceeding simultaneously. The tail end, appearing as a straight line, represents the elimination phase. To obtain the plot of α phase alone, the initial segment of the plot has to be resolved into its components. Resolution is achieved by extrapolating the line representing β phase to zero time and subsequently subtracting the data points on the extrapolated segment from the data points on the composite curve. The plot of resulting values versus time yields a straight line representing the α phase.

The volume of distribution can be calculated by using equation 2.5a:

$$\mathrm{VD} = \mathrm{Am}/\mathrm{AUC} \times k_{\beta} \tag{2.5a}$$

where Am is the total amount (g) of the compound administered, and AUC is the area under the curve from time 0 to infinity. AUC is expressed by

[5]The first-order reactions are characterized by a linear plot of the logarithm of concentration vs. time. The derivation of this plot is as follows. According to the first-order kinetics, $-dC/dt = kC$, where C is concentration, t is time, k is the rate constant, and $-dC/dt$ is the change of concentration over time. Rearrangement of the equation gives $-dC/C = k\,dt$, or $d \ln C = -k\,dt$. Integration yields the linear equation $\ln C = -kt + \text{constant}$, or $\log C = (-k/2.303)t + \text{constant}$.

Table 2.2. Mechanisms of Cellular Uptake and Their Characteristics

Mechanism	*Compound*	*Kinetics*	C_o *vs.* C_i	*Energy*
Diffusion through pores	<0.4 nm	$v_i = c_d A(C_o - C_i)/d$	$C_o > C_i$	None
Diffusion through lipid layer	Lipophilic	$v_i = c_d A(C_o - C_i)/d$	$C_o > C_i$	None
Facilitated diffusion	Miscellaneous	$v_i = v_m C_s/(K_M + C_s)$	$C_o > C_i$	None
Active transport	Miscellaneous	$v_i = v_m C_s/(K_M + C_s)$	$C_o > C_i$ or $C_o < C_i$	Required

SYMBOLS: C_o and C_i are concentration outside and inside the cell, respectively; v_i is uptake velocity (initial velocity); C_d is the diffusion coefficient; A and d are area and thickness of the membrane, respectively; v_m is maximal velocity; C_s is substrate concentration; and K_M is the Michaelis–Menten constant.

$$AUC = C_\alpha/k_\alpha + C_\beta/k_\beta \quad (2.5b)$$

where the reaction rates, k_α and k_β, are slopes of the a and β phase, respectively, multiplied by 2.303, and C_α and C_β are ordinate intercepts of the distribution and elimination phase, respectively (Figure 2.8, A).

Another case to consider is when the equilibration between tissue and plasma is much faster than the elimination of a compound. In such a case, a distribution equilibrium will be established promptly and no α phase will be apparent. A plot of the logarithm of concentration versus time will give a straight line, corresponding to the β phase (Figure 2.8, B). Because there is no α phase, AUC in equation 2.5b is reduced to C_β/k_β, and equation 2.5a becomes

$$VD = Am/C_\beta \quad (2.5c)$$

Because Am is given in mass units and C in concentration units, VD has dimensions of a volume and is always given in liters. The reaction rates, k_α and k_β, can be easily calculated from the relationship between the rate constant and the half-life, $t_{1/2}$, where $k = 0.693/t_{1/2}$.

An easy-to-use program called Lagran, which can be used with desktop or laptop computers, is now available for computation of pharmacokinetic parameters, such as k_β, $t_{1/2}$ of β phase, AUC, and VD (*13*). Table 2.3 shows the interpretation of the relationship between VD and body weight (BW).

The entry of toxins into the brain and central nervous system (CNS) is frequently more difficult than into other tissues. The function of this *blood–brain barrier* is related to impaired permeability of the blood capillaries in brain tissue, the necessity for toxins to penetrate glial cells, and the low protein content of the CNS interstitial fluid (*7*). Lipid solubility of a toxin is an important factor in the penetration of the blood–brain barrier.

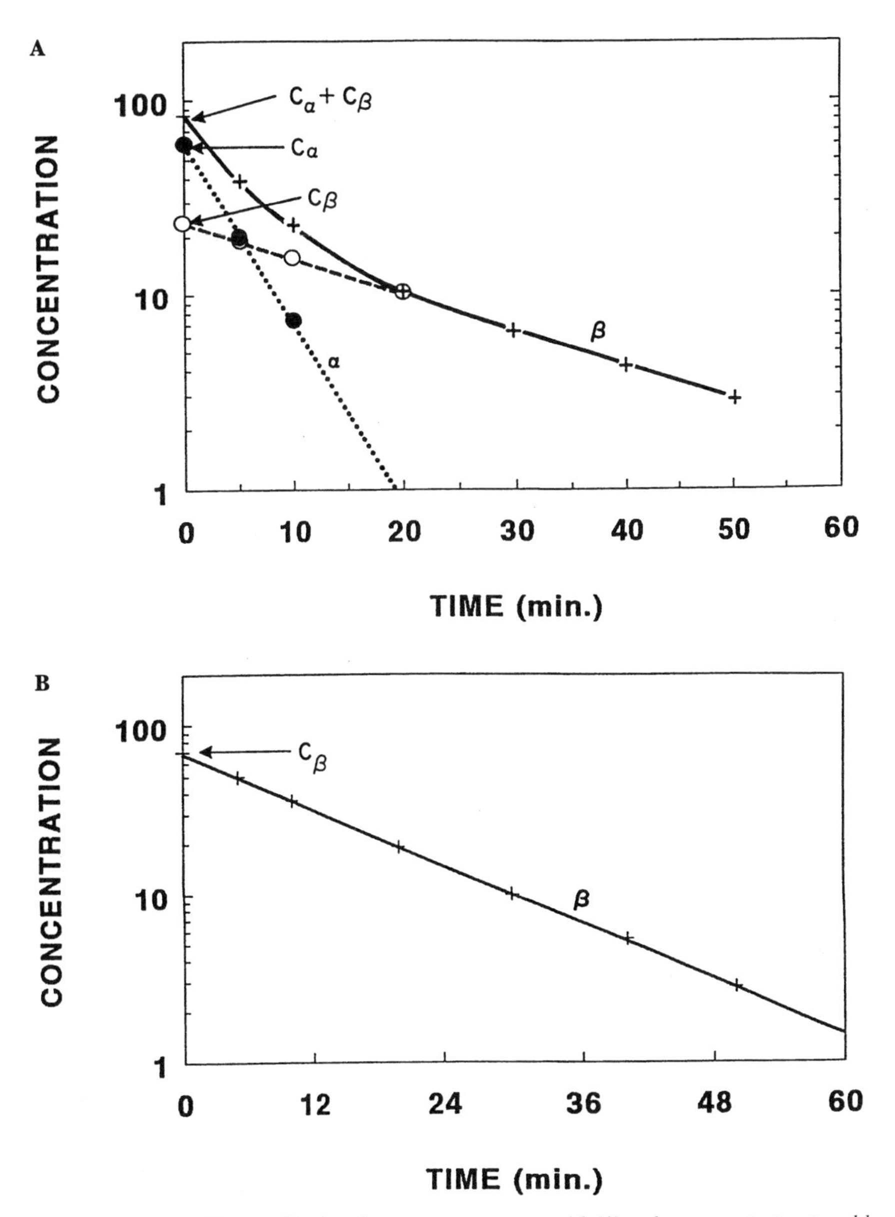

Figure 2.8. Pharmacokinetics of a two-compartment model (A) and a one-compartment model (B). Key: α, distribution phase; β, elimination phase.

Table 2.3. Interpretation of the Relationship Between Volume of Distribution and Body Weight

VD vs. BW	*Meaning*	*Possible Interpretation*
VD > BW	$C_t > C_p$	High lipophilicity or strong receptor binding or deposition in fat
VD < BW	$C_p > C_t$	Hydrophilic compounds with poor transport or binding to plasma protein

SYMBOLS: VD is volume of distribution; BW is body weight; C_p and C_t are concentrations in plasma and tissue, respectively.

Storage of Chemicals in the Body

An important factor to be considered is the capability of certain chemicals or their metabolites to be stored in the body. In general, a compound will accumulate in the body after repeated intake if its elimination or biotransformation is slower than the frequency of uptake. The best example of this phenomenon is the accumulation and persistence of alcohol in the blood after prolonged drinking. The human body metabolizes, on the average, one drink (a 12-oz can of beer, a 5-oz glass of wine, or one shot of 86-proof liquor) per hour. For a person weighing 140–160 pounds, the blood alcohol level rises 20 mg% per drink per hour. Accumulation of alcohol in blood after consuming one drink per hour or two drinks per hour, respectively, is shown in Figure 2.9. When two drinks per hour are consumed, the uptake of alcohol is much faster than its metabolism, so the alcohol levels build up rapidly. To maintain legally safe levels of alcohol in

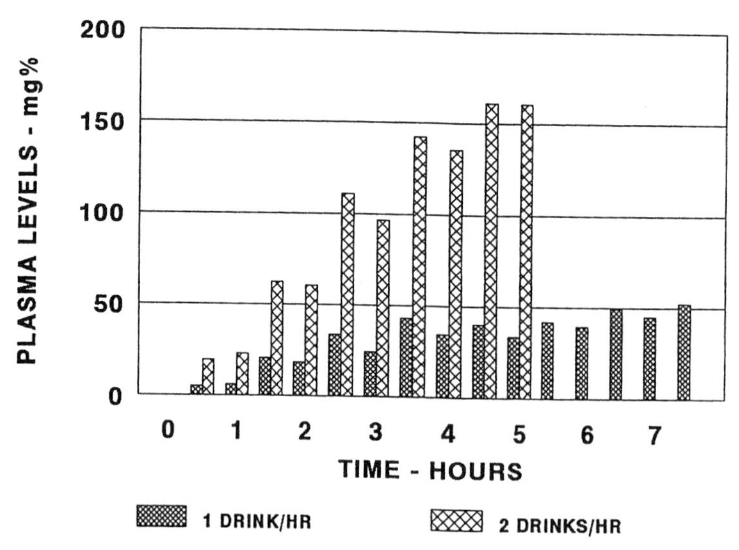

Figure 2.9. Accumulation of alcohol in humans after prolonged drinking. (1 drink = 1 oz of 100-proof whiskey.) (Based on data in reference 14.)

the blood while driving (less than 50 mg%), it is recommended that one consume no more than one drink per hour.

Some compounds are stored in the body in specific tissues. Such storage effectively removes the material from circulation and thus decreases the toxicity of the compound. Repeated doses of a toxic substance may be taken up and subsequently stored without apparent toxicity until the storage receptors become saturated; then toxicity suddenly occurs. In some cases, the stored compound may be displaced from its storage receptor by another compound that has an affinity for the same receptor. Examples of this phenomenon are the displacement of antidiabetic sulfonylureas by sulfonamides and the ability of antimalarial drugs such as quinacrine (Atabrine) and primaquine to displace each other (*15*) (Chart 2.1). A special danger in such cases is that compounds may have escaped detoxifying metabolism while stored in the body, and that their toxicity may be potent and prolonged when they are released.

Lipophilic compounds (such as halogenated hydrocarbons, DDT (dichlorodiphenyltrichloroethane), PCBs (polychlorinated biphenyls), etc.) may be stored in fat without apparent harm to the exposed organism. However, these toxins tend to accumulate in the food chain. Eventually the storage capacity of an organism at the end of the food chain may be exceeded, and the toxin may be re-

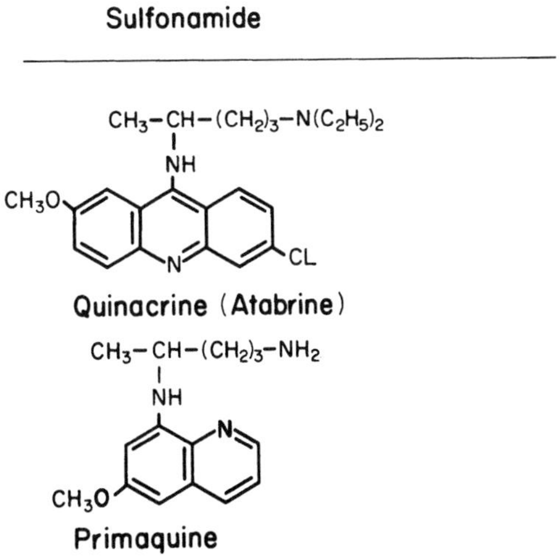

Sulfonylurea (Tolbutamide)

Sulfonamide

Quinacrine (Atabrine)

Primaquine

Chart 2.1. Chemical structures of sulfonylurea vs. sulfonamide, and quinacrine vs. primaquine.

leased into circulation and into the milk. Another danger is that during a period of starvation, as frequently happens to wild animals in winter, fat deposits are mobilized for energy. Stored toxins are then released, causing sickness or death.

In addition to possible lasting inactivation of xenobiotics due to storage in various tissues, living organisms are partially protected by their reserve functional capacity. Some organs (such as the lungs, liver, and kidney) may withstand a certain amount of injury without any demonstrable symptoms. In such cases, the injury can be demonstrated only histologically.

References

1. Bliss, C. I. *Ann. Appl. Biol.* **1935,** *22,* 134.
2. Lichfield, J. T., Jr.; Wilcoxon, F. *J. Pharmacol. Exp. Ther.* **1949,** *96,* 99.
3. The following programs are applicable: PCNONLIN (1986, Statistical Consultants, Inc., Carl M. Metzler and Daniel L. Weiner); SAS procedure PROBIT (SAS Institute, Inc., SAS Circle, Box 8000, Cary, NC 27512).
4. Ehrlich, P. *Lancet* **1913,** *2,* 445.
5. Baxter, J. D.; Forsham, P. H. *Am. J. Med.* **1972,** *53,* 573.
6. Loomis, T. A. *Essentials of Toxicology;* Lea & Febiger: Philadelphia, PA, 1978; Chapter 5, p 68.
7. Klaassen, C. D. In *Cassarett and Doull's Toxicology;* Klaassen, C. D.; Amdur, M. O.; Doull, J., Eds.; MacMillan: New York, 1986; Chapter 3, p 33.
8. *Review of Physiological Chemistry;* Harper, H. A.; Rodwell, V. W.; Mayers, P. A., Eds.; Lange Medical: Los Altos, CA, 1979; Chapter 15, p 218.
9. Worden, A. N.; Harper, K. H. *Proc. Eur. Soc. Study Drug Toxic.* **1963,** *2,* 15.
10. Goldstein, A.; Aronow, L.; Kalman, S. M. *Principles of Drug Action;* John Wiley: New York, 1974; Chapter 2, p 129.
11. *Review of Physiological Chemistry;* Harper, H. A.; Rodwell, V. W.; Mayers, P. A., Eds.; Lange Medical: Los Altos, CA, 1979; Chapter 9, p 112.
12. Greenblatt, D. J.; Koch-Weser, J. *N. Engl. J. Med.* **1975,** Oct. 2, 702.
13. Gibaldi, M. In *Biopharmaceutics and Clinical Pharmacokinetics;* Lea & Febiger: Philadelphia, PA, 1977; Chapter 1, p 1.
14. Forney, R. B.; Hughes, F. W. *Clin. Pharm. Ther.* **1963,** *4,* 619.
15. Loomis, T. A. *Essentials of Toxicology;* Lee & Febiger: Philadelphia, PA, 1978; Chapter 3, p 36.

3 Metabolism of Xenobiotics

Phases of Metabolism

The action of most xenobiotics ends in either excretion or metabolic inactivation. Some compounds, on the other hand, require metabolic activation before they can exert any biological action. In most cases these biotransformations, activations as well as inactivations, are carried out by specialized enzyme systems. The essential role of these enzymes is to facilitate elimination of xenobiotics. Water-soluble compounds usually do not need to be metabolized, as they can be excreted in their original forms. Lipophilic compounds can be disposed of through biliary excretion, or they may undergo metabolism to become more polar and thus more water-soluble so that they can be disposed of through the kidneys.

The metabolism of xenobiotics is usually carried out in two phases. Phase 1 involves oxidative reactions in most cases, whereas phase 2 involves conjugation (combination) with highly water-soluble moieties. Occasionally the products of biotransformation are unstable and decompose to release highly reactive compounds such as free radicals, strong electrophiles, or highly stressed three-member rings (epoxides, azaridines, episulfides, and diazomethane; Chart 3.1) that have a tendency toward nucleophilic ring opening.

For order to be retained within the cells, the chemical reactions have to occur through enzymatic processes in which the substrate is activated while bound to the enzyme. Only after the desired reaction takes place is a stable product released. Freely roaming reactive compounds are not welcome in a living organism because they react randomly with macromolecules such as DNA, RNA, and proteins. Alteration of DNA leads to faulty replication and transcription. Alteration

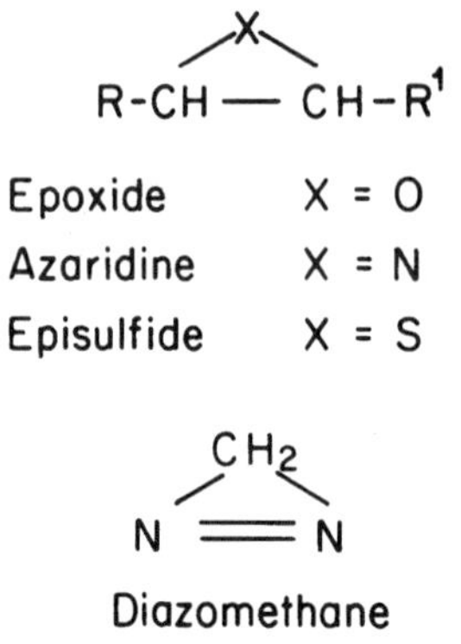

$$R\text{-}CH \overset{X}{—} CH\text{-}R^1$$

Epoxide	X = O
Azaridine	X = N
Episulfide	X = S

$$N \overset{CH_2}{\equiv} N$$

Diazomethane

Chart 3.1. Unstable three-member rings.

of RNA causes faulty messages that, in turn, lead to the synthesis of abnormal proteins and thus alter enzymatic and regulatory activity.

Phase 1 Biotransformations

Phase 1 processes are carried out by a series of similar enzymes (commonly designated as mixed-function monooxidases) or cytochrome P-450.[1] The basic reactions catalyzed by cytochrome P-450 enzymes involve introduction of oxygen into a molecule. In most cases the oxygen is retained, but sometimes it is removed from the end product. The oxygen carrier is a prosthetic group containing porphyrin-bound iron (Scheme 3.1, center). The overall reaction catalyzed by these enzymes is hydroxylation.

$$RH + O_2 + H_2 \rightarrow ROH + H_2O \tag{3.1}$$

Its flow diagram is presented in Scheme 3.1 (*1*).

Although some authors propose slightly different schemes, the crux of the matter is that two single electrons are transferred to the P-450–substrate complex in two separate reactions. These electrons originate from reduced nicotinamine–adenine dinucleotide phosphate (NADPH). The reductions carried out by NADPH involve the transfer of a hydride ion (i.e., a hydrogen atom carrying two electrons) (Scheme 3.2) (*2*). Because both electrons would be transferred simultaneously, a step-down mechanism is needed for transfer of a single electron. This single-electron transfer is achieved by coupling cytochrome P-450 with another enzyme called cytochrome P-450 reductase, which has two prosthetic

[1]The name P-450 comes from the observation that, when exposed to CO, the enzyme exhibits a characteristic light absorption with a maximum at 450 nm.

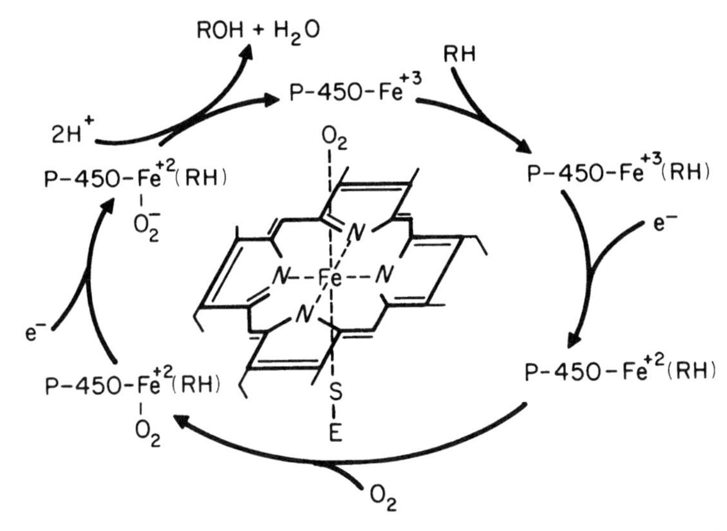

Scheme 3.1. Outside: suggested sequence of hydroxylation reactions carried out by cytochrome P-450. Inside: schematic presentation of the configuration of the P-450 prosthetic group.

groups: flavin mononucleotide (FMN) and flavin–adenine dinucleotide (FAD) (Scheme 3.3). Both FMN and FAD are capable of a two-stage single-electron transfer involving a semiquinone free-radical intermediate (*2, 3*). The electron flow between NADPH and the substrate, via cytochrome P-450 reductase and cytochrome P-450, is presented in Scheme 3.4.

The reactions catalyzed by cytochrome P-450 are listed in Chart 3.2. The last three reactions in Chart 3.2 deserve comment. They involve reductive, rather than oxidative, transformation. In this case the substrate, not oxygen, accepts electrons and is reduced (*4*).

Both enzymes, cytochrome P-450 and cytochrome P-450 reductase, are bound inside the cell to the endoplasmic reticulum (ER). The ER, a network of

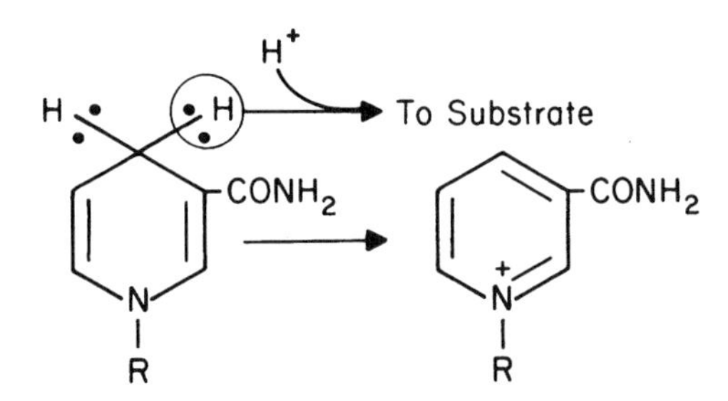

Scheme 3.2. Mechanism of reduction by NADPH, which is itself oxidized. R is ADP-(2′-phosphate)ribosyl.

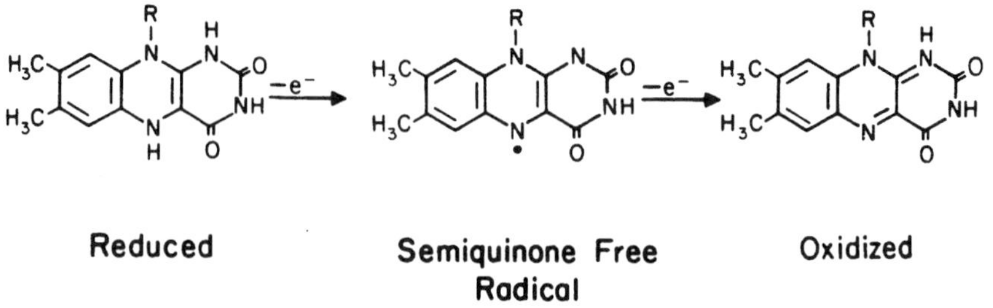

Reduced — Semiquinone Free Radical — Oxidized

Scheme 3.3. Mechanism of reduction of 6,7-dimethylisoalloxazine by single-electron transfer. R is D*-1′-ribityl-5′-phosphate (in FMN) or ADP-*D*-1′-ribityl (in FAD).*

membranes within the cell, is continuous with the outer nuclear membrane. When cells are homogenized, the ER is degraded to small vesicles called microsomes, which can be isolated by fractional centrifugation. Cytochrome P-450 can be solubilized by treatment of microsomal preparation with sodium dodecyl sulfate (*5*). Both cytochrome P-450 and its reductase are predominantly located in the liver. However, measurable quantities of these enzymes are also found in the kidney, lungs, intestine, brain, and skin (*6*).

Endoplasmic reticulum contains still another oxidizing enzyme system that competes with cytochrome P-450 for oxidation of amines. Enzymes of this group, historically referred to as mixed-function amine oxidases, contain FAD as a prosthetic group. Although it was originally thought that this system was specific for amines only, it now appears that it also metabolizes sulfur-containing xenobiotics. Mixed-function amine oxidases convert primary amines into hydroxylamines and oximes, secondary amines into hydroxylamines and nitrones, and tertiary amines into amine oxides. They also oxidize thioethers to sulfoxides and sulfones and thiols to RS–SR compounds (*4*).

Mammalian systems also contain soluble xenobiotic-reducing enzymes that carry out the reduction of carbonyl, nitro, and azo groups, and esterases that hydrolyze esters and amides to the corresponding carboxylic acids and alcohols or amines, respectively. An in-depth treatment of soluble xenobiotic-metabolizing enzymes is available in *Burger's Medicinal Chemistry* (*7*).

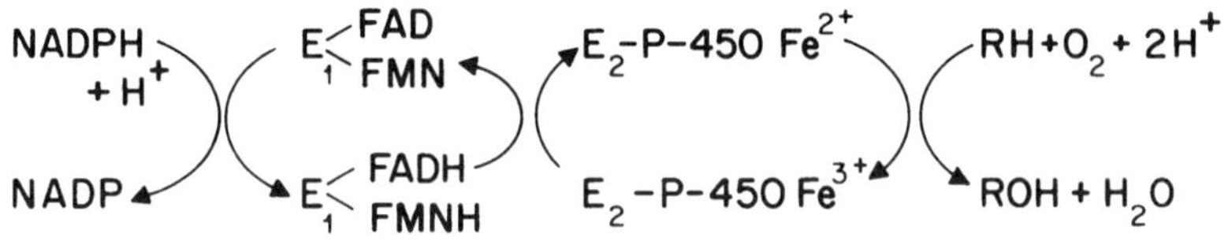

NADPH + H^+ → NADP; E_1<FAD/FMN ⇄ E_1<FADH/FMNH; E_2-P-450 Fe^{2+} ⇄ E_2-P-450 Fe^{3+}; RH + O_2 + $2H^+$ → ROH + H_2O

Scheme 3.4. Electron flow between NADPH and a substrate in the cytochrome P-450 catalyzed reactions. E_1 *is cytochrome P-450 reductase apoenzyme;* E_2 *is cytochrome P-450 apoenzyme.*

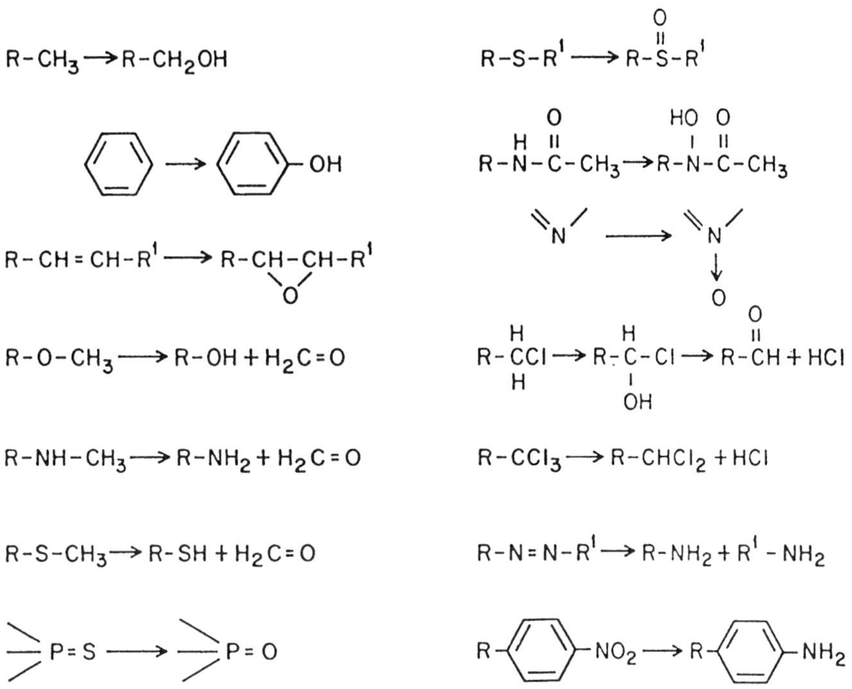

$$R{-}CH_3 \rightarrow R{-}CH_2OH$$

$$C_6H_6 \rightarrow C_6H_5{-}OH$$

$$R{-}CH{=}CH{-}R' \rightarrow R{-}\underset{\diagdown O \diagup}{CH{-}CH}{-}R'$$

$$R{-}O{-}CH_3 \rightarrow R{-}OH + H_2C{=}O$$

$$R{-}NH{-}CH_3 \rightarrow R{-}NH_2 + H_2C{=}O$$

$$R{-}S{-}CH_3 \rightarrow R{-}SH + H_2C{=}O$$

$$>P{=}S \rightarrow >P{=}O$$

$$R{-}S{-}R' \rightarrow R{-}\overset{O}{\overset{\|}{S}}{-}R'$$

$$R{-}\overset{H}{N}{-}\overset{O}{\overset{\|}{C}}{-}CH_3 \rightarrow R{-}\overset{HO}{\overset{|}{N}}{-}\overset{O}{\overset{\|}{C}}{-}CH_3$$

$$=N{-} \rightarrow =\underset{\underset{O}{\downarrow}}{N}{-}$$

$$R{-}\overset{H}{\underset{H}{C}}Cl \rightarrow R{-}\overset{H}{\underset{\underset{OH}{|}}{C}}{-}Cl \rightarrow R{-}\overset{O}{\overset{\|}{C}}H + HCl$$

$$R{-}CCl_3 \rightarrow R{-}CHCl_2 + HCl$$

$$R{-}N{=}N{-}R' \rightarrow R{-}NH_2 + R'{-}NH_2$$

$$R{-}C_6H_4{-}NO_2 \rightarrow R{-}C_6H_4{-}NH_2$$

Chart 3.2. Reactions catalyzed by cytochrome P-450.

Disposition of Epoxides

Epoxides are frequent intermediates or end products of cytochrome P-450 catalyzed reactions. Because they are inherently unstable, they are liable to react in the cell with macromolecules (specifically with DNA); these reactions lead to mutations or carcinogenic changes. Whether they react with macromolecules or not depends on the stability of the epoxide and its suitability as a substrate for epoxide-metabolizing enzymes. Extremely unstable epoxides, with a half-life of a couple of minutes or less, do not represent much of a danger because they will be decomposed before they have an opportunity to react with DNA. The extremely stable epoxides will react with DNA only slowly, if at all, and will probably be transformed enzymatically to harmless compounds.

Two enzymatic and two nonenzymatic reactions dispose of epoxides. An enzyme bound to ER called epoxide hydrolase (also called epoxide hydrase) converts epoxides to *trans*-diols (Chart 3.3, A). Then the *trans*-diols can be conjugated as described in the following section. The other reaction involves glutathione and an enzyme, glutathione *S*-transferase (Chart 3.3, B). The end product, a *trans*-(hydroxy)glutathione conjugate, is eventually split to a corresponding derivative of mercapturic acid.

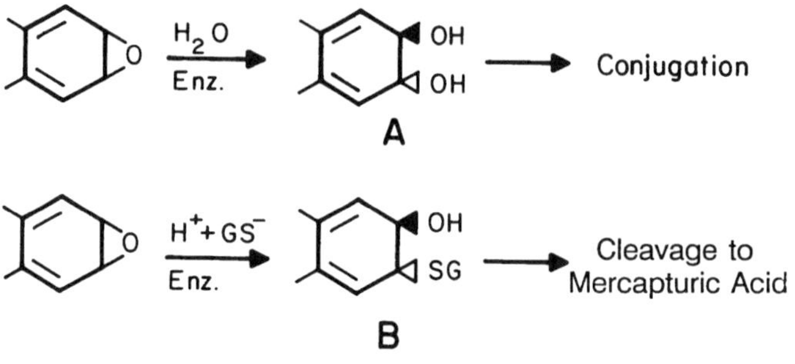

Chart 3.3. *Enzymatic disposition of epoxides by epoxide hydrolase (A) and glutathione transferase (B). Black triangles indicate valences directed above the plane; white triangles indicate valences directed below the plane.*

The two nonenzymatic reactions are the S_N2-type addition of water,[2] resulting in the formation of a *trans*-diol, and the S_N1-type rearrangement referred to as the NIH shift (*8*), resulting in the formation of a phenol (or arenol) (Scheme 3.5).

Conjugations (Phase 2)

The lipophilic compounds that are converted by phase 1 processes into polar, somewhat more hydrophilic, products may undergo further transformation into highly water-soluble materials by different types of conjugations. From the chemical point of view, conjugations may be divided into *electrophilic conjugations* (the conjugating agent is an electrophile) and *nucleophilic conjugations* (the conjugating agent is a nucleophile). Electrophilic conjugations involve glucuronide, sulfate, acetate, glycine, glutamine, and methyl transfer; the first three types are the most common. Nucleophilic conjugation involves glutathione only.

Electrophilic conjugations proceed through the S_N2 mechanism, which is

---

[2]First-order nucleophilic substitution (S_N1) proceeds as follows:

$$RCl \rightarrow R^+ + Cl^- \quad \text{(slow)}$$

$$R^+ + X^- \rightarrow RX \quad \text{(fast)}$$

$$RCl + X^- \rightarrow RX + Cl^-$$

Because the first step is rate-limiting, the reaction exhibits first-order kinetics. Second-order nucleophilic substitution (S_N2) proceeds as follows:

$$X{:} + R{:}Cl \rightarrow X{:}R + {:}Cl^- \quad \text{(slow)}$$

S_N2 reactions proceed with the reversal of the stereo configuration.

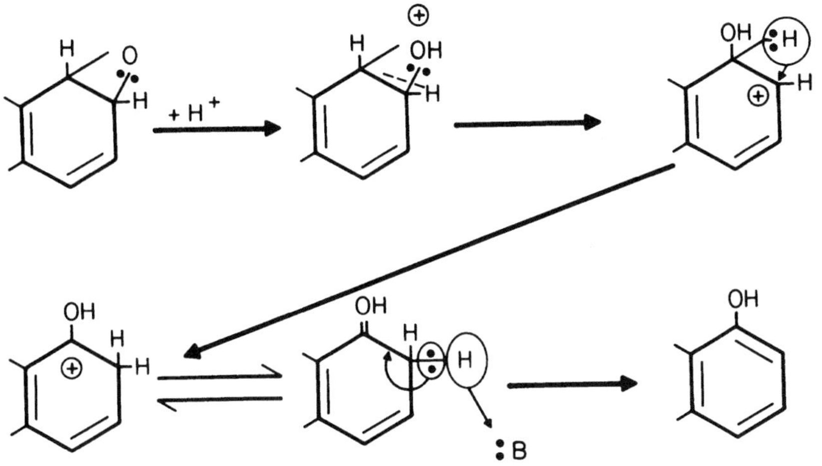

Scheme 3.5. *Conversion of an epoxide to an arenol by NIH rearrangement; :B stands for base.*

characterized by a stereospecific attack of the xenobiotic on the electrophilic atom of the conjugating agent as shown in eq 3.2.

$$R\text{–}X\!: + {}^{+}Y\!:\!Z^{-} \rightarrow R\text{–}X\text{–}Y + :\!Z \tag{3.2}$$

where X is O, N, or S; R–X is a nucleophilic xenobiotic; and Y:Z is an electrophilic conjugating agent.

Glucuronidation is carried out by the ER-bound glucuronyl transferase, an enzyme of 200,000–300,000 molecular weight, consisting of 3–6% glycoprotein. The substrates are phenols, alcohols, carboxylic acids, amines, hydroxylamines, and mercaptans. The glucuronic acid group is donated by uridine diphosphate glucuronic acid (UDPGA). This cofactor is formed from uridine diphosphate glucose (UDPG) by oxidation. The structure of the cofactor and the scheme of the reaction are presented in Chart 3.4. The α configuration on the 1′ carbon of the cofactor is reversed to β in the conjugated product. The glucuronide conjugates are hydrolyzed to aglycons by β-glucuronidase, an enzyme occurring in lysosomes and in intestinal bacteria.

Phenols (arenols), steroids, and *N*-hydroxy species undergo conjugation with sulfate. The enzymes in these reactions are cytoplasmic sulfotransferases, and the cofactor is a mixed anhydride between sulfuric and phosphoric acid, 3′-phosphoadenosine 5′-phosphosulfate (PAPS). The sulfate conjugation is presented in Scheme 3.6. The sulfate conjugates are sensitive to attack by sulfatases, which split them back to the starting materials.

Conjugation with acetate is restricted to amines and is carried out by a cytoplasmic enzyme, *N*-acetyltransferase. Oxygen and sulfur acetylation occurs in

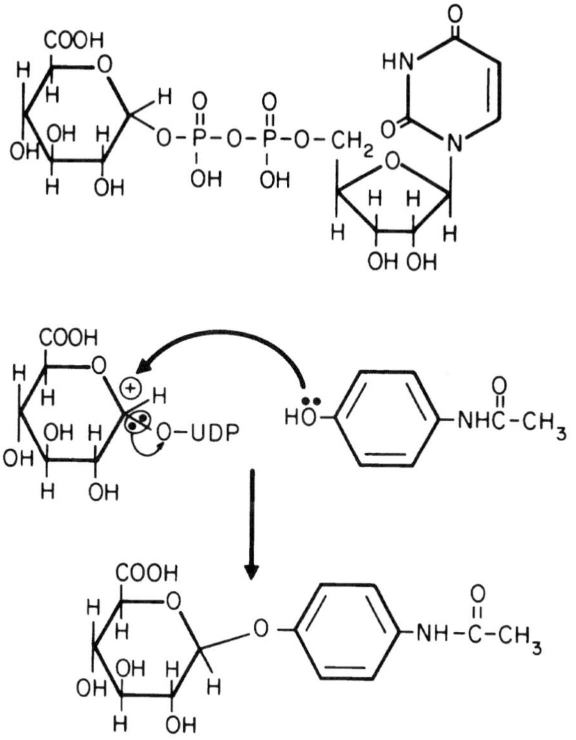

Chart 3.4. Uridine 5′-diphospho-D-glucuronic acid (UDPGA) (top). Mechanism of conjugation of p-*hydroxyacetylalanine with glucuronic acid (bottom).*

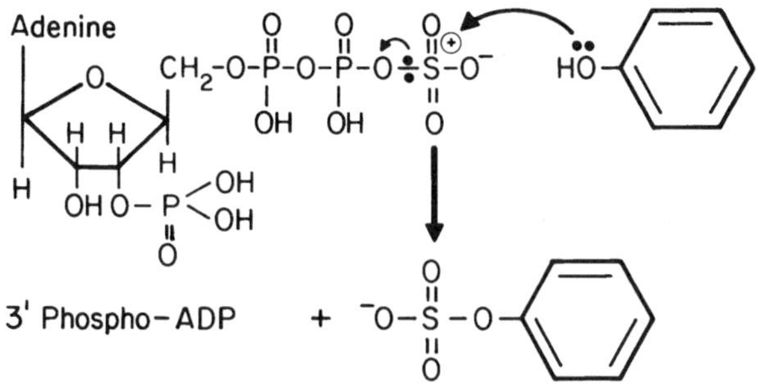

Scheme 3.6. Mechanism of the reaction of 3′-phosphoadenosine 5′-phosphosulfate (PAPS) with phenol.

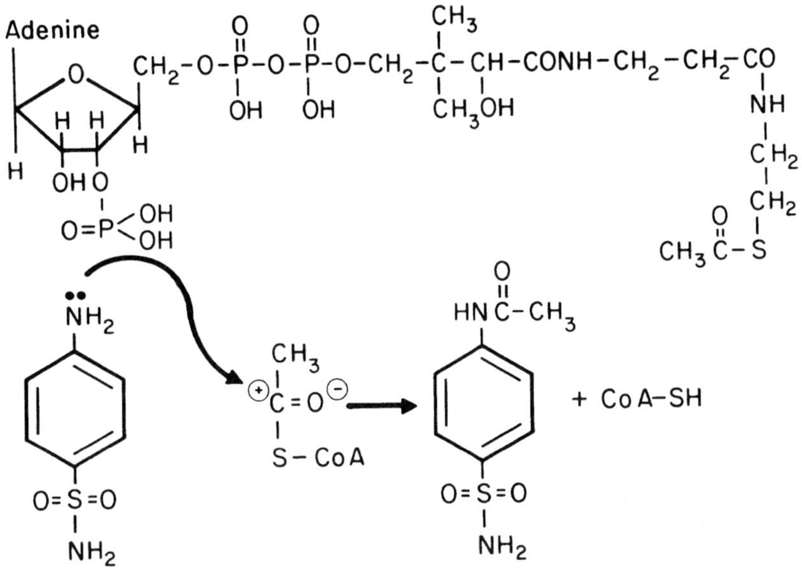

Chart 3.5. Structure of acetyl coenzyme A (top). Reaction of sulfanilamide with acetyl coenzyme A (bottom).

normal primary metabolism but not in the metabolism of xenobiotics. The acetyl donor is *S*-acetyl coenzyme A (Chart 3.5).

Conjugation with amino acids (glycine and glutamine) is carried out by mitochondrial enzymes (*N*-acetyltransferases), and is restricted to carboxylic acids, especially aromatic ones. The carboxylic acid requires activation with adenosine 5′-triphosphate (ATP) and coenzyme A before being conjugated (*7*). Methylations are catalyzed by a cytoplasmic enzyme, methyltransferase, which utilizes *S*-adenosylmethionine (SAM) as a cofactor.

Glutathione

Glutathione is a γ-glutamyl–cysteinyl–glycine tripeptide (Structure 3.1) that occurs in most tissues, but especially in the liver (100 g of liver tissue contains 170 mg of reduced glutathione). Glutathione plays many important roles in cell metabolism. As far as the metabolism of xenobiotics is concerned, it is involved in enzymatic as well as nonenzymatic reactions. Nonenzymatically, it acts as a low-molecular-weight scavenger of reactive electrophilic xenobiotics. As long as its concentration remains high enough, it is likely to outcompete DNA, RNA, and proteins in capturing electrophiles.

Enzymatic reactions involving glutathione are catalyzed by a series of

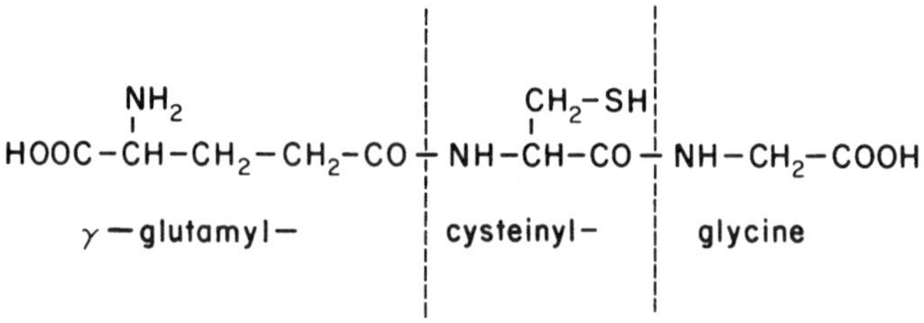

$$HOOC-CH(NH_2)-CH_2-CH_2-CO \;|\; NH-CH(CH_2-SH)-CO \;|\; NH-CH_2-COOH$$

γ-glutamyl- | cysteinyl- | glycine

Structure 3.1. Glutathione.

isozymes, known under the common name of glutathione *S*-transferase, with broad specificity for electrophilic substrates. (*Isozymes* are enzymes with different chemical compositions but performing the same catalytic functions.) At least five isozymes together comprise 10% of soluble liver protein. Glutathione *S*-transferase catalyzes the reaction between glutathione and aliphatic and aromatic epoxides, as well as aromatic and aliphatic halides (Chart 3.6). The conjugated product is further hydrolyzed with the removal of glutamyl and glycyl residues, followed by *N*-acetylation by acetyltransferase. The end product is mercapturic acid, which is highly water-soluble and easily excreted in urine.

Glutathione *S*-transferase also catalyzes reactions of organic nitrates with glutathione. These reactions, however, do not proceed through the mercapturic acid pathway. They lead instead to reduction of the organic nitrate to inorganic nitrite and oxidation of glutathione to its S–S dimer (Chart 3.7). This reaction is

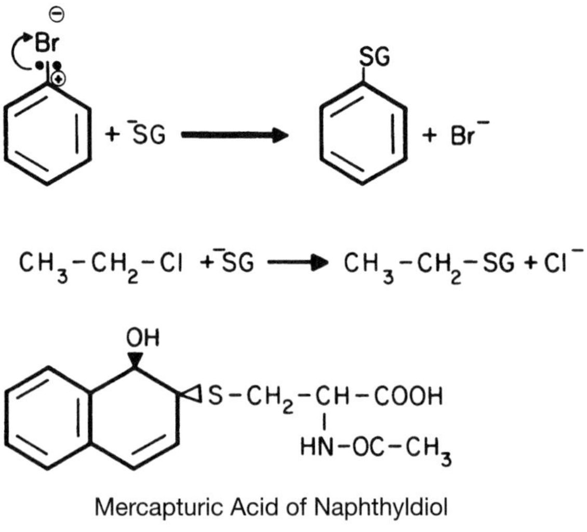

$$C_6H_5Br + {}^{-}SG \longrightarrow C_6H_5SG + Br^-$$

$$CH_3-CH_2-Cl + {}^{-}SG \longrightarrow CH_3-CH_2-SG + Cl^-$$

OH

S-CH₂-CH-COOH
HN-OC-CH₃

Mercapturic Acid of Naphthyldiol

Chart 3.6. Mechanism of the reaction between aromatic (top) and aliphatic (middle) halides and glutathione. Structure of mercapturic acid (bottom).

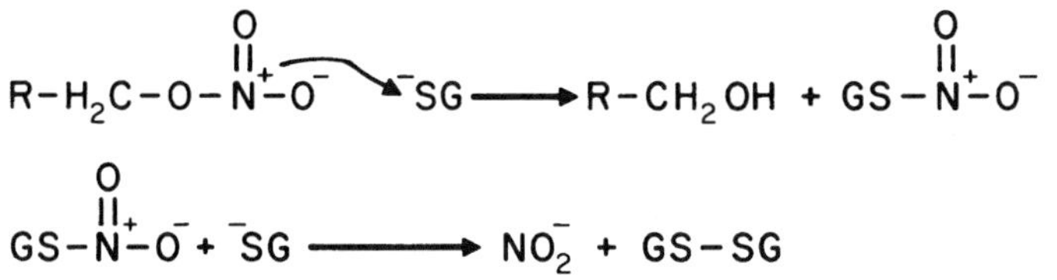

$$R{-}H_2C{-}O{-}\overset{O}{\overset{\|}{N^+}}{-}O^- \; {}^-SG \longrightarrow R{-}CH_2OH + GS{-}\overset{O}{\overset{\|}{N^+}}{-}O^-$$

$$GS{-}\overset{O}{\overset{\|}{N^+}}{-}O^- + {}^-SG \longrightarrow NO_2^- + GS{-}SG$$

Chart 3.7. Mechanism of the reaction between an organic nitro compound and glutathione.

responsible for the rapid inactivation of nitroglycerin, a vasodilator used in the treatment of myocardial ischemia. The nitrites formed in such reactions may interact with amines and thus lead to the formation of carcinogenic nitrosamines.

Another reaction that does not proceed through the mercapturic acid pathway is catalyzed by glutathione peroxidase. In this reaction, highly reactive peroxides are reduced to alcohols, whereas glutathione is oxidized.

The importance of glutathione as a detoxifying agent is obvious. Its depletion, either by genetic predisposition or by persistent heavy loads of xenobiotics, predisposes to hepatotoxicity and mutagenicity by other external agents. Some examples of compounds that cause depletion of liver glutathione in rats are given in Table 3.1.

Induction and Inhibition of P-450 Isozymes

Cytochrome P-450 is a mixture of at least 10 isozymes. Although all of them perform essentially the same catalytic functions and utilize the same substrates, they exhibit quantitative substrate preferences. They also vary in their molecular weight and in their electrophoretic mobility. In addition, these isozymes differ in their response to specific inducers.

Enzyme induction is a phenomenon in which a xenobiotic causes an increase in the biosynthesis of an enzyme. It was first observed in studies involving *N*-

Table 3.1. Compounds That Cause Depletion of Liver Glutathione

Compound	*Dose (mg/kg)*	*Time After Dose (h)*	*Remaining GSH[a] (percent of control)*
Methyl iodide	70	2	17
Benzyl chloride	500	6	18
Naphthalene	500	6	10

[a]GSH is reduced glutathione.

demethylation of aminoazo dyes in rat livers. Dietary factors, or pretreatment of the animals with various chemicals, enhanced the liver's ability to demethylate the dyes (*9*). The phenomenon of induction proceeds via a cytoplasmic receptor–inducer complex (*10*), which in turn interacts with an appropriate gene to cause an increase in production of the enzyme.

Laboratory Studies with P-450 Inducers

Haugen et al. (*11*) purified cytochrome P-450 from rabbit liver microsomes and presented evidence for the occurrence of at least four forms. The mixture of isozymes could be separated by gel electrophoresis into distinct bands. Two of them were purified to homogeneity and were designated as LM_2 and LM_4 (LM stands for liver microsomes, and the subscript designates the sequential number of the band). LM_2, which has been shown to be inducible by phenobarbital (PB), has a molecular weight of 50,000. LM_2 is inducible by β-naphthoflavone and has a molecular weight of 54,000 (Chart 3.8). LM_4 can also be induced by 3-methylcholanthrene (3MC) and has been shown to have substrate preference for aromatic hydrocarbons (*12*); it is therefore referred to as aromatic hydrocarbon hydroxylase (AHH). Furthermore, when combined with CO, this isozyme's peak light absorption is at 448 nm, not at 450 nm as is the case with the other isozymes.

In addition to the increase in the activity of specific isozymes, pretreatment of animals with PB causes a marked proliferation of smooth endoplasmic reticulum and an increase in liver weight. Pretreatment with 3MC, on the other hand, causes liver weight gain but has only a slight effect on endoplasmic reticulum. PB

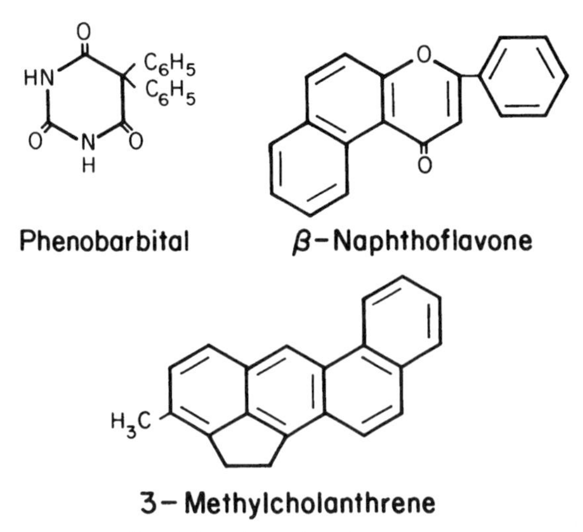

Chart 3.8. Inducers of cytochrome P-450.

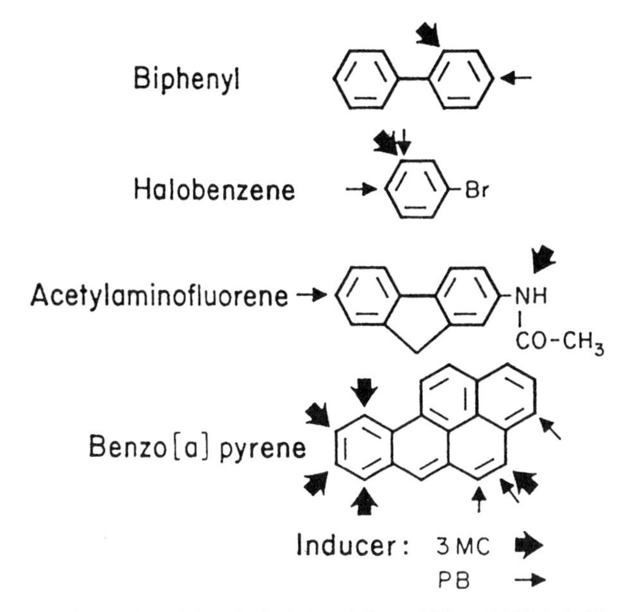

Chart 3.9. Comparison of site-selective hydroxylating activities of 3MC-inducible vs. PB-inducible cytochrome P-450.

does not induce extrahepatic cytochrome P-450, whereas 3MC induces hepatic as well as extrahepatic P-450 enzymes (*6*).

Cytochrome P-450 isozymes differ not only in their substrate preference; they also exhibit site- and stereoselective activities. The site selectivity is illustrated in Chart 3.9 (*12*).

Hydroxylation of the rodenticide warfarin (Structure 3.2) is a good example of stereoselective activity (*5*). Because of the asymmetric carbon (marked with an asterisk), warfarin has two stereoisomers, (*R*) and (*S*). Table 3.2 shows the relative amounts of warfarin hydroxylation (*R/S*) after induction of P-450 in rats with 3MC and PB.

*Structure 3.2. Warfarin; * indicates asymmetric carbon.*

Table 3.2. Stereoselective Hydroxylation of the (*R*) and (*S*) Isomers of Warfarin

Inducer	*6*	*7*	*8*	*4'*	*Benzylic*
3MC	+305/+165	−50/0	+1040/+315	−60/−60	−65/−10
PB	+95/+95	+130/+295	+110/+200	+135/+75	+50/+750

NOTE: The values shown are the percent increase (+) or decrease (−) of hydroxylation at the indicated positions, as compared to untreated control after induction of P-450 in rats with 3MC and PB. The first value is for the (*R*) isomer; the second is for the (*S*) isomer.
SOURCE: Adapted from data in reference 5.

As will become evident later in this chapter, knowledge of site selectivity is vital in assessing the risk of exposure to potential mutagens and carcinogens.

Inhibitors

Inhibitors of cytochrome P-450 can be reversible or irreversible. Frequently the *reversible inhibitors* are slowly metabolized substrates of P-450. They occupy the active site of the enzyme and thus retard the processing of other xenobiotics. A typical example of a reversible inhibitor is 2-diethylaminoethyl 2,2-diphenylvalerate, known as SKF 525-A (Structure 3.3).

This compound is bound relatively tightly to LM_2 isozyme (inhibition constant $K_i = 10^{-6}$) and is slowly metabolized by hydroxylation of the benzene rings and dealkylation of nitrogen. Another example of a reversible inhibitor is α-naphthoflavone (Structure 3.4). A similar compound, β-naphthoflavone, is an inducer of LM_4.

An example of an irreversible inhibitor of P-450 is carbon tetrachloride (CCl_4). It acts by causing peroxidation of lipids, which in turn destroys cell membrane integrity, with a subsequent loss of P-450.

The effect of an inhibitor can be assessed by measuring the increase in sleeping time of animals anesthetized with hexobarbital. Because hexobarbital is inactivated by cytochrome P-450, inhibitors of P-450 prolong sleeping time, whereas inducers shorten it.

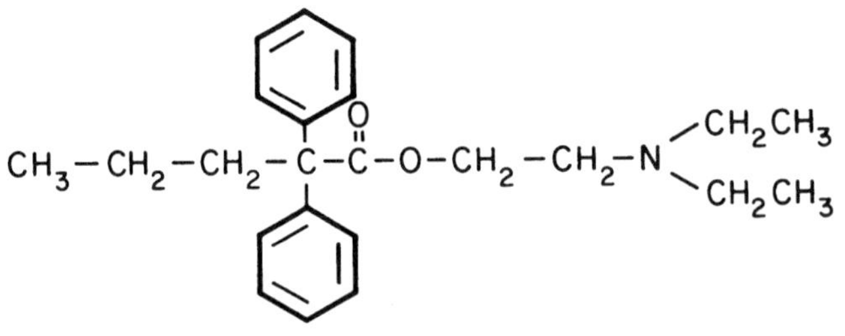

$CH_3-CH_2-CH_2-C(C_6H_5)_2-C(=O)-O-CH_2-CH_2-N(CH_2CH_3)_2$

Structure 3.3. 2-Diethylaminoethyl 2,2-diphenylvalerate.

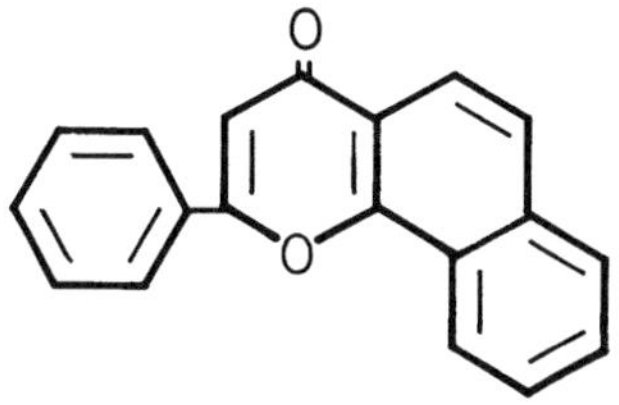

Structure 3.4. α-Naphthoflavone.

Environmental Inducers of P-450

A number of environmental agents affect cytochrome P-450. It has been reported (*13*) that the insecticide DDT (1,1,1-trichloro-2,2-bis(*p*-chlorophenyl)ethane) (Chart 3.10, top), when fed to rats at 50 mg kg^{-1} day^{-1}, decreased the sleeping time of animals anesthetized with hexobarbital. This change indicates induction of P-450. DDT also reduced the number of mammary tumors produced by dimethylbenzanthracene (*14*). This result may be due to the induction of the P-450

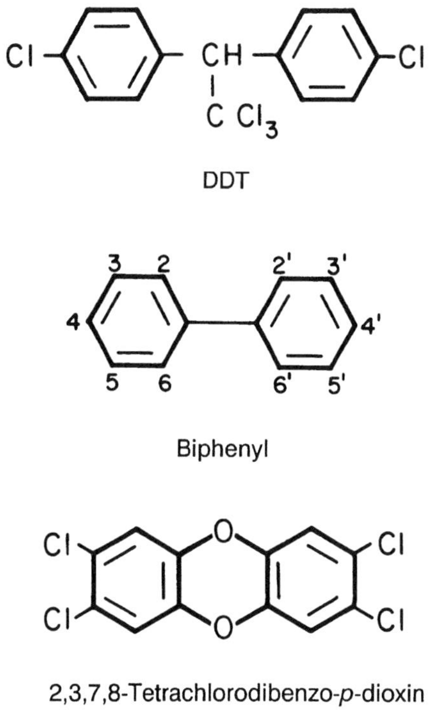

DDT

Biphenyl

2,3,7,8-Tetrachlorodibenzo-*p*-dioxin

*Chart 3.10. Structures of DDT, biphenyl, and tetrachlorodibenzo-*p*-dioxin.*

isozyme responsible for noncarcinogenic hydroxylation of dimethylbenzanthracene, or to the induction of epoxide hydrolase (*see* the following section of this chapter) or glutathione *S*-transferase, or any combination of these effects. Indeed, evidence has been presented (*6*) that both epoxide hydrolase and glutathione *S*-transferase are inducible. Other chlorinated hydrocarbon pesticides (such as aldrin, dieldrin, hexachlorobenzene, and hexachlorohexane) also act as P-450 inducers.

Monsanto arochlors are mixtures of polychlorinated biphenyls (PCBs) (Chart 3.10, middle). They are named by using four-digit numbers. The first two digits (1,2) indicate a biphenyl structure; the remaining two digits indicate the average percentage of chlorine. (For example, Arochlor 1254 is a mixture of chlorinated biphenyls with an average chlorine content of 54% by weight.) PCBs were widely used as insulating fluids in capacitors, transformers, vacuum pumps, and gas transmission turbines. Their biological activity varies somewhat, depending on the position of the chlorine atoms. Generally they exert a number of effects, such as induction of P-450 and of *p*-nitrophenol and testosterone glucuronyl transferases. In addition, they cause an increase in liver weight and in microsomal protein (*14*).

Another environmental contaminant of great concern is TCDD (2,3,7,8-tetrachlorodibenzo-*p*-dioxin) (Chart 3.10, bottom). This extremely toxic compound has no practical application and is not being manufactured deliberately. However, it is present in the environment. It is formed on incineration of chlorinated organic substances and thus is found in exhaust and in ash from municipal incinerators. It is also formed in the process of pulp bleaching in paper manufacturing and as a byproduct of the manufacturing of a herbicide, 2,4,5-T [(2,4,5-trichlorophenoxy) acetic acid], and a wood preservative, pentachlorophenol. TCDD is 30,000 times more potent an inducer of AHH than 3MC.

The inducers discussed so far are specific for cytochrome P-450 isozymes, although some of them may also have inducing activity for other xenobiotics' metabolizing enzymes. Inducers specific for phase 2 metabolizing enzymes occur in cruciferous vegetables (broccoli, cauliflower, mustard, cress, and other cabbage-related plants). They specifically induce glutathione *S*-transferases and quinone reductase. One representative of this class has been isolated from broccoli and identified as (–)-1-isothiocyanato-(4*R*)-(methylsulfinyl)butane, which is known as sulforaphane (Structure 3.5) (*15*).

Because a diet rich in green and yellow vegetables lowers the risk of cancer in humans (*15*), it is assumed that this protection against cancer is due to the in-

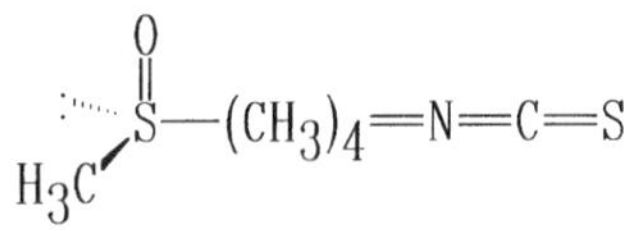

$$H_3C-S(=O)-(CH_3)_4=N=C=S$$

Structure 3.5. Sulforaphane.

duction of phase 2 enzymes that detoxify the carcinogens. The role of glutathione and of glutathione *S*-transferases as detoxifying agents has been discussed. It will be shown in the following section that one pathway of carcinogenic activation of benzo[*a*]pyrene involves the formation of quinones. Thus, quinone reductase may prevent this pathway of activation.

Activation of Precarcinogens

As mentioned earlier, in some cases the metabolism of xenobiotics leads to the formation of unstable intermediates that react with cellular macromolecules. This reaction leads to mutagenic or carcinogenic transformation. In the following pages, activation of the most typical precarcinogens will be discussed.

2-Naphthylamine, a compound used in dye manufacturing, has been found to produce bladder cancer among workers employed in dye manufacturing. Injected 2-naphthylamine and other aromatic amines do not produce tumors at the site of injection. Rather, they produce tumors in distant organs such as the liver and urinary bladder. The tumor location indicates that these chemicals are not carcinogens per se, but that metabolism of the chemical is required to produce the carcinogenic insult (*16*). It was proposed (*4*) that 2-naphthylamine becomes a carcinogen upon *N*-hydroxylation by cytochrome P-450. When the hydroxylamine is stabilized by conjugation with glucuronide, it becomes harmless. However, the conjugated compound can be hydrolyzed back to the carcinogenic hydroxylamine, either by the action of β-glucuronidase in the kidney or by acidic pH in the urine (Scheme 3.7).

Aminofluorene was developed as an insecticide. However, because of its carcinogenicity it was not released for commercial application. This compound is

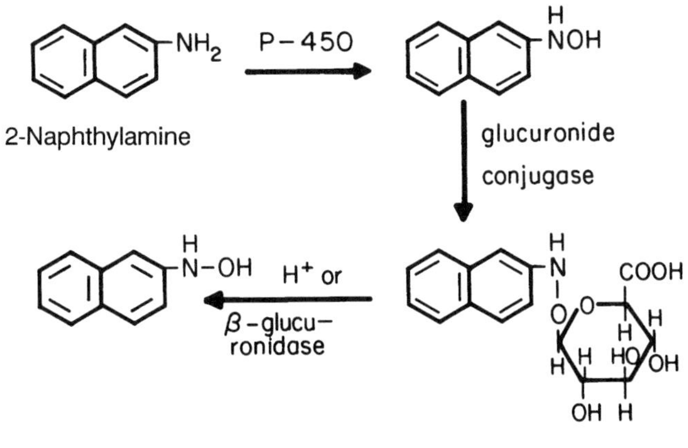

Scheme 3.7. Carcinogenic activation of 2-naphthylamine.

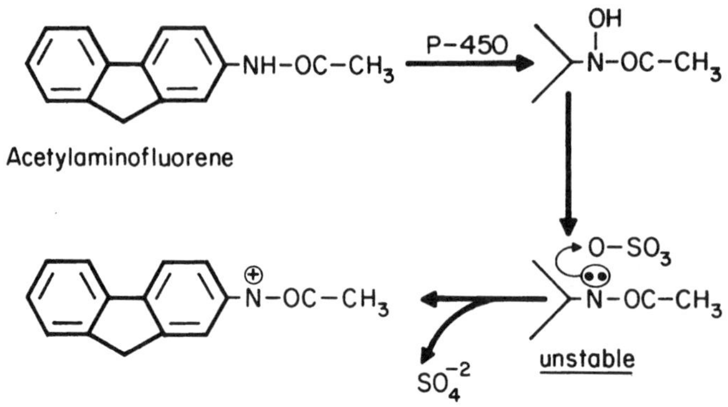

Acetylaminofluorene

P-450

unstable

Scheme 3.8. Carcinogenic activation of acetylaminofluorene.

acetylated, *N*-hydroxylated, and subsequently conjugated with sulfate, which is unstable and breaks down to a powerful electrophile (Scheme 3.8) (*7*).

Dichloroethane is a waste product of vinyl chloride production and also a laboratory solvent. Its analog, dibromoethane, is used as a gasoline additive and as an insecticide. Both are carcinogens and mutagens. They may be metabolized by conjugation with glutathione to produce haloethyl-*S*-glutathione, a compound structurally similar to sulfur mustard, which was used as a war gas during World War I (Yperite). Haloethyl-*S*-glutathione acts by spontaneous formation of an unstable three-member ring that, upon ring opening, reacts with cellular macromolecules (Scheme 3.9) (*17*).

Vinyl chloride is a starting material in the manufacture of poly(vinyl chloride) plastics. Epidemiological studies of workers exposed to vinyl chloride revealed an unusually high frequency of angiosarcoma, an otherwise rare liver cancer. The proposed mechanism of carcinogenic activation involves epoxide formation. This epoxide, however, may be further metabolized, as presented in Scheme 3.10 (*18*).

A group of compounds designated as aflatoxins is produced by a mold, *Aspergillus flavus*. Under favorable conditions it contaminates crops such as corn and peanuts. The compound of major concern is aflatoxin B_1 (AFB_1); in human and

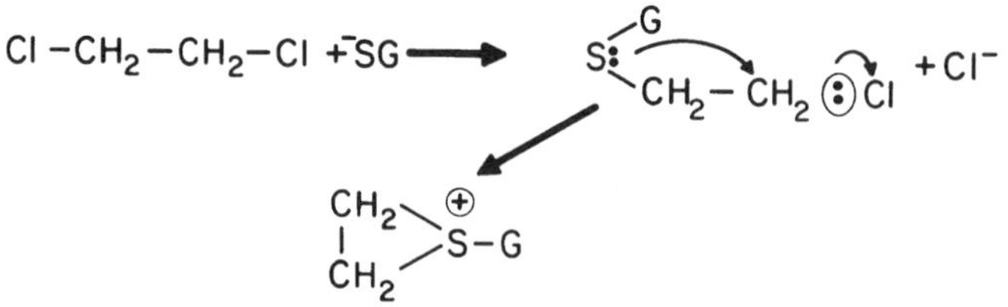

$$Cl-CH_2-CH_2-Cl + {}^{-}SG \longrightarrow$$

Scheme 3.9. Carcinogenic activation of 1,2-dichloroethane.

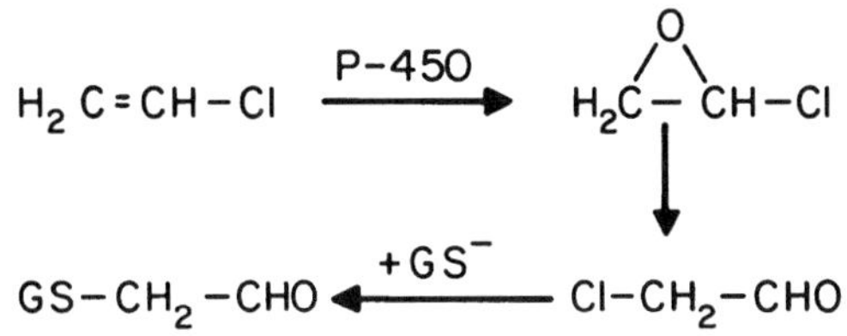

Scheme 3.10. Carcinogenic activation and further metabolism of vinyl chloride.

animal species it may be activated to a powerful hepatocarcinogen. AFB_1 is metabolized by cytochrome P-450 isozymes in multiple ways, one of which (2,3-epoxidation) leads to the formation of a carcinogen (Scheme 3.11) (*19*). Although this reaction is catalyzed by the 3MC-inducible enzyme, this enzyme is distinctly separate from AHH and is controlled by a different gene (*20*).

Benzo[*a*]pyrene is a major polycyclic hydrocarbon carcinogen in the environment. It is formed by the pyrolysis of hydrocarbons and thus occurs in industrial smoke, cigarette smoke and tar, and in fried, broiled, or smoked food. Benzo[*a*]pyrene in its native state is harmless, but it is metabolized by cytochrome P-450. The complete metabolism is rather complicated because of the many available positions. Oxygen can be introduced by cytochrome P-450 at all positions except C-11 (Scheme 3.12). These reactions lead to the formation of epoxides. The epoxides are then converted to *trans*-diols by epoxide hydrolase, to glutathione conjugates by glutathione transferase, or to arenols by nonenzymatic NIH rearrangement.

The velocity of these conversions depends on the chemical stability of the epoxides and on their substrate suitability for the enzymatic reactions involved. These two factors, in turn, depend on the position of the epoxides in the molecule. The diols and arenols can be conjugated with glucuronic acid or with sulfate, respectively. The products of the initial conversion can be reprocessed over and over again with the formation of new epoxides. The critical conversion that activates benzo[*a*]pyrene and other polycyclic hydrocarbons to carcinogens depends on the presence of the bay region and proceeds as presented in Scheme 3.12 (*7*).

The first step in the carcinogenic activation of benzo[*a*]pyrene is the forma-

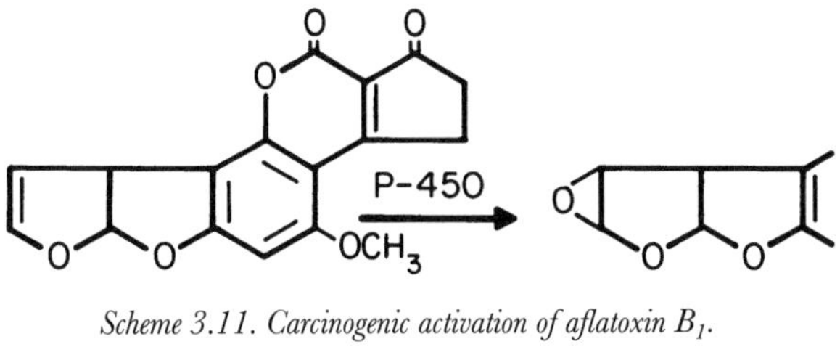

Scheme 3.11. Carcinogenic activation of aflatoxin B_1.

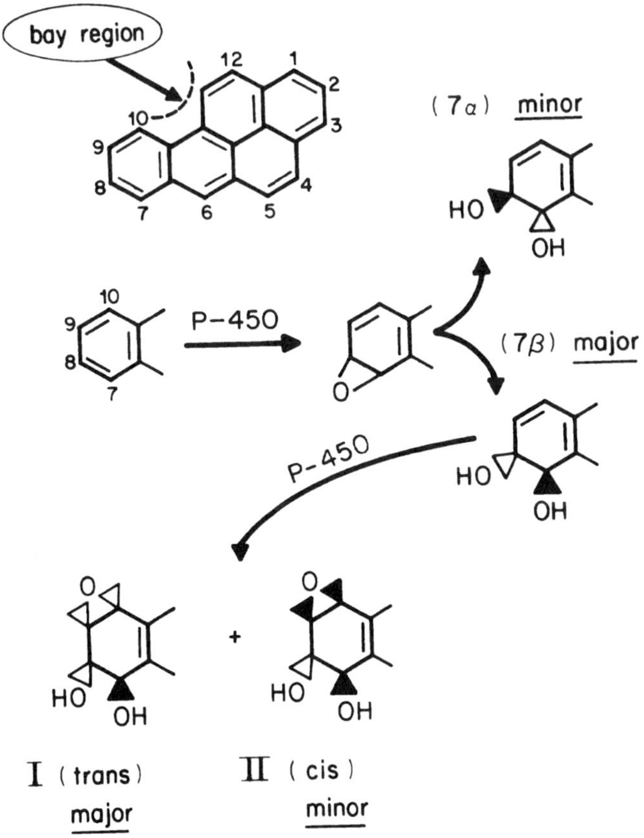

*Scheme 3.12. Carcinogenic activation of benzo[*a*]pyrene.*

tion of 7,8-epoxide. This substance is converted by epoxide hydrolase to two *trans*-diols, of which 7β is the major form. The diol formation activates the 9,10 double bond and thus facilitates formation of two 7,8-diol-9,10-epoxides, the major component being the *trans* form (I in Scheme 3.12), and the minor the *cis* form (II in Scheme 3.12). Both compounds are poor substrates for epoxide hydrolase. The 7,8-dihydrodiol-9,10-*trans*-epoxide is the carcinogenic form of benzo[*a*]pyrene. Its half-life is 8 min, which is probably long enough to react with DNA. In contrast, its *cis* analog has a half-life of only 0.5 min and thus is too unstable to damage the cells (*21*).

Concurrent with the reactions described, activation of benzo[*a*]pyrene may involve formation of 1,6-, 3,6-, and 6,12-quinones. In turn, these compounds may form carcinogenic 6-phenoxy radicals (*16*) (*see* Chapter 5.)

Another class of precarcinogenic compounds that require P-450 activation is the nitrosamines. They are formed by the reaction of nitrite ions (NO_2^-) with secondary and, to a lesser extent, tertiary amines (Scheme 3.13). Nitrite originates, directly or indirectly, from food. It is added directly to meat products as a preservative, to protect them from bacterial contamination and to preserve the fresh color. Indirectly, it comes from nitrate (NO_3^-), which occurs in drinking water and in vegetables (*see* Chapter 10). Nitrate is reduced to nitrite by salivary enzymes.

Dimethylamine is an important industrial material used in rubber, leather, and soap manufacturing. It reacts with nitrite to form dimethylnitrosamine. The course of its activation to alkylating electrophiles is presented in Scheme 3.13.

The formation of carcinogenic nitrosamines can frequently be prevented by compounds that compete with secondary and tertiary amines for the nitrite ion, such as the primary amines, ascorbic acid, and tocopherol. Ascorbic acid is especially useful; if present in twice the concentration of nitrite, it will completely inhibit formation of nitrosamines. Ascorbic acid reacts with nitrite by forming dehydroascorbic acid and NO. However, NO reenters the circulation by oxidation to nitrate.

These examples represent the most typical, and perhaps the best studied, cases of the failure of nature's detoxifying system. The factors that influence the metabolism of xenobiotics will be discussed in the next chapter.

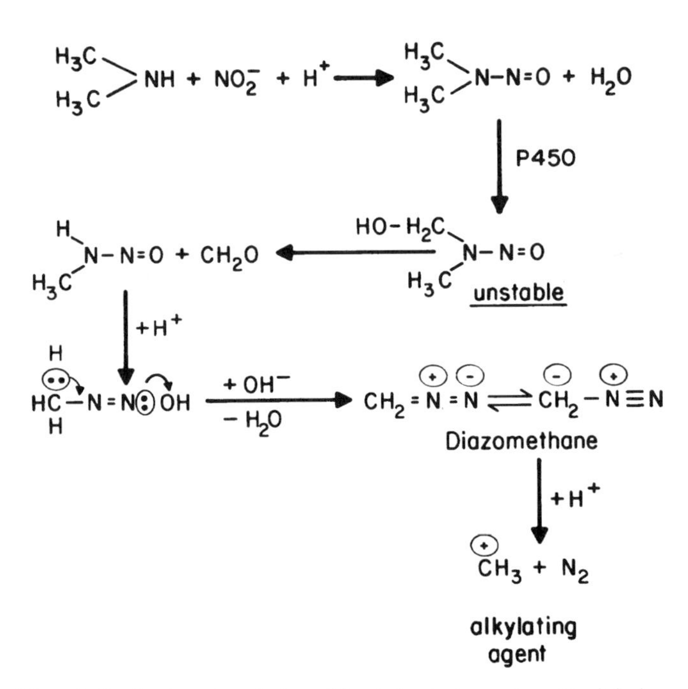

Scheme 3.13. Carcinogenic activation of dimethylamine by reaction with nitrite ions.

References

1. Harper, H. A.; Rodwell, V. W.; Mayers, P. A. *Review of Physiological Chemistry;* Lange Medical: Los Altos, CA, 1979; Chapter 19, p 266.
2. Mahler, H. R.; Cordes, E. H. *Biological Chemistry;* Harper and Row: New York, 1966; Chapter 8, p 322.
3. Beinert, H.; Sands, R. H. In *Free Radicals in Biological Systems;* Blois, M. S., Jr.; Brown, H. W.; Lemmon, R. M.; Lindblom, R. O.; Weissbluth, M., Eds.; Academic: New York, 1961; p 17.
4. Sipes, G.; Gandolfi, A. J. In *Cassarett and Doull's Toxicology,* 3rd ed.; Klaassen, C. D.; Amdur, M. O.; Doull, J., Eds.; MacMillan: New York, 1986; Chapter 4, p 64.
5. Fasco, M. J.; Vatsis, K. P.; Kaminsky, L. S.; Coon, M. J. *J. Biol. Chem.* **1978**, *253*, 823.
6. Nelson, S. D. In *Burger's Medicinal Chemistry*, 4th ed.; Wolff, M. E., Ed.; Wiley: New York, 1980; Part I, Chapter 4, p 227.
7. Low, L. K.; Castagnoly, N., Jr. In *Burger's Medicinal Chemistry,* 4th ed.; Wolff, M. E., Ed.; Wiley: New York, 1980; Part I, Chapter 3, p 107.
8. Guroff, G.; Daly, J. W.; Jerina, D. M.; Renson, J.; Witkop, B.; Undenfriend, S. *Science (Washington, D.C.)* **1967**, *157*, 1524.
9. Brown, R. R.; Miller, J. A.; Miller, E. C. *J. Biol. Chem.* **1954**, *209*, 211.
10. Poland, A.; Glover, E.; Kende, A. S. *J. Biol. Chem.* **1976**, *251*, 4936.
11. Haugen, D. A.; van der Hoeven, T. A.; Coon, M. J. *J. Biol. Chem.* **1975**, *250*, 3567.
12. Conney, A. H.; Lu, A. Y. H.; Levin, W.; West, S.; Smogyi, A.; Jacobson, M.; Ryan, D.; Kunzman, R. *Drug Metab. Dispos.* **1973**, *1*, 1.
13. Okey, A. B. *Life Sci.* **1972**, *11*, 833.
14. Goldstein, J. A.; McKinney, J. D.; Lucier, G. W.; Hickman, P.; Bergman, H.; Moore, J. A. *Toxicol. Appl. Pharmacol.* **1976**, *36*, 81.
15. Yuesheng, Z.; Talalay, P.; Cheon-Gyu, Cho.; Posner, G. H. *Proc. Natl. Acad. Sci. U.S.A.* **1992**, *89*, 2399.
16. Selkirk, J. K. In *Burger's Medicinal Chemistry*, 4th ed.; Wolff, M. E., Ed.; Wiley: New York, 1980; Part I, Chapter 12, p 455.
17. Williams, G. M.; Weisburger, J. H. In *Cassarett and Doull's Toxicology*, 3rd ed.; Klaassen, C. D.; Amdur, M. O.; Doull, J., Eds.; MacMillan: New York, 1986; Chapter 5, p 99.
18. Andrews, L. S.; Snyder, R. In *Cassarett and Doull's Toxicology,* 3rd ed.; Klaassen, C. D.; Amdur, M. O.; Doull, J., Eds.; MacMillan: New York, 1986; Chapter 20, p 636.
19. Hayes, J. R.; Campbell, T. C. In *Cassarett and Doull's Toxicology,* 3rd ed.; Klaassen, C. D.; Amdur, M. O.; Doull, J., Eds.; MacMillan: New York, 1986; Chapter 24, p 771.
20. Koser, P. L.; Faletto, M. B.; Maccubbin, A. E.; Gurtoo, H. *J. Biol. Chem.* **1988**, *263*, 12584.
21. Walsh, C. Lecture notes on metabolic processing of drugs, toxins, and other xenobiotics, presented in MIT Course 20.610, 1979, "Principles of Toxicology".

4 Factors That Influence Toxicity

Selective Toxicity

The more species are removed from each other in evolutionary development, the greater is the likelihood of differences in response to toxic agents. One obvious difference that affects toxicity is the size of the organisms. Much less toxin is needed to kill a small insect than a considerably larger mammal (everything else being equal). In addition, there is an inverse relationship between the weight of an animal and its surface area; the smaller the animal, the larger its surface area per gram of weight.

Thus, the weight ratio of a human being (70 kg) to a rat (200 g) is 350, but the surface area ratio of a human being to a rat is only 55. Roughly, the surface area of an animal (S) can be calculated as follows: $S(\mathrm{m}^2) = \text{weight (kg)}^{2/3}/10$. This type of calculation is important when one is considering the selective eradication of an uneconomical species, such as certain insects, by spraying an area with insecticide. The goal is to control the insects without harming wildlife, livestock, and human beings.

Other factors, such as the rate of percutaneous absorption, also have to be considered. For instance, it has been shown that DDT (dichlorodiphenyltrichloroethane) is about equally toxic to insects and mammals when given by injection, yet when applied externally it is considerably more toxic to insects. This toxicity is due not only to the difference of the surface area:body weight ratio, but also to the fact that the chitinous exoskeleton of the insect is more permeable to DDT than unprotected mammalian skin (*1*). Of course, in real-life situations (i.e., outside the laboratory), most mammalian skin is covered by fur, which gives the animals additional protection.

The foregoing discussion is not meant to imply that unrestricted spraying with pesticides (especially chlorinated hydrocarbons, which are fat-soluble and poorly biodegradable) is environmentally sound. Problems with their use include lack of selectivity among insect species; leaching into watersheds and groundwater; and bioaccumulation in the food chain. These problems will be discussed in detail in Chapter 10.

Metabolic Pathways

Metabolic-pathway differences among species may provide another rationale for achieving selective toxicity. A good example of this type of selectivity is the chemotherapeutic use of sulfonamides. Human beings and, as far as we know, most vertebrates require an exogenous supply of folic acid. Folic acid is converted in the organism to tetrahydrofolic acid, an important cofactor involved in the de novo biosynthesis of purine and pyrimidine nucleotides.

Certain gram-negative bacteria, on the other hand, are unable to assimilate preformed folic acid. Instead, they have the capacity to synthesize a precursor of tetrahydrofolic acid (namely, dihydropteroic acid) from 6-hydroxymethyl-7,8-dihydropteridine and *p*-aminobenzoic acid (Chart 4.1) (*2*). Sulfonamides, because of their structural similarity to *p*-aminobenzoic acid (*see* Chart 2.1 in Chapter 2), inhibit this reaction (*3*). Thus, these bacteria are deprived of tetrahydrofolic acid cofactors. In turn, this deprivation results in bacterial-growth inhibition. Humans are not affected because they are not capable of carrying on this synthetic reaction.

Although sulfonamides have toxic side effects in humans, this toxicity is not related to their biochemical mode of action. Instead, their low solubility in urine makes them tend to precipitate in the kidney.

Enzyme Activity

In some cases metabolic pathways may be the same for several species, but the enzymes that carry out certain reactions may differ. Hitchings and Burchall (*4*) compared the inhibitory activity of two compounds toward the enzyme dihydrofolate reductase (*see* Chart 4.1) obtained from different species. The results of this experiment are summarized in Table 4.1.

The high sensitivity of the enzyme from the two bacterial strains to trimethoprim and its lack of sensitivity to pyrimethamine, as compared to the relative insensitivity of the mammalian enzymes to both compounds, are evident. Even so, pyrimethamine is not selective for bacteria; it was found to be effective against plasmodia, the parasites that cause malaria. Trimethoprim is used selectively against bacterial infections. The structures of both compounds are presented in Chart 4.2.

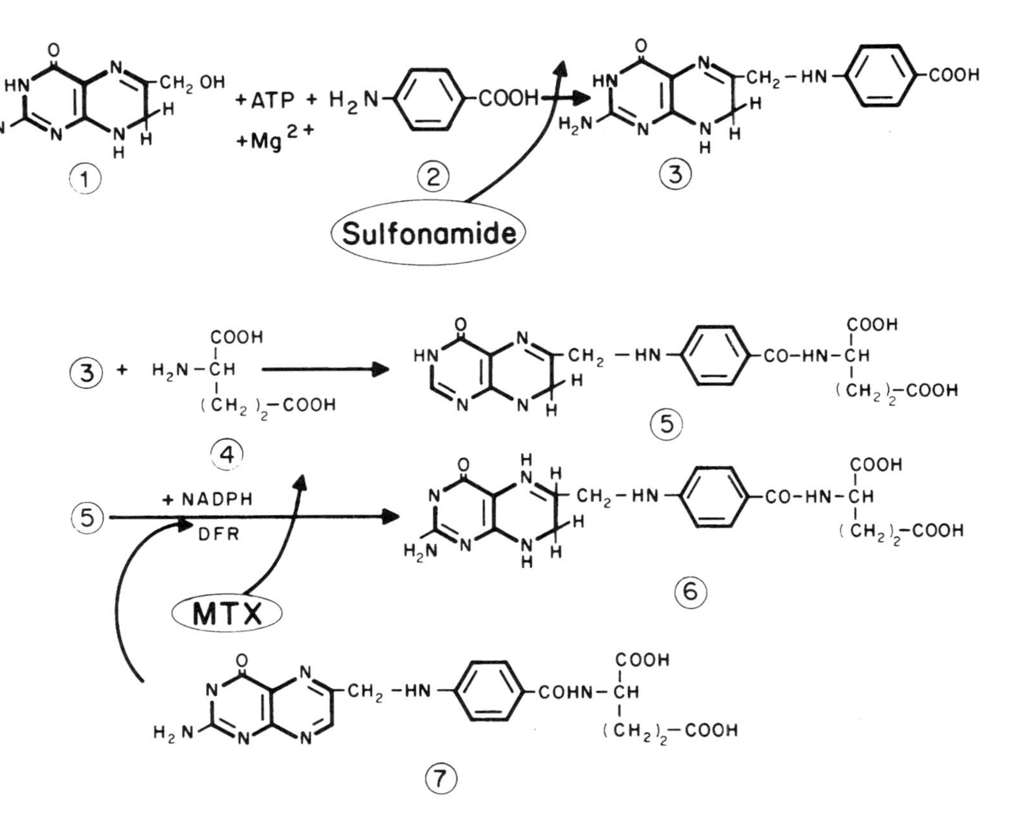

Chart 4.1. Synthetic pathways leading to the formation of tetrahydrofolic acid. 1: 6-hydroxymethyl-7,8-dihydropteridine; 2: p*-aminobenzoic acid; 3: 7,8-dihydropteroic acid; 4:* L*-glutamic acid; 5: 7,8-dihydrofolic acid; 6: 5,6,7,8,-tetrahydrofolic acid; 7: folic acid. MTX is methotrexate.*

Table 4.1. Inhibitory Activity of Pyrimethamine and Trimethoprim Toward Dihydrofolate Reductase

Source of Enzyme	*Pyrimethamine*	*Trimethoprim*
Human liver	180	30,000
Escherichia coli	2,500	0.5
Proteus vulgaris	1,500	0.4
Rat liver	70	26,000

NOTE: All values are IC_{50} in units of molarity × 10^{-8}.

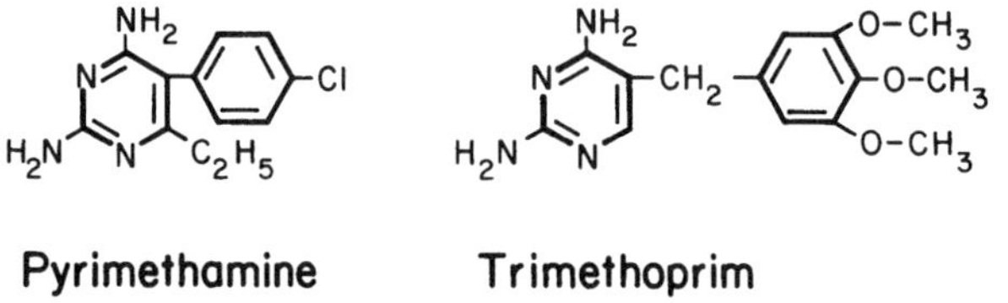

Pyrimethamine Trimethoprim

Chart 4.2. Structures of pyrimethamine (Daraprim) and trimethoprim.

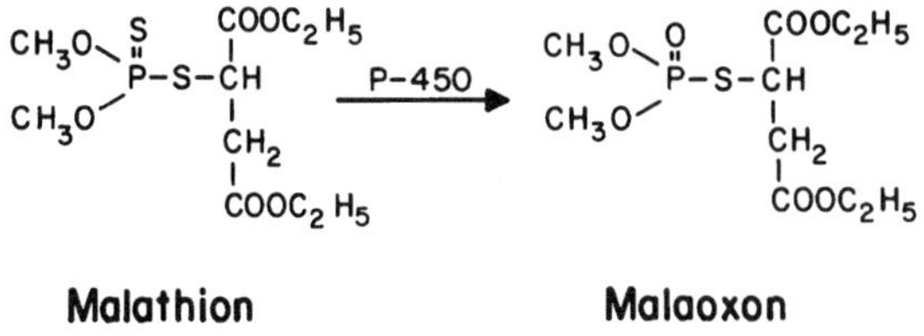

Malathion Malaoxon

Scheme 4.1. Conversion of malathion to malaoxon.

Xenobiotic-Metabolizing Systems

Selective toxicity also may be based on differences in xenobiotic-metabolizing systems. For instance, the insecticide malathion (Scheme 4.1), upon being converted by cytochrome P-450 to malaoxon, becomes an inhibitor of acetylcholinesterase. It is nearly 38 times less toxic when given orally to rats than when applied topically to houseflies (*5*). The explanation is that mammals possess very active esterases that inactivate malaoxon by hydrolyzing the ester groups. Insects also contain esterases, but they act much more slowly than the mammalian enzymes.

An interesting case of selective toxicity is the use of synthetic pyrethroids as insecticides. This group of compounds is derived from the naturally occurring toxins called pyrethrins (Structure 4.1) that are isolated from chrysanthemum flowers. The pyrethroids are highly selective in their toxicity toward insects. For instance, one member of this group, permethrin, has an LD_{50} 1400 times larger for rats than for the desert locust (*6*). Possibly because the toxicity of pyrethroids increases with decreasing temperature, they seem to be more toxic to cold-blooded than to warm-blooded species. Thus, temperature dependence may be the reason for their selective toxicity toward insects (*7*). This concept is supported by the observation that pyrethroids are extremely toxic to fish in the laboratory. Another possibility is that pyrethroids undergo rapid bioinactivation, namely, hydrolysis of the ester bond, in mammals but not in insects (*8*).

Toxicity Tests in Animals

The three types of toxicity studies in animals are acute toxicity determination, subchronic toxicity determination, and chronic toxicity determination. The chronic toxicity determination, which usually concerns carcinogens, is discussed in Chapter 5.

Acute toxicity studies involve determination of LD_{50}. Groups of animals (5–10 males and an equal number of females per group) are treated with a chemical at three to six different dose levels. The number of animals that die within 14 days is tabulated. The weight of the animals and any changes in their behavior are noted. At the end of the experiment the survivors are sacrificed and all animals (including the control group) are examined for pathological changes.

Subchronic toxicity studies involve daily administration of the compound to be tested to groups of males and females at three dose levels: the maximum tolerated dose (MTD), lowest observable adverse effect level (LOAEL), and no observable adverse effect level (NOAEL). MTD is chosen so that it does not exceed LD_{10}. Usually two species and frequently two routes of exposure are tested, one being the same as the expected human exposure. The duration of the tests vary between 5 and 90 days. Mortality, weight, and behavioral changes are noted. Blood chemistry measurements are performed prior to, halfway through, and at the end of the experiment. Subsequently, all the animals are sacrificed for pathologic study.

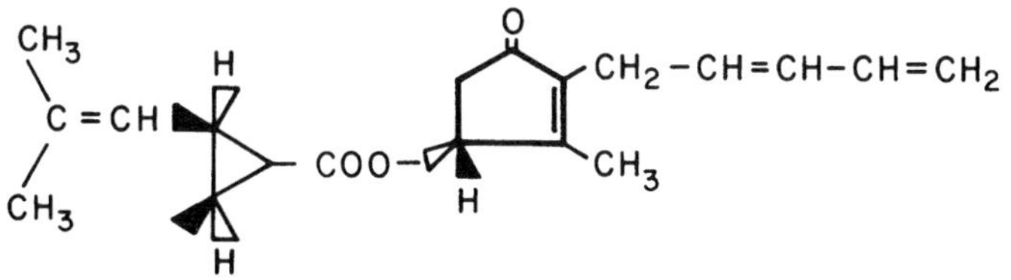

Structure 4.1. Pyrethrin I.

Species Differences

When using animal assay data for predicting human toxicity, the goal is to minimize species differences. Unfortunately, this is frequently difficult to achieve. Even within a single class, such as mammals, metabolic differences among species may be considerable. In most cases the differences are quantitative, although occasionally qualitative differences are encountered.

For instance, only primates, guinea pigs, and fruit-eating bats and birds have a need for vitamin C. Somewhere during evolutionary development, these particular species lost the synthetic pathway for ascorbic acid; other mammals and birds can synthesize it. Another example is the toxic response to the anticancer drug methotrexate. Although methotrexate is very toxic to humans, mice, rats, and dogs, it is not toxic to guinea pigs and rabbits. These examples indicate the importance of an appropriate choice of an animal model.

Because of the relative ease of maintenance and availability, most toxicity evaluation is done with mice or rats. Dogs, cats, or primates are sometimes used in limited quantities, especially for the study of pathology. Whatever the animal model, extrapolation of the results to humans has to be done with caution because considerable quantitative differences between humans and the model may be encountered. For this reason the Food and Drug Administration (FDA) requires a toxicity study in two unrelated species (usually rats or mice and dogs) before an approval of phase I clinical trials is granted. (Phase I clinical trials are designed to test the toxicity of a new drug in human patients.)

The variability of response to toxic agents may be further illustrated by an analysis of the NCI (National Cancer Institute) carcinogenicity assay data from 190 compounds that were tested in two species, mice and rats. Of these, only 44 were found to be carcinogenic in both species, whereas 54 were carcinogenic in either mice or rats, but not in both (*9*).

Exposure Mode

In any evaluation of the toxicity of environmental and industrial compounds, it is important that the test animals be exposed to the presumed toxin in a manner similar to the anticipated human exposure. This point assumes special importance when a judicial battle threatens to ban or restrict the use of a toxic substance. For example, early demonstrations of the carcinogenicity of tobacco tar were dismissed by the tobacco industry as invalid because the tar was painted on the skin of the test animals. This application is not comparable to human exposure.

Carcinogenicity tests in animal models present a special problem. To obtain a significant number of tumors during the life span of mice or rats, within practical limits of the size of the population tested, it is necessary to use relatively large doses of the suspected carcinogen. This high dosage may, or may not, simulate the actual conditions of occupational exposure to carcinogens. In any case, it does not faithfully reproduce the chronic exposure of the population at large to the very small amounts of environmental carcinogens. Thus, although the dose–response curve for large doses can be traced, its extrapolation for small doses remains purely hypothetical. For these reasons, risk assessment of exposure to environmental carcinogens is difficult. (Further discussion of this topic is presented in Chapter 6.)

The current U.S. government's policy is that, as far as carcinogens are concerned, there is no threshold dose (a dose below which there is no cancer risk); any exposure, no matter how small the dose, is considered to be harmful. In 1958 the U.S. Congress passed an amendment to the Food and Cosmetic Act of 1938, known as the Delaney Clause, which states: "... no additive shall be deemed to be safe if it is found to induce cancer when ingested by man or animal, or if it is found, after tests which are appropriate for the evaluation of safety of food additives, to induce cancer in man or animal. ..." In practical terms the Delaney amendment concerns mainly residues of cancer-causing pesticides in processed food. Since early 1993, both the federal administration and the U.S. Congress began a push for replacement of the Delaney amendment with risk assessment, that is, allowing residues of carcinogenic pesticides in processed food as long as they present negligible risk only; negligible risk was defined as no more than one additional cancer per one million people over a 70-year lifetime. The justification for the change of policy was that modern analytical methods allow detection of much smaller residues than was possible in 1958 when the Delaney Clause was formulated. Thus, strict application of the Delaney Clause imposes unnecessary hardship on the agricultural and food-processing industries, without providing much protection for the public. The revision of the Delaney Clause remains controversial. The replacement of the Delaney Clause with risk assessment is supported by the Agricultural Chemical Manufacturers Association and by the food-processing industry, but is opposed by many environmental organizations.[1]

The risk assessment of carcinogens and the problems encountered with pesticides are discussed in more detail in Chapters 6 and 10.

Individual Variations in Response to Xenobiotics

Variations among individuals within a species in response to xenobiotics may be due to environmental causes, to the genetic makeup of an individual or a group of individuals, and to the age of the individuals.

[1]While this book was in production, President Clinton signed the Food Quality Protection Act on August 3, 1996, in which the Delaney Clause was replaced with a new standard of "reasonable certainty that no harm will result from aggregate exposure to the pesticide chemical residue" (*see* Chapter 14).

Environmental and Endocrine Factors

It has been demonstrated that the metabolism of a xenobiotic may be influenced by diet (*see* reference 9 in Chapter 3). Another factor may be concurrent exposure to other xenobiotics, such as drugs or environmental toxins. Induction and inhibition of xenobiotic-metabolizing enzymes were discussed in the preceding chapter. Metabolism of one chemical may be accelerated or retarded by exposure to another one that happens to be an inducer or an inhibitor of cytochrome P-450 or any of the conjugating enzymes.

There is ample evidence that the hormonal status of an individual also affects response to toxins. This condition is manifested not only in different responses between males and females, but also in different responses within an individual, depending on the time of the day. These variations are due to fluctuating levels of serum corticosterone, which in turn depend on the light cycle, often referred to as circadian rhythm (*10*).

Genetic Factors

As discussed in Chapter 2, any apparently homogeneous biological system, even where all individuals are maintained under identical conditions and fed an identical diet, is in fact heterogeneous. The quantal dose–response curve (Figure 4.1) shows that most of the individuals in a system respond to a chemical injury in a similar way. However, there is always a small fraction of individuals on either end of the curve who are either exceptionally sensitive or exceptionally resistant to the insult. These individuals are endowed with genetic characteristics designated as hypersensitivity (left end of the curve) and hyposensitivity (right end of the curve). The hyper- and hyposensitivities are not considered to result from genetic mutation. They merely represent normal genetic deviation within a population.

In some cases, when a large population sample is screened for certain traits

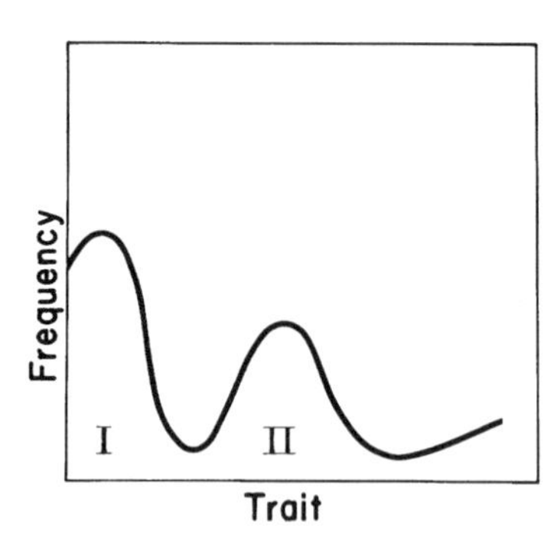

Figure 4.1. Quantal dose–response curve indicating the presence of a mutated population (peak II).

$$\text{C}_5\text{H}_4\text{N}\text{-CO-NH-NH}_2 \xrightarrow{\text{CH}_3\text{-CO-S-CoA}} \text{C}_5\text{H}_4\text{N}\text{-CO-NH-NH-OC-CH}_3$$

Scheme 4.2. Acetylation of isoniazid (INH).

and the data are presented as a quantal dose–response plot, a multiphasic curve is obtained. In the hypothetical plot depicted in Figure 4.1, the main peak represents the "normal" population and the minor peak the mutated population.

An example of genetic mutation is the so-called acetylation polymorphism. The action of the antitubercular drug isoniazid (INH) is terminated by acetylation (Scheme 4.2), a reaction that is carried out by *N*-acetyltransferase (*see* Chapter 3). A genetic deficiency of this enzyme is encountered among certain groups in the population, both in humans and animals.

When blood levels of INH are determined in a large sample of population 6 h after administration of a standard dose of the drug and the results are plotted as a quantal dose–response relation, a triphasic curve is obtained. Thus, there are three populations: the population under the first (major) peak are the fast acetylators who had none, or very little, of the INH in their blood; the population under the second peak are the slow acetylators, who had considerably higher levels of the drug remaining; and the population under the third peak are the very slow acetylators, who had the largest levels of the drug remaining (*11*).

It appears that deficiency of *N*-acetyltransferase is a genetic trait; it runs in families. The predisposition for this characteristic is related to race; frequency of occurrence is highest among blacks and Caucasians, lesser among Japanese and Chinese, and lowest among Eskimos.

Acetylation polymorphism is but one example of genetic mutations expressed by altered capacity to metabolize xenobiotics. A more extensive treatment of this subject may be found in Ted Loomis's *Essentials of Toxicology* (*11*).

Genetically altered populations develop when genetic mutations occur in reproductive cells. If the mutation results in the deficiency of an enzyme that is indispensable for normal metabolism, the offspring will not survive. Therefore, the only observable mutated populations are those in which the deficient enzyme is not essential for survival. These individuals lead a normal life, but injury may occur when they are challenged with a drug or a xenobiotic.

Influence of Age

In general, both developing and aging organisms are more susceptible to the toxic effects of xenobiotics than are young adults. This increased susceptibility is probably due to the fact that very young individuals have not fully developed sufficient levels of detoxifying enzymes and the levels of these enzymes have decreased in aging individuals. An insufficiently developed immune system in children and depressed immunity in aged organisms may also play a role (*see also* the section "Lead Pollution" in Chapter 10 and "Radiosensitivity" in Chapter 12.)

References

1. Klaassen, C. D. In *Cassarett and Doull's Toxicology*, 3rd ed.; Klaassen, C. D.; Amdur, M. O.; Doull, J., Eds.; MacMillan: New York, 1986; Chapter 3, p 33.
2. Brown, G. M. In *Chemistry and Biology of Pteridines;* Iwai, K.; Akino, M.; Goto, M.; Iwashami, Y., Eds.; International Academic: Tokyo, Japan, 1970.
3. Brown, G. M. *J. Biol. Chem.* **1962**, *237*, 536.
4. Hitchings, G.; Burchall, J. *Mol. Pharmacol.* **1965**, *1*, 126.
5. Metcalf, R. L. In *Pest Control Strategies for the Future;* Agricultural Board Division of Biology and Agriculture, National Research Council, National Academy of Sciences: Washington, DC, 1972; p 137.
6. Elliot, M. *Environ. Health Perspect.* **1976**, *14*, 3.
7. Norton, S. In *Cassarett and Doull's Toxicology*, 3rd ed.; Klaassen, C. D.; Amdur, M. O.; Doull, J., Eds.; MacMillan: New York, 1986; Chapter 13, p 35.
8. Murphy, S. D. In *Cassarett and Doull's Toxicology*, 3rd ed.; Klaassen, C. D.; Amdur, M. O.; Doull, J., Eds.; MacMillan: New York, 1986; Chapter 18, p 519.
9. Office of Technology Assessment. *Cancer Risk. Assessing and Reducing the Dangers in Our Society. Summary;* Westview: Boulder, CO, 1982; p 3.
10. Nelson, S. D. In *Burger's Medicinal Chemistry*, 4th ed.; Wolff, M. E., Ed.; Wiley: New York, 1980; Part I, Chapter 4, p 227.
11. Loomis, T. *Essentials of Toxicology;* Lee & Febiger: Philadelphia, PA, 1978; Chapter 6, p 81.

5 Chemical Carcinogenesis and Mutagenesis

Environment and Cancer

Cancer is a common name for about 200 diseases characterized by abnormal cell growth. According to Kundson (*1*), the causes of cancer may be classified into the following groups:

1. genetic predisposition,
2. environmental factors,
3. environmental factors superimposed on genetic predisposition, and
4. unknown factors.

Typical examples of the first group are childhood cancers such as retinoblastoma (a genetically predisposed malignancy of the retina), neuroblastoma (a malignancy of the brain), and Wilms' tumor (a malignancy of the kidney). In adults, an example is polyposis of the colon, a genetic condition that frequently leads to colon cancer.

The third group is represented by xeroderma pigmentosum, a genetic condition characterized by a deficient DNA excision repair mechanism (*see* the discussion later in this chapter). Individuals so predisposed develop skin cancer when exposed to ultraviolet light. The variable susceptibility of the population to the carcinogenic effects of cigarette smoke may also reflect genetic predisposition.

Very little can be said about the fourth group because the causes of this group of cancers are not known.

Groups 2 and 3 combined (i.e., cancer attributable to environmental causes, with or without genetic predisposition) probably account for 60–90% of all cancers (*2*). The environment, in this context, involves not only air, water, and soil, but also food, drink, living habits, occupational exposure, drugs, and practically all aspects of human interaction with the surroundings. This definition implies that a great majority of cancers could be prevented by avoiding exposure to potential carcinogens and by changing living habits. It is therefore not surprising that the study of chemical carcinogenesis represents a major aspect of environmental toxicology.

Table 5.1 gives an overview of estimated environmentally associated cancer mortality or incidence in the United States. The data presented in this table have to be considered as rough estimates only. There are great variations in the estimates, depending on the investigators and their methods of collecting the pertinent statistics. The Office of Technology Assessment report on cancer risk offers a more in-depth treatment of this subject (*2*). Because of cancer's long latency period (*see* the next section in this chapter), such statistics refer to the situation of two decades ago, rather than to the present. Data to be published 20 years from now may present a completely different picture. For instance, the National Cancer Institute reported that in 1988 the incidence of lung cancer among American males declined for the first time in several decades. Yet the smoking habit, the principal cause of lung cancer, was decreasing steadily since the 1960s.

As shown in Table 5.1, tobacco smoking is the main single cause of environmentally induced cancer. It has been estimated that in 1992 there were 168,000 new cases of lung cancer (*3*) and that the medical expenses and lost wages due to tobacco use were, on the average, $52 billion annually. Most of the lung cancer was caused by smoking; however, passive smoking (exposure to the tobacco smoke of others), occupational exposure to industrial carcinogens, and residential exposure to the radioactive gas radon also contributed to the cancer incidence.

Table 5.1. Cancer Mortality (Incidence) Associated with Environmental Exposure in the United States

Factor	*Percent of Total Cancer*	*Year Estimated*
Tobacco	30 (mortality)	1977
	76 (mortality)	1980
Alcohol	4–5 (mortality)	1978
Diet	35 (mortality)	1977
Asbestos	13–18 (incidence)	Near term and future
	3 (incidence)	Now or future
Air pollution	2 (mortality)	Future

SOURCE: Reproduced from reference 2.

The statistics on cancer mortality due to air pollution may be misleading; though the mortality due to direct inhalation of carcinogens may be low, the indirect effect of air pollution may be quite significant. Many air pollutants, such as polycyclic aromatic hydrocarbons (PAHs), deposited on land or water, enter the food chain and thus are classified as cancer caused by food and not by air pollution. In addition, inhaled carcinogens may also find their way, via the mucociliary escalator, to the digestive tract. The highest incidence of cancer caused directly by inhalation of air pollutants occurs in highly industrialized areas and affects mostly people in certain occupations such as coke-oven and coal-tar pitch workers; such occupational exposure to PAH may be 30,000 times higher than the exposure of the public at large.

The relatively high cancer mortality associated with diet deserves comment. Except for the correlation between liver cancer and the consumption of crops contaminated with aflatoxin, no direct epidemiological evidence linking any specific food or food contaminant to human cancer has been presented. However, many carcinogens have been found in foods.

Nitrites, which are added to meats as preservatives, are precarcinogens. Nitrates occur in vegetables, fruits, and drinking water, usually as a result of the leaching of nitrate fertilizers into groundwater; although not carcinogenic in their own right, they are reduced to nitrites by salivary enzymes. PAHs are produced when meat or fish is broiled, fried, or smoked. In addition, fish or shellfish from polluted waters may contain chlorinated hydrocarbon pesticides, PAHs, polychlorinated biphenyls (PCBs), and other organic contaminants. The fact that no correlation between cancer and consumption of specific foods has been found does not imply that none exists.

A relationship between obesity and cancer mortality has been found. Whether this effect is due to obesity itself or the obesity is a reflection of a certain lifestyle conducive to cancer is not known.

Multistage Development of Cancer

The concept of a multistage development of cancer goes back to the experiments of Berenblum and Shubik (*4*). These investigators studied the carcinogenicity of 9,10-dimethylbenzanthracene (DMBA) and benzo[*a*]pyrene (BP) in mice. When a 1.5% solution of DMBA in liquid paraffin was applied only once to the skin of 45 mice, only one mouse developed a tumor. However, when the single application of DMBA was followed by the application of 5% croton oil in liquid paraffin twice weekly for 20 weeks, 20 out of 45 mice developed tumors. No tumors were observed when croton oil was applied twice weekly for 2 weeks prior to the DMBA treatment.

Further evidence, provided by epidemiological and laboratory studies, led to the development of the present concept of cancer initiation, promotion, and progression. *Initiation* is caused by the interaction of a genotoxic (*see* the definition later in this chapter) compound with cellular DNA. Once the injury to the DNA has occurred and is not repaired, the cell is permanently mutated. Such a latently premalignant cell can remain in an animal for most of its natural life without ever developing into a cancerous growth. In humans the latent period may be 20 years or longer. According to some investigators (*5*), the latent period is inversely related to the dose of the initiator. The validity of this assumption is being questioned by others.

Exposure of the premalignant cell to a *promoter*, even after a delay of as long as 1 year (*6*), converts the cell to an irreversibly malignant state. Promotion is a slow process, and exposure to the promoter must be sustained for a certain period of time. This requirement explains why the risk of cancer diminishes rapidly after one quits the cigarette smoking habit; both initiators and promoters appear to be contained in tobacco smoke.

To date many promoters have been identified. The most extensively studied examples are the phorbol esters (Structure 5.1), a family of diterpenes isolated from croton oil. Bile acids have been shown to be promoters in colon carcinogenesis. Alcohol acts as a promoter in people exposed to the carcinogens in tobacco smoke. Smokers seldom develop cancer in the upper gastrointestinal tract or in the oral cavity; however, smokers who also drink alcohol frequently develop malignancies there. Certain inducers of cytochrome P-450, such as phenobarbital, DDT, and butylated hydroxytoluene (BUT, a food-additive antioxidant) have been identified as promoters; so are some hormones, if they are present in excessive amounts.

The mode of action of promoters is not well understood. To a certain degree, their activity may be accounted for by their action on cellular membranes. Recent findings with phorbol esters indicate that they may be involved in gene repression and derepression (*7*). Another concept, supported by more recent experimental evidence, is that cells are able to "communicate" with each other by transmitting small growth-regulating molecules through the so-called gap junc-

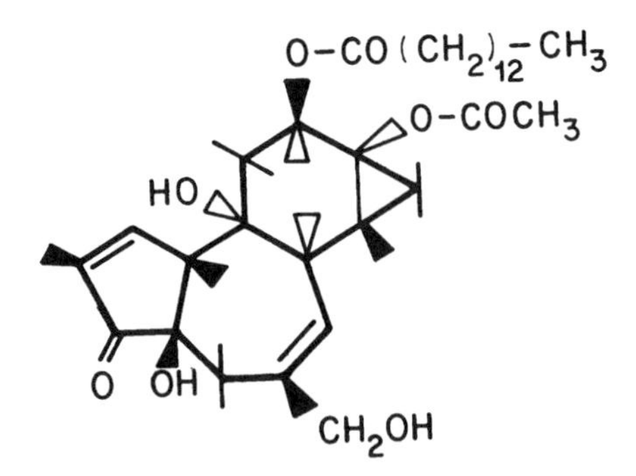

Structure 5.1. 12-O-Tetradecanoylphorbol-13-acetate (TPA), the most active tumor-promoting constituent of croton oil.

tion. Studies in cell culture have demonstrated that promoters are capable of inhibiting this intercellular communication. Such interference may release a latently premalignant cell from these growth-inhibiting restraints, to result later in cancerous growth (*7*).

Some compounds, although not necessarily carcinogenic by themselves when administered prior to or with a carcinogen, potentiate its activity. Such compounds are referred to as *cocarcinogens*. Some promoters, such as phorbol esters (*8*, *9*), are also cocarcinogens.

The distinction between these two classes is sometimes vague. The main difference is that cocarcinogens potentiate the neoplastic conversion, whereas promoters are involved in events following this conversion. Typical examples of cocarcinogens are catechols. As components of tobacco smoke, catechols potentiate the action of PAHs, the principal carcinogens of tobacco. Similarly, asbestos potentiates the carcinogenicity of tobacco smoke. Exposure to asbestos alone causes pleural and peritoneal mesotheliomas,[1] but not lung cancer. However, in smokers, exposure to asbestos greatly increases the incidence of lung cancer.

Types of Carcinogens

Carcinogens are divided into two categories: genotoxic and epigenetic. Compounds that interact directly or indirectly with DNA are, in most cases, mutagens. They are designated as genotoxic because they have the potential to alter the genetic code. The directly acting *genotoxic carcinogens* are either strong electrophiles, or consist of, or contain in the molecule highly stressed heterocyclic three- or four-member rings such as epoxides, azaridines, episulfides (*see* Chapter 3), and lactones. These cyclic compounds have a tendency to nucleophilic ring opening. As discussed in Chapter 3, many xenobiotics enter the body as innocuous compounds and become carcinogens after metabolic activation. Such xenobiotics are referred to as precarcinogens.

The indirectly acting genotoxic carcinogens occur less frequently than the directly acting ones. They react with non-DNA targets, releasing oxygen or hydroxy radicals such as $O^{\cdot-}$ (superoxide) or $\cdot OH$, as well as H_2O_2 and $^1O_2$ (singlet oxygen[2]). These activated species interact with DNA to cause strand breaks or damage the purine or pyrimidine bases. This sequence is essentially the mode of carcinogenic activity of ionizing radiation. However, certain types of compounds

---

[1]Mesotheliomas are tumors of the mesothelium, an outermost monolayer of flat epithelial cells that covers the lining of coelomic cavities (such as pericordial, pleural, and peritoneal).

[2]Singlet oxygen is molecular oxygen with one of its valence electrons elevated to a higher energy level; thus it is highly reactive.

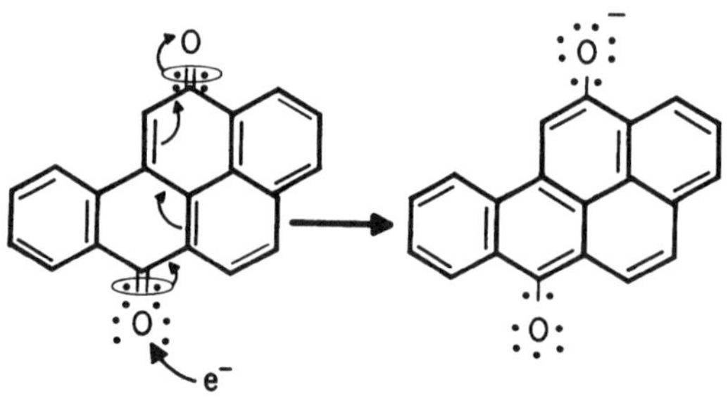

Scheme 5.1. Formation of 6-phenoxy radical from benzo[a]pyrene-6,12-quinone (see Chapter 3).

that contain the quinoid structure or are activated to form quinoids are postulated to act through free-radical formation, either directly or indirectly, via oxygen or hydroxy radicals (*10, 11*) (Scheme 5.1).

The mode of action of genotoxic carcinogens on the molecular level has been studied extensively. There is a wealth of information concerning their interaction with DNA. This subject will be discussed in more detail later in this chapter.

Much less is known about the mode of action of *epigenetic carcinogens*. Because the designation "epigenetic" includes all carcinogens that are not classified as genotoxic, a multitude of mechanisms may be involved. The epigenetic carcinogens comprise a wide variety of compounds, such as metal ions (nickel, beryllium, chromium, lead, cobalt, manganese, and titanium); solid-state carcinogens (asbestos and silica); immunosuppressors (azathioprine and 6-mercaptopurine); promoters; and the recently discovered xenoestrogens.

Promoters deserve special attention. In addition to known promoters such as tetradecanoylphorbol acetate (TPA) and phenobarbital, some environmental contaminants belong to this group. These are PCBs, tetrachlorodibenzodioxin (TCDD), and chlorinated hydrocarbon pesticides (DDT, aldrin, chlordane, etc.), all of which have been shown to produce liver cancer in rodents (*7*).

Review of DNA and Chromosomal Structure

Before discussing mutagenesis and the interaction of chemicals with DNA, a brief review of DNA and chromosomal structure is in order. The three main components of DNA are purine and pyrimidine bases, sugar (deoxyribose), and phosphate.

The three related pyrimidines are cytosine, thymine, and uracil, and the two related purines are guanine and adenine (Chart 5.1). Of the three pyrimidines, only thymine and cytosine occur in DNA, whereas only cytosine and uracil occur in RNA. Each of the bases can exist in two tautomeric forms, lactim or lactam

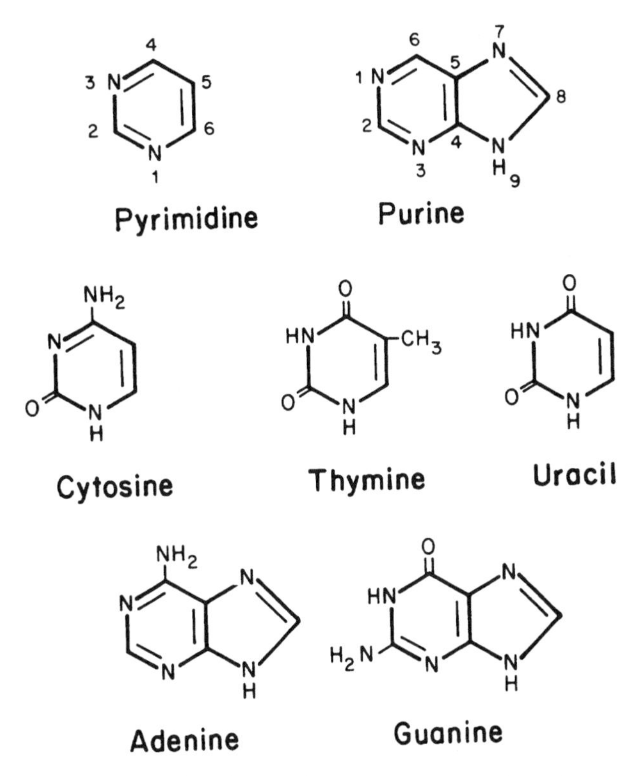

Chart 5.1. Purine and pyrimidine bases occurring in nucleic acids.

(Chart 5.2). Under physiological conditions the tautomeric form of each base is that depicted in Chart 5.1. Because of the pi electron clouds, the bases are planar. Both of these conditions are important prerequisites for the structure of the DNA double helix. Only planarity will allow stacking of the bases on top of each other, and only the proper tautomeric configurations will allow proper pairing of the bases.

Lactim ⇌ Lactam Lactim ⇌ Lactam

Chart 5.2. Tautomeric forms of purine and pyrimidine bases.

The next higher order of organization in DNA is the nucleosides (Chart 5.3), in which purine or pyrimidine bases are connected by a glycosidic linkage to the C-1′ of deoxyribose or ribose, in DNA or RNA, respectively. In pyrimidines the sugar is attached at N-1, in purines at N-9.

The glycosidic linkage is relatively acid-labile. Depending on the type of sugar, the nucleosides are called, collectively, ribosides or deoxyribosides. Individually they are called adenosine (A) or deoxyadenosine (dA), guanosine (G) or deoxyguanosine (dG), cytidine (C) or deoxycytidine (dC), thymidine (T) (no "d" prefix is needed because it occurs only as a deoxyriboside), and uridine (U), which occurs only as a riboside.

The free rotation around N-9 or N-1, as the case may be, and C-1′ of the sugar, is restricted by steric hindrance; thus two conformations, *syn* and *anti*, are possible. In the naturally occurring nucleosides, the *anti* conformation is favored (Chart 5.3).

The esterification of the 3′ or 5′ hydroxyl of the sugar with phosphoric acid leads to the formation of nucleotides. Individually they are designated as adenosine monophosphate (adenylate) (AMP) or deoxyadenosine monophosphate (dAMP), and so on. In accordance with the nomenclature used with nucleosides, deoxythymidilate is designated as TMP.

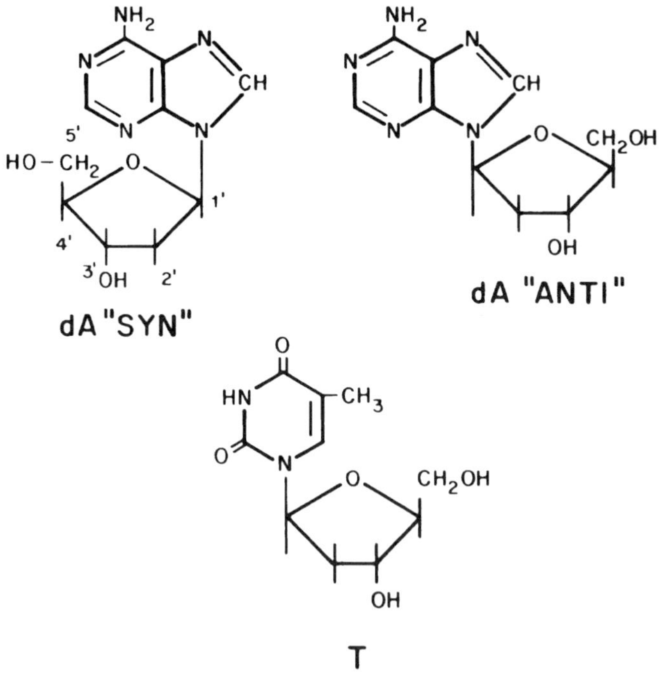

Chart 5.3. Possible conformation of nucleosides.

DNA is a polymer consisting of a chain of 2′ deoxyriboses connected by a 3′,5′ phosphodiester linkage, with the purine and pyrimidine bases projecting outward from the C-1′ of each deoxyribose (Structure 5.2). A chain of this sort has polarity; one end terminates in 5′-OH and the other one in 3′-OH.

In the late 1940s Chargaff and co-workers observed that, although the content of different nucleotides varied in different DNA species, the amount of dA was always equal to that of T, and the amount of dG was always equal to that of dC (*12*).

This observation, as well as X-ray diffraction data from the DNA molecule, led Watson and Crick (*13*) to postulate the model of double-stranded DNA (Chart 5.4). In this model, the two chains of DNA possess opposite polarity (i.e., one runs in the 5′–3′ direction and the other runs in the 3′–5′ direction). The chains are held together by hydrogen bonds between the bases. Because of the predominant tautomeric forms of the bases and the *anti* configuration of the deoxyribose, dA can pair only with T, and dG only with dC.

Two hydrogen bonds are present in the dA–T pair and three in the dG–dC pair; thus the binding force between dG and dC is 50% stronger than that between dA and T. Therefore, the dG–dC combination is more compact than the dA–T combination. The higher the dG–dC content, the greater the buoyant density of DNA. The bases in the helix are stacked on top of each other. The normal, B-form, DNA contains 10 base pairs per turn; this corresponds to a length of 3.4 nm.

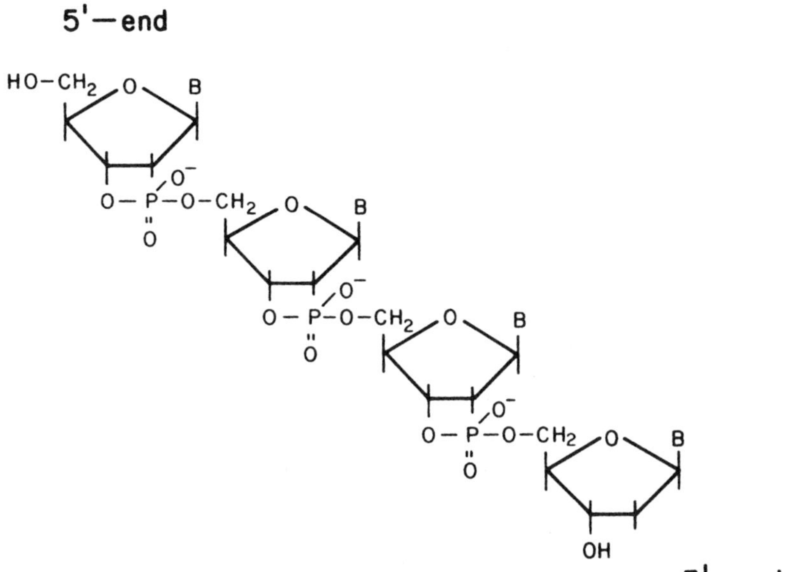

Structure 5.2. A single strand of DNA.

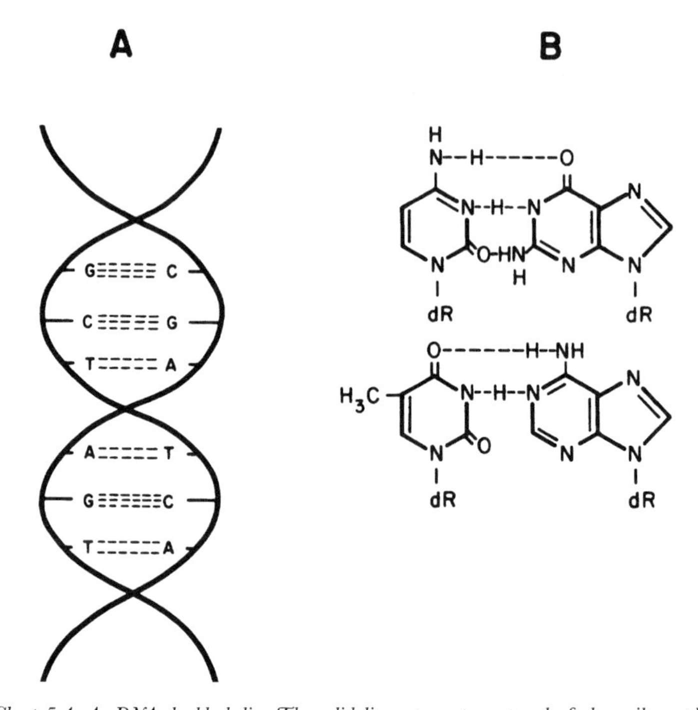

Chart 5.4. A, DNA double helix. The solid line represents a strand of deoxyribose–phosphate units; A, G, T, and C are the bases, and the broken lines represent hydrogen bonds. B, Hydrogen bonding between dA–T and dG–dC.

Increasing the temperature or decreasing the salt concentration results in melting or *denaturation* of DNA. In this process the two chains pull apart. This pulling apart is accompanied by an increase in the optical density of DNA, referred to as *hyperchromicity of denaturation.* The three-dimensional structure of the double helix reveals two grooves, referred to as the *major groove* and *minor groove.* In these grooves, specific proteins interact with DNA.

Only one of the DNA strands in the double helix, the so-called *sense* strand, contains genetic information. The other strand, which serves only as a template for replication, is called the *antisense* strand. During replication, the strands are pulled apart as the synthesis of the new strands, complementary to the old strands, proceeds in the 5′ to 3′ direction (Figure 5.1) (*14*).

The sense strand serves as a template for transcription of a specific sequence of nucleotides to form messenger RNA. The message contained in mRNA is, in turn, translated into a specific sequence of amino acids in proteins. A sequence of three nucleotides in the DNA is termed a codon; each codon codes for a specific amino acid. With four bases available and with three bases in each codon, there are 64 possible messages ($4^3 = 64$) to provide for 20 amino acids. Three codons do not code for any amino acid and are called nonsense codons; at least

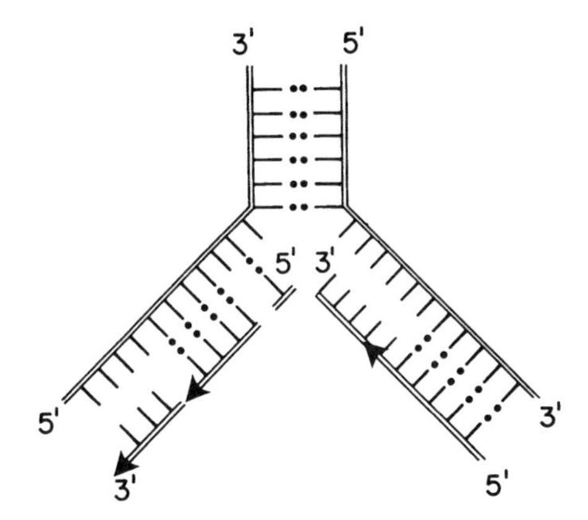

Figure 5.1. Schematic representation of the DNA replication process. The strand growing toward the outside of the fork replicates in segments; the gaps are closed later by ligases.

two of these code for termination of the amino acid chain. Because the remaining 61 triplets code for 20 amino acids, it appears that several different triplets code for the same amino acid. This phenomenon is referred to as *degeneracy* of the genetic code.

A chain of codons of about 1000 base pairs responsible for the synthesis of a specific protein is called a *gene*. Genes are assembled into *chromosomes*. A chromosome consists of about 10^8 base pairs. In addition to DNA, it contains a considerable amount of protein.

Chromosomal material extracted from the nuclei of eukaryotic organisms is called *chromatin*. It consists of double-stranded DNA and about an equal mass of basic proteins (called *histones*), a smaller amount of acidic proteins (called *nonhistones*), and a small amount of RNA.

The five types of histones are the lysine-rich H1, slightly lysine-rich H2A and H2B, and arginine-rich H3 and H4. Histones are involved in the folding ("superpacking") of DNA strands. The initial electron microscopic study of chromatin revealed that it consists of spherical particles about 12.5 nm in diameter (*nucleosomes*) connected by DNA filaments (*14*).

Further investigation of nucleosome structure disclosed that the double-stranded DNA is wound, in two complete turns, around a core consisting of an octamer of two of each: H2A, H2B, H3, and H4. There are 140 base pairs in this supercoiled arrangement. At each end of the coil there are straight segments of DNA (usually 20 base pairs or more). These segments, referred to as linker DNA, connect the nucleosomal particles. H1 is located at the entrance and at the exit of the coil (Figure 5.2) (*15*). Histone H1 is least tightly bound; when it is removed the chromatin becomes soluble. The histones are the same, or nearly so, for most eukaryotic species. When histones are mixed with DNA, chromatin is spontaneously formed, regardless of the origin of the various components. This

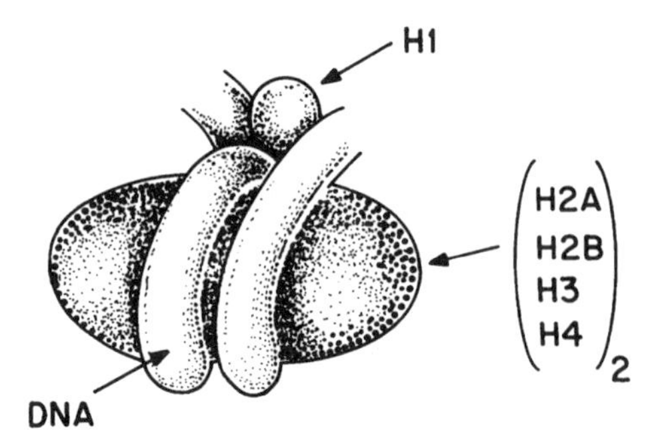

Figure 5.2. Conceptual image of a nucleosome.

chromatin formation results in folding of the double-stranded DNA to 1/7 of its original length.

Whereas histones are related to packing of nuclear material at the lower structural level of chromosomes, the nonhistone proteins appear to be involved in regulatory functions of gene expression. They cover or uncover specific areas of DNA as needed for transcription. The nonhistone proteins may also be involved in a higher level of organization of chromosomal DNA as scaffolding proteins (*16*).

Although the exact folding of the secondary structures (i.e., of the chains of nucleosomes in a chromosome) is not known, electron microscopic study indicates the folding of a thin fiber 5–10 nm in diameter into a heavier fiber 25–30 nm in diameter.

Chromosomal structure can be studied with light microscopy. At the point during cell division called metaphase, mammalian chromosomes appear as X-shaped objects. The two sides of the X are referred to as *sister chromatids*, and the connecting point as the *centromere*. The position of the centromere is characteristic for each chromosome. The long arms of the chromatids are designated as "q" and the short ones as "p". When chromosomes are stained with quinacrine or Giemsa stain, a characteristic pattern of horizontal bands appears. This banding is highly reproducible within species but varies among species.

Mutagenesis

A large number of environmental chemicals interact with the human and animal genetic systems. Three types of observable genetic lesions are

1. changes in DNA known as point mutations,
2. changes in chromosomal structure, such as breaking off of a part of a chromosome or translocation of an arm, known as clastogenesis, and

3. uneven separation of chromosomes during cell division, known as aneuploidization.

Point Mutation

Point mutation may involve either base substitution or frameshift mutation. Two types of *base substitution* are transition (when a purine is replaced by a purine, such as A by G or vice versa, or a pyrimidine is replaced by a pyrimidine, such as C by T or vice versa) and transversion (when a purine is replaced by a pyrimidine or vice versa). Altogether there are six possible base substitutions: two transitions (AT–GC; GC–AT) and four transversions (AT–TA; AT–CG; GC–CG; GC–AT).

A single base substitution is expected to be of little consequence. First of all, because of the degeneracy of the genetic code, misincorporation of a base into DNA may not affect the incorporation of the proper amino acid into a protein at all. Second, even if the wrong amino acid should be incorporated, unless it happens to be positioned in the active site of an enzyme, the activity of the enzyme will not be affected.

Base substitutions that do not produce changes in amino acids of proteins, or that produce changes that do not alter enzyme activity, are termed *cryptic mutations*. However, it may happen that the base substitution will lead to the formation of a nonsense codon, one that codes for termination of protein synthesis. In this case an incomplete enzyme will be synthesized, which may have serious consequences.

Frameshift mutation occurs when base pairs are added or deleted and their number is other than three or a multiple of three. In this case the triplet code is misread entirely (Chart 5.5), and the result is a radical change of the protein structure.

Point mutations cannot be detected by morphological examination of chromosomes. If a point mutation occurs in reproductive cells, a mutated offspring may result. Heritable disorders due to point mutations may have their origin from ei-

A

ATC AAT GCG TTA
TAG TTA CGC AAT

B

TCA ATG CGT TA
AGT TAC GCA AT

Chart 5.5. Schematic representation of the process of frameshift mutation. Key: A, original sequence of codons; B, sequence upon deletion of one base.

ther the paternal or maternal side. In contrast to the chromosomal aberrations to be discussed, the frequency of point mutations increases with paternal age.

Clastogenesis

The normal human carries 46 chromosomes: 22 pairs called autosomes, which are designated by consecutive numbers from 1 to 22, and two sex chromosomes, XX in females and XY in males. This composition is referred to as the normal human karyotype. The study of chromosomes and their abnormalities can be done in cell culture, in bone marrow, or in peripheral lymphocytes.

The chromosomes are best characterized at mitosis, because during this period they are visible by light microscope. The banding that appears upon staining allows identification of chromosomal fragments. Thus breaks, gaps, unstained segments, sister chromatid exchanges, and combinations of two chromosomes or their fragments can be determined.

Some evidence suggests that, at least in some cases, *clastogenesis* is a result of chemical injury. A correlation between intercalator-induced DNA strand breaks and sister chromatid exchanges has been presented (*17*).

Aneuploidization

Aneuploidization is a term for uneven distribution of chromosomes during cell division. Although many hereditary disorders are caused by this phenomenon, the causes and mechanism of aneuploidization are still largely unknown. Except for the effects of X-rays, no other causative factor has been found.

The following code is used to designate the type of chromosomal abnormality. The first number indicates the total number of chromosomes in the karyotype, and the second one designates the additional or missing chromosome, followed by + or 0, respectively. According to this code, Down's syndrome is designated as (47, 21+) and Turner syndrome as (45, X0). In the former case there is trisomy of chromosome 21, and in the latter case one sex chromosome is missing.

The chances of aneuploidy increase with maternal age, but in general the frequency of live births with abnormal chromosomal patterns is relatively low (23–30%), as compared to the frequency of occurrence. Most abnormal fetuses are spontaneously aborted. An in-depth treatment of this subject can be found in the review by Thilly and Call (*18*).

Interaction of Chemicals with DNA

Alkylations

The susceptibility of DNA to nucleophilic substitution results from its large content of hetero atoms, such as nitrogen and oxygen, which carry pairs of free elec-

Table 5.2. Positions in DNA Susceptible to Electrophilic Attack

Position	*Guanine*	*Adenine*	*Cytosine*	*Thymine*
N-1	Yes	Yes	—	—
N-3	Yes	Yes	Yes	Yes
N-7	Yes	Yes	—	—
Exocyclic atoms	O^6	N^6	O^2, N^4	O^4
C-8	Yes	Yes	—	—

NOTE: — indicates not applicable.

trons. Practically all endo- and exocyclic nitrogens, except N-9 in purine and N-1 in pyrimidine bases, are subject to electrophilic attack. So are the oxygens in the bases and the nonesterified phosphate oxygens of the backbone of the DNA strands. In addition, the acidic C-8 of purines assumes nucleophilic properties by dissociating its hydrogen as a proton. Table 5.2 lists the positions of each base that are susceptible to electrophilic attack. In position notation, the superscript indicates an exocyclic atom.

The preferred substitution site in the base molecule depends on the nucleophilicity of the atom undergoing substitution, accessibility of the site, and the size of the alkylating agent. For small alkylating agents, where steric hindrance is not a factor, the rate of reaction depends on electrophilicity of the alkylating agent and nucleophilicity of the site of substitution, as related by the Swain–Scott equation.[3] Alkylating agents with a large Swain–Scott *s* parameter react via the S_N2 mechanism, and only with the strongest nucleophiles. Those with a small *s* react via the S_N1 mechanism, with strong and weak nucleophiles alike (*see* Chapter 3, footnote 2).

In general (but not always), the bulky electrophiles show a preference for N-7 and C-8 of guanine and are preferentially incorporated into the linker, rather than into the core, DNA (*10*). With small alkylating agents (such as *N*-nitroso compounds) that react via carbonium ions ($R–CH_2^+$) (Scheme 3.13 in Chapter 3), every N and O in purine and pyrimidine bases, as well as nonesterified O in the phosphates, are potential subjects for interaction (*11, 19*). The relative extent of alkylation of adenine and guanine by methylnitrosourea (MNU), presented in Table 5.3, indicates the relative nucleophilicity of the hetero atoms of the purine bases. The extraordinarily strong nucleophilicity of N-7 of guanine is worth noting.

[3]C. G. Swain and C. B. Scott developed the following two-parameter equation to correlate the relative rates of reaction of nucleophilic agents with various organic substrates (electrophiles): $\log(k/k_0) = sn$, where k_0 and k are rate constants for reactions with water and any other nucleophile, respectively; s (substrate constant) is the electrophilic parameter, which is equal to 1.0 for the reference compound, methyl bromide; and n is the nucleophilic parameter, which is equal to 0.00 for water (*35*).

Table 5.3. Relative Extent of Alkylation of Adenine and Guanine by Methylnitrosourea

Position	*Adenine*	*Guanine*
N-3	8.2	0.6
N-1	2.7	—
N-7	1.2	65.6
O^6	—	6.7

NOTE: The values in this table represent ratios.
SOURCE: Adapted from reference 20.

Some alkylations result in the formation of altered but stable products that persist permanently or until excised in the process of repair. However, other alkylations lead to unstable adducts that subsequently undergo a series of rearrangements. Consequences of alkylation may vary, depending on the type of substituent and position of alkylation, from a relatively innocuous base substitution to a very injurious DNA strand break or removal of a base.

Methylation or ethylation at O^6 of guanine causes a change in its tautomeric form so that it will resemble adenine (Scheme 5.2). Thus, during replication of DNA or transcription of messenger RNA, 6-methylguanine will pair with thymine (or uracil) instead of cytosine. Such base substitution, as explained earlier, may cause a perceivable or a cryptic mutation.

The consequences of substitution on N-7 or N-3 of purines are much more serious. Table 5.3 shows that N-7 is the most reactive atom in guanine, whereas in adenine N-3 is the most reactive. Aflatoxin B_1, upon metabolic activation to 2,3-epoxide, reacts with N-7 of guanine (*21*). N-7-substituted purines are unstable and may decompose in two ways, as depicted in Chart 5.6 (*22*). The pyrazine ring opening (reaction I) leads to a rather stable product that distorts the fidelity of the genetic code. The depurination (reaction II), which may also occur with N-3-substituted purines, leaves a gap in the sequence of nucleotides leading to a frameshift mutation.

Free deoxyribose (like other sugars) exists in two forms that are in equilibrium with each other: the cyclic furanose and the open aldose. In DNA, because of

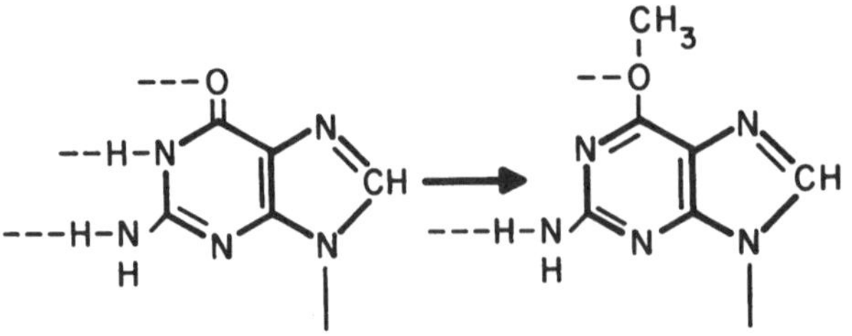

Scheme 5.2. Consequences of methylation on O^6 of guanine.

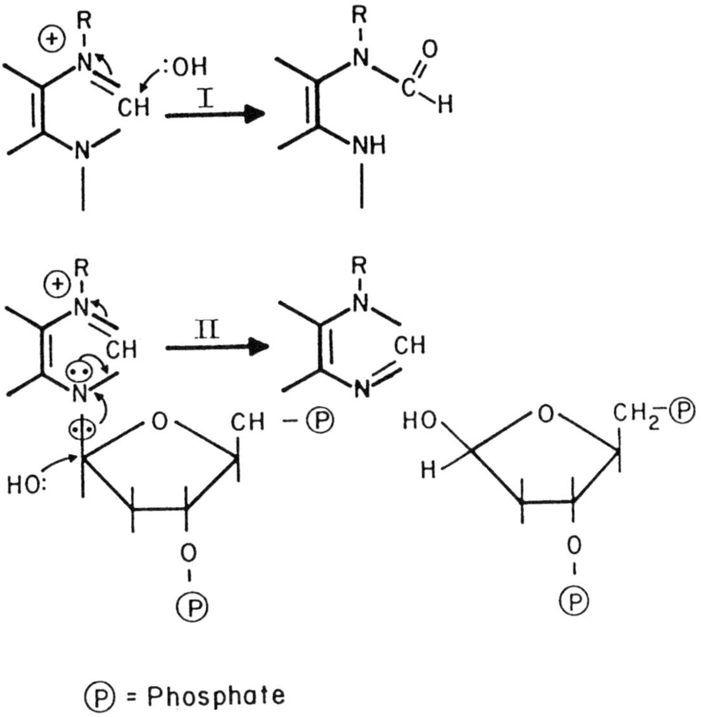

Ⓟ = Phosphate

Chart 5.6. Consequences of alkylation on N-7 of guanine.

the glycosidic linkage with the bases, there is only the furanose form. Upon depurination, equilibrium between furanose and aldose is established (Scheme 5.3). Aldose is susceptible to base-catalyzed rearrangement leading to a strand break at the 3′ position. Alternately, the free aldehyde may cross-link, via Schiff base formation, with a nearby amino group. Both of these reactions will cause additional distortion of the DNA.

Acetylaminofluorene (AAF) is activated, as described in Chapter 3, to a strong electrophile. The positively charged nitrogen reacts with the nucleophilic C-8 of guanosine (Scheme 5.4), and forces rotation of guanine around its glycoside bond. The planar AAF intercalates between stacked bases, and guanine slips out so that it projects to the outside of the helix. This movement is referred to as *base displacement* (*19*). A gap in the nucleotide sequence is thus created and results in a frameshift mutation.

The ubiquitous environmental carcinogen benzo[*a*]pyrene, upon activation to 7,8-dihydrodiol-9,10-epoxide (Chapter 3), forms a covalent bond between C-10 of the hydrocarbon and the exocyclic nitrogen of guanine (Scheme 5.5). Both stereoisomers of 7,8-dihydrodiol-9,10-epoxide, *cis*- and *trans*-epoxy (with respect to 7-hydroxy), react with DNA in vitro, but only the *trans* isomer reacts in vivo

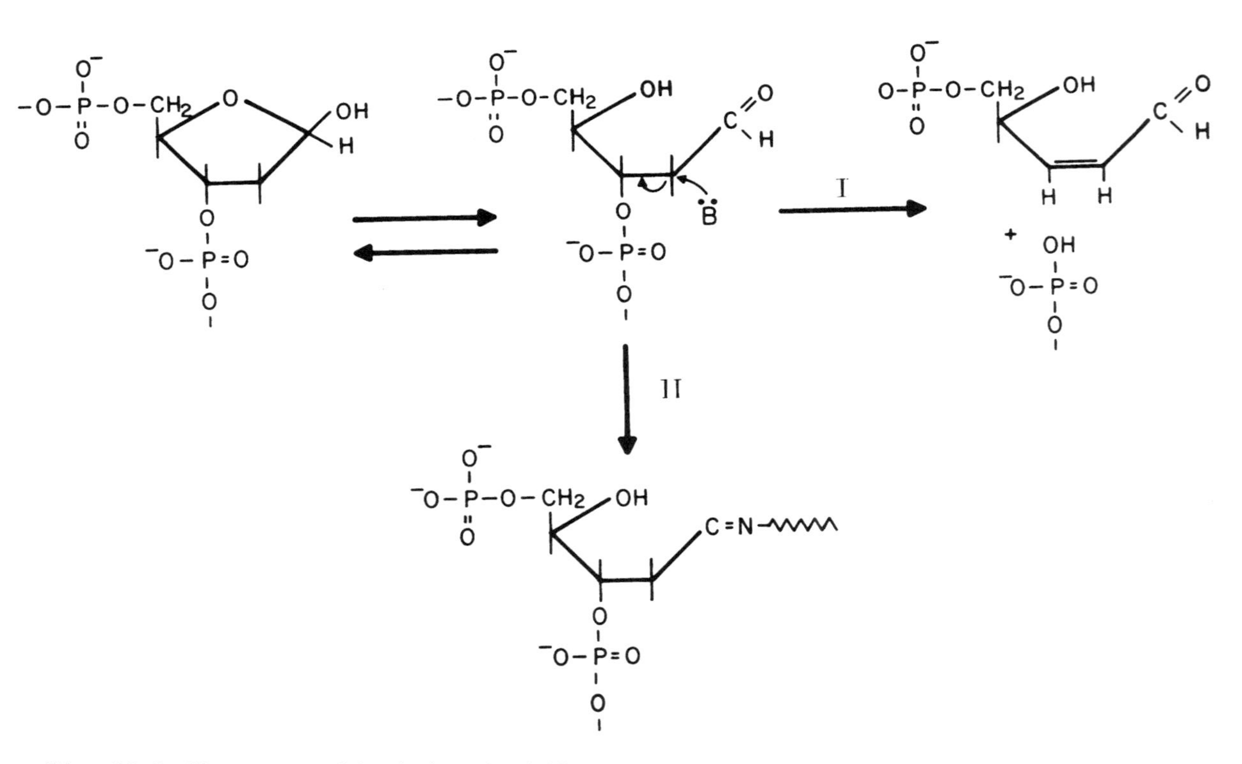

Scheme 5.3. Possible consequences of depurination or depyrimidination. Reaction I is a strand break. Reaction II is a Schiff base cross-link.

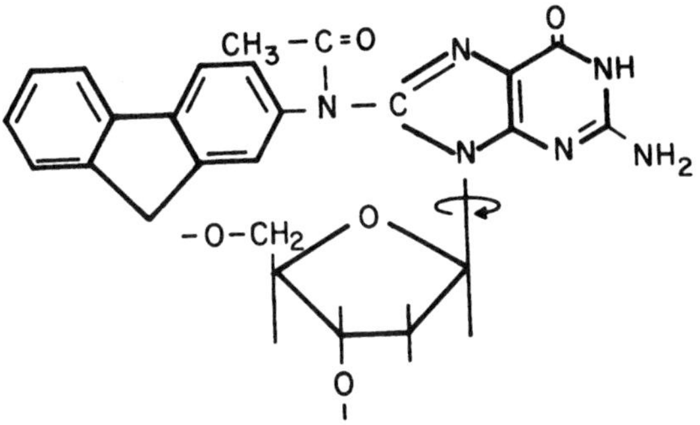

Scheme 5.4. Interaction of activated AAF with C-8 of guanosine.

(*23, 24*). This effect may be due to the instability of the *cis* isomer, as postulated in Chapter 3. Alkylation by benzo[*a*]pyrene was reported (*19*) to cause frameshift mutation. Whether this mutation results from its interaction with guanine or from the alleged alkylation of phosphate has not been established. Table 5.4 compares relative reactivities of the N-7 of guanine to that of phosphate oxygens, with four alkylating agents.

An initial attack on the OH group of phosphate is difficult because of the resonance between the two free oxygens (Chart 5.7). However, once the alkylation takes place, the positions of the electrons are fixed and the subsequent alky-

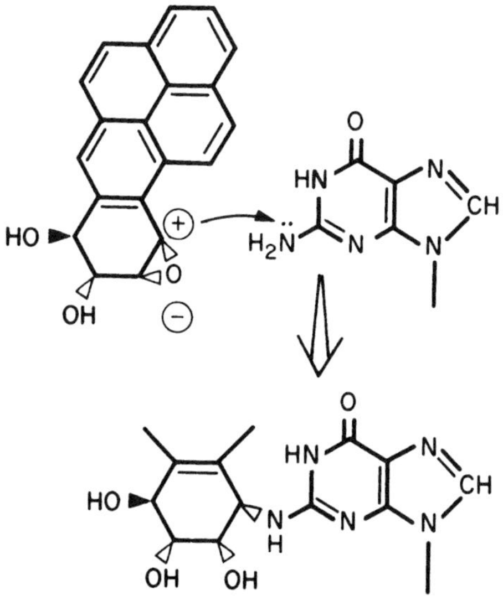

Scheme 5.5. Interaction of 7,8-dihydrodiol-9,10-epoxide of benzo[a]pyrene with the amino group of guanine.

Table 5.4. Relative Reactivity of N-7 of Guanine and Phosphate Oxygens with Four Alkylating Agents

Alkylating Agent	*N-7*	*Phosphate*
Methyl methanesulfonate	81.4	0.82
Ethyl methanesulfonate	58.4	12.00
N-Methyl-*N*-nitrosourea	66.4	12.10
N-Ethyl-*N*-nitrosourea	11.0	55.40

NOTE: Values are nondimensional and relative.
SOURCE: Adapted from reference 20.

lation of the triester is greatly facilitated. This second attack is followed either by removal of the first alkyl group (thus retaining the status quo) or by a strand break between the phosphate and the 3′-OH of deoxyribose. The phosphotriester may be subject to alkali-catalyzed hydrolysis, which likewise results in either removal of the alkyl group or strand scission. This type of scission cannot be repaired because the ligases designed to mend strand breaks can join only 3′-phosphate with 5′-OH, but not 3′-OH with 5′-phosphate.

Intercalating Agents

Certain aromatic or heterocyclic planar compounds are able to insert themselves between stacked bases of DNA. This type of interaction is called *intercalation.* Intercalation results in local spreading and distortion of the helix so that the length of the helix per turn is increased (*25*). Some examples of intercalating agents are presented in Chart 5.8. All these compounds are characterized by their dimen-

$$1.\quad 3'\text{–O–P}(\text{O}^-)(=\text{O})\text{–O–5}' \rightleftharpoons 3'\text{–O–P}(=\text{O})(\text{O}^-)\text{–O–5}'$$

$$2.\quad 3'\text{–O–P}(=\text{O})(\text{O}^-)\text{–O–5}' + \text{X}^{\oplus} \longrightarrow 3'\text{–O–P}(=\text{O})(\text{OX})\text{–O–5}'$$

$$3.\quad 3'\text{–O–P}(=\text{O})(\text{OX})\text{–O–5}' + \text{X}'^{\oplus} \longrightarrow \text{X}'\text{O–P}(=\text{O})(\text{OX})\text{–O–5}'$$

Chart 5.7. Alkylation on phosphate (triester formation). Key: 1, resonance between nonesterified oxygens; 2, first alkylation; 3, second alkylation and strand break.

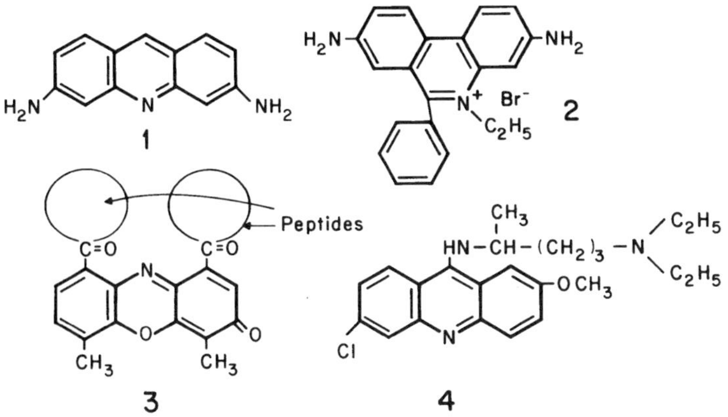

Chart 5.8. Examples of intercalating agents. Key: 1, acriflavine; 2, ethidium bromide; 3, actinomycin; 4, quinacrine.

sions, which correspond to three condensed aromatic (or heterocyclic) rings, about the same as the diameter of the DNA double helix.

One study (*17*) presents evidence that intercalators interfere with the action of topoisomerase II. Topoisomerase II catalyzes transient double-strand breaks of DNA for purposes such as replication and transcription. Although strand scission occurs in the presence of intercalators, the topoisomerase II remains firmly bound to the nicked DNA and thus prevents ligation of the strand.

Effect of Ultraviolet Radiation

X-rays and shortwave ultraviolet radiation cause strand breaks via free-radical formation. However, ultraviolet light of wavelengths around 290 nm, which is in the range of light absorption of pyrimidines, causes dimerization of neighboring pyrimidines (Scheme 5.6). Such dimerization results in unwinding of the DNA

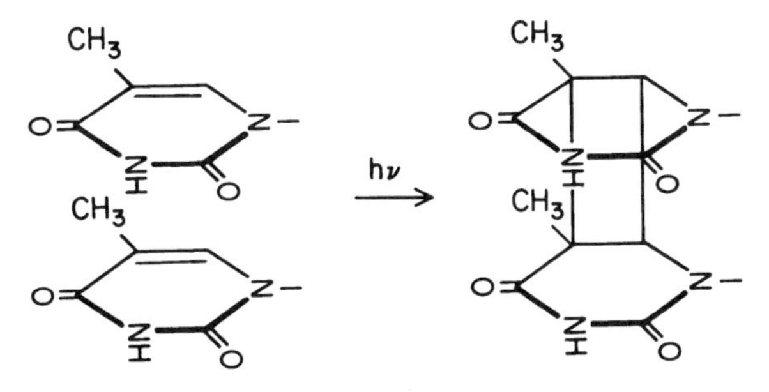

Scheme 5.6. Dimerization of thymidine.

helix and disruption of hydrogen bonds between the dimerized pyrimidines and their complementary purines.

DNA Repair Mechanism

Although no direct evidence proves that mutations lead to cancer, many indications suggest that this is indeed the case. Nearly all chemicals that are mutagenic (in vitro as well as in vivo) cause cancer. Tumors originate from a single cell, and the cancer cell phenotype is heritable; presumably a mutated cell is the origin of the tumor. Certain genetic abnormalities predispose to cancer and so do defects in the DNA repair mechanism, as exemplified by xeroderma pigmentosum.

Chemical or radiation-induced DNA damage will lead to mutation only if it is not *properly* repaired before, or immediately after, replication of the genome. The emphasis is on properly, because misrepair may cause mutation in itself. The original alteration of a DNA base, caused by alkylation or dimerization, is referred to as *premutagenic change*. The mutation is fixed only if the damage is misrepaired or not repaired at all.

Several types of DNA repair occur. The best-elucidated type is excision repair. Excision repair may involve two different mechanisms. In the case of thymidine dimers, a nick is produced in the DNA strand near the damaged area, the nucleotides are released, and the lesion is repaired with new nucleotides by using the undamaged strand as a template. If a single base is damaged, the repair involves removal of the base, followed by scission of the strand and resynthesis of the damaged area as in the former case (*18*). Excision repair usually functions with high fidelity. A positive correlation has been found between the DNA excision repair potential of a species and its longevity (*26*).

Other modes of repair that frequently occur following DNA replication are not well understood. Some of them are error-prone and may be responsible for establishment of mutations and the development of cancer (*17*). According to some sources, certain repairs, such as demethylation of O^6-methylguanine, can be performed by a special methyltransferase (*27*).

Regardless of how good or how bad the repair mechanism is, mutation will occur if the frequency and extent of injuries exceeds the capability of the system or if the repair mechanism is deficient or suppressed.

Xenoestrogens

Epigenetic carcinogens that recently attracted attention are the female hormones, estrogen and progesterone. It is estimated that about 40% of all cancers in women are hormonally mediated (*28*). The mode of action of these hormones as carcinogens is not understood; however, it appears that the length and the timing of exposure play a large part in determining breast cancer risk. It seems that the longer the period in the life of a woman between the onset of the menstrual cycle and menopause, the greater the likelihood that she will develop breast cancer (*29*). This fact may explain the difference in the breast cancer incidence and mortality rates between races. For instance, the rate of mortality due to breast cancer in the United States was 22.4 per 100,000 people during 1986–1988, whereas in China it was only 4.7. Correspondingly, American girls reach menarche on average at the age of 12.8, while Chinese girls reach it at the age of 17 (*29*).

Incidence rates of breast cancer in the United States increased by about 3% a year between 1980 and 1988, from 84.8 per 100,000 in 1980 to 109.5 per 100,000 in 1988 (*3*). A similar increase has been observed in other industrialized countries. The improved detection methods (mammography) may account in part for the observed rise, but they cannot entirely explain the pattern. A recent study showed a correlation between concentration of DDE (1,1-dichloro-2,2-bis(*p*-chlorophenyl)ethylene) (the principal metabolite of the pesticide DDT) in women's serum and the incidence of breast cancer (*30*).

Lately, several structurally unrelated synthetic compounds that bind to estrogen receptors have been identified. They are designated by a common name, "xenoestrogens". Most of them are either pesticides, such as DDT, DDE, Kepone, and dieldrin, or industrial byproducts, such as some PCBs, alkyl phenols, and PAH. They either mimic the natural hormone, or they inhibit its action. In either case they create havoc in women's endocrine systems (*31*).

Devra Lee Davis, scientific adviser at the U.S. Department of Health and Human Services, and H. Leon Bradlow at the Strang Cornell Cancer Research Laboratory, postulated a mechanism of action for xenoestrogens (*31*). The natural estrogen, estradiol, is metabolized via two pathways: conversion to 2-hydroxyestrone and to 16-hydroxyestrone (Scheme 5.7). Whereas the former has a weak estrogenic activity and is not carcinogenic, the latter has a powerful estrogenic activity and damages DNA. According to these scientists, xenoestrogens inhibit the pathway leading to the formation of 2-hydroxyestrone and shift the metabolism toward the formation of 16-hydroxyestrone.

Recent research in humans and animals indicates that prenatal exposure to minute amounts of xenoestrogens disturbs the entire endocrine balance of the fetuses, resulting in abnormal sexual and mental development. Thus, not only the levels and activities of estrogens are affected but also those of androgens and thyroid hormone. The name *endocrine disrupters* has been coined for compounds with such activity (*32*).

Carcinogenic Effect of Low-Frequency Electromagnetic Fields

Since the past decade there has been concern about the health effects of low-frequency electromagnetic fields, such as those produced by power lines, home ap-

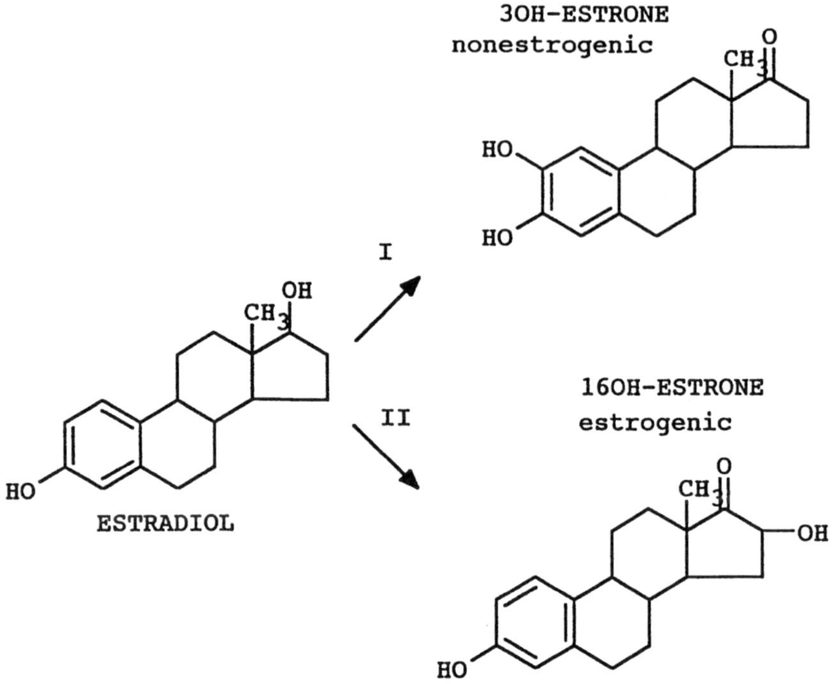

Scheme 5.7. Metabolism of estradiol.

pliances, and electric gadgetry. This concern was precipitated by reports of clusters of elevated cancer incidence, especially childhood leukemia, among people residing in the vicinity of power lines. In response to this concern several epidemiological studies were undertaken. Whereas some of them showed a weak association between exposure to low-frequency electromagnetic fields and childhood leukemia and other types of cancer, others did not. Similarly, animal experiments gave contradictory results. The animal studies were complicated because there was no clear dose–response effect, and because the effect depended on the frequency, the waveform, and the angles between the applied field and that of the earth's magnetic field (*33*).

The most recent epidemiological study examining records of women who died of breast cancer indicated 38% higher mortality among electrical workers as compared to women employed in other occupations (*34*). The connection between exposure to a low-frequency electromagnetic field and breast cancer has its theoretical bases. It has been observed that electromagnetic fields reduce the production, by the pineal gland, of the nocturnal hormone melatonin. Melatonin is an antagonist of estrogen and as such suppresses the tumor-enhancing activity of this hormone. Although the study quoted above lends support to the melatonin theory, the authors caution that their study had serious limitations and that more research is needed to prove definitively that a connection between electromagnetic-field exposure and breast cancer really exists. An extensive review of the health effects of electromagnetic fields has been published (*33*).

References

1. Kundson, A. G. In *Origins of Human Cancer;* Hiatt, H. H.; Watson, J. D.; Winsten, J. A., Eds.; Cold Spring Harbor Laboratory: Cold Spring Harbor, NY, 1977; Cold Spring Harbor Conference on Cell Proliferation, Vol. 4, Book A, p 45.
2. Office of Technology Assessment. *Cancer Risk: Assessing and Reducing the Dangers in Our Society;* Westview: Boulder, CO, 1982.
3. *Cancer Facts & Figures–1992;* American Cancer Society: Atlanta, GA, 1992.
4. Berenblum, I.; Shubik, P. *Br. J. Cancer* **1974,** *1,* 379.
5. Chand, N.; Hoel, D. G. *A Comparison of Models Determining Safe Levels of Environmental Agents in Reliability and Biometry; Statistical Analysis of Lifelength;* Society for Industrial and Applied Mathematics: Philadelphia, PA, 1974.
6. Miller, E. C. *Cancer Res.* **1978,** *38,* 1479.
7. Williams, G. M.; Weisburger, J. H. In *Cassarett and Doull's Toxicology;* Klaassen, C. D.; Amdur, M. O.; Doull, J., Eds.; MacMillan: New York, 1986; Chapter 5, p 99.
8. Berenblum, I. *Carcinogenesis as a Biological Problem;* Frontiers of Biology, Vol. 34; North-Holland: Amsterdam, 1974.
9. Hecker, E. In *Carcinogenesis, A Comprehensive Survey;* Slaga, T. J.; Sivak, A.; Boutwell, R. K., Eds.; Raven: New York, 1978; Vol. 2, Mechanism of Tumor Promotion and Cocarcinogenesis, p 11.
10. Bachur, N. R.; Gordon, S. L.; Gee, M. V. *Cancer Res.* **1978,** *38,* 1795.
11. Ceretti, P. A. In *Chemical Carcinogenesis;* Nicolini, C., Ed.; Plenum: New York, 1982.
12. Zamenhof, S.; Brawerman, G.; Chargaff, E. *Biochim. Biophys. Acta* **1952,** *9,* 402.
13. Watson, J. D.; Crick, F. H. C. *Nature (London)* **1953,** *171,* 737.
14. Harper, H. A.; Rodwell, V. W.; Mayers, P. A. *Review of Physiological Chemistry;* Lange Medical: Los Altos, CA, 1979; Chapter 30, p 460.
15. Allan, J.; Hartman, P. G.; Crane-Robinson, C.; Aviles, F. X. *Nature (London)* **1980,** *288,* 675.
16. Laemmli, U. K. *Pharmacol. Rev.* **1979,** *30*(4), 469.
17. Pommier, Y.; Zwelling, L. A.; Kao-Shan, C. S.; Whang-Peng, J.; Bradley, M. O. *Cancer Res.* **1985,** *45,* 3143.
18. Thilly, W. G.; Call, M. K. In *Cassarett and Doull's Toxicology;* Klaassen, C. D.; Amdur, M. O.; Doull, J., Eds.; MacMillan: New York, 1986; Chapter 6, p 174.
19. Weinstein, I. B. *Bull. N. Y. Acad. Med.* **1978,** *54*(4), 366.
20. Lawley, P. D. In *Chemical Carcinogenesis,* 2nd ed.; Searle, C. E., Ed.; ACS Monograph 182; American Chemical Society: Washington, DC, 1984; Vol. 1, Chapter 7, p 325.
21. Essigman, J. M.; Croy, R. G.; Nadzan, A. M.; Busby, W. F., Jr.; Reinhold, V. N.; Buechi, G.; Wogan, G. N. *Proc. Natl. Acad. Sci. U.S.A.* **1977,** *74*(5), 1870.
22. Wogan, G. N.; Croy, R. G.; Essigman, J. M.; Groopman, J. D.; Thilly, W. G.; Skopek, T. R.; Liber, H. L. In *Environmental Carcinogenesis;* Emmelot, P.; Kriek, E., Eds.; Elsevier North-Holland Biomedical: Amsterdam, Netherlands, 1979; p 97.
23. Weinstein, I. B.; Jeffrey, A. M.; Jennette, K. W.; Blobstein, S. H.; Harvey, R. G.; Harris, C.; Antrup, H.; Kasai, H.; Nakanishi, K. *Science (Washington, D.C.)* **1976,** *193,* 592.
24. Jeffrey, A. M.; Jennette, K. W.; Blobstein, S. H.; Weinstein, I. B.; Beland, F. A.; Harvey, R. G.; Kasai, H.; Miura, I.; Nakanishi, K. *J. Am. Chem. Soc.* **1976,** *98,* 5714.
25. Lerman, L. S. *J. Mol. Biol.* **1967,** *3,* 18.

26. Hart, R. W.; Setlow, R. B. *Proc. Natl. Acad. Sci. U.S.A.* **1974,** *71*(6), 2169.
27. Yarosh, B. D. *Mutat. Res.* **1985,** *145,* 1.
28. Davis, D. L.; Bradlow, H. L.; Woodruff, T.; Hoel, D. G.; Anton-Culver, H. *Environ. Health Perspect.* **1993,** *101*(5), 372.
29. Marshall, E. *Science (Washington, D.C.)* **1993,** *259,* 618.
30. Wolff, M. S.; Toniolo, P. G.; Lee, E. W.; Rivera, M.; Dubin, N. *J. Natl. Cancer Inst.* **1993,** *85*(8), 648.
31. Hileman, B. *Chem. Eng. News* January 31, 1994, p 19.
32. Colborn, T.; Dumanoski, D.; Peterson Myers, J. *Our Stolen Future;* Penguin Books: New York, 1996.
33. Hileman, B. *Chem. Eng. News* November 8, 1993, p 15.
34. Loomis, D. P.; Savitz, D. A.; Ananth, C. V. *J. Natl. Cancer Inst.* **1994,** *86*(12), 921.
35. Swain, C. G.; Scott, C. B. *J. Am. Chem. Soc.* **1953,** *75,* 142.

6

Risk Assessment

The purpose of risk assessment is estimation of the severity of harmful effects to human health and the environment that may result from exposure to chemicals present in the environment. The Environmental Protection Agency (EPA) procedure of risk assessment, whether related to human health or to the environment, involves four steps:

1. hazard assessment,
2. dose–response assessment,
3. exposure assessment, and
4. risk characterization.

Hazard Assessment

The quantity of chemicals in use today is staggering. According to the data compiled by Hodgson and Guthrie in 1980 (*1*), there were then 1500 active ingredients of pesticides, 4000 active ingredients of therapeutic drugs, 2000 drug additives to improve stability, 2500 food additives with nutritional value, 3000 food additives to promote product life, and 50,000 additional chemicals in common use. Considering the growth of the chemical and pharmaceutical industries, these amounts must now be considerably larger.

Past experience has shown that some of these chemicals, although not toxic unless ingested in large quantities, may be mutagenic and carcinogenic with chronic exposure to minute doses, or may interfere with the reproductive or immune systems of humans and animals. To protect human health it is necessary to determine that compounds to which people are exposed daily or periodically in their daily lives (such as cosmetics, foods, and pesticides) will not cause harm upon long-term exposure.

The discussion in this chapter will focus on carcinogenicity and mutagenicity. The carcinogenicity of some chemicals was established through epidemiological studies. However, because of the long latency period of cancer, epidemiological studies require many years before any conclusions can be reached. In addition, they are very expensive.

Another method that could be used is bioassay in animals. Such bioassays, although quite useful in predicting human cancer hazard, may take as long as 2 years or more and require at least 600 animals per assay. This method is also too costly in terms of time and money to be considered for large-scale screening. For these reasons an inexpensive, short-term assay system is needed for preliminary evaluation of potential mutagens and carcinogens.

Bacterial Mutagenesis Test

Several versions of the bacterial mutagenesis test exist, but by far the most commonly used is the Ames test (*2*). This test uses genetically engineered strains of *Salmonella typhimurium* that are incapable of synthesizing the amino acid histidine and thus require histidine for growth. The test measures the frequency of back mutations to a histidine-independent parent strain.

The bacteria are seeded on agar plates with minimum growth medium that contains just enough histidine to produce a background growth, and with the compound to be tested. The back-mutated organisms produce colonies that are counted. Control plates are set to score for spontaneous mutations. A dose–response curve can be traced with increasing doses of the mutagen. Because many potential mutagens–carcinogens require metabolic activation and bacteria do not have such an activating system, a liver microsomal preparation (postmitochondrial supernatant, PMS) is added to the plates.

Several mutated strains, differing in their genetic makeup, have been developed. This variety shows a distinction between base substitution and frameshift mutation. In addition, supersensitive strains lack a DNA repair system or lipopolysaccharide coating. Thus, they are more vulnerable to exogenous chemicals.

The predictive reliability of the Ames assay has been tested experimentally, and 85% of the known carcinogens tested positive. Among compounds classified as noncarcinogens, fewer than 10% tested positive. A newer study (*3*) indicated that the predictability of the *Salmonella* test depended greatly on the chemical class of compounds tested. Thus only 40% of the chlorinated carcinogens were identified as mutagens, whereas 75 and 100% of the carcinogenic amines and nitro compounds, respectively, tested positively as mutagens. Also, 29% of the compounds that were not mutagens were positive in this test.

Another bacterial assay involves *Escherichia coli* that is deficient in a DNA repair mechanism. Mutagens that produce DNA lesions are more toxic to the genetically altered strain than to the parent strain (*4*).

DNA Repair Assay

This assay, performed in mammalian cell culture, is designed to detect compounds injurious to DNA. The test presupposes that the injury to DNA stimulates the repair mechanism. DNA repair is measured by the increase in the incorporation of ^{3}H-thymidine into DNA above that of the control. The radioactivity is determined either by scintillation counting or by autoradiography. PMS is added to the cultures to activate precarcinogens.

A modification of this procedure, the hepatocyte primary culture–DNA repair assay, uses freshly isolated, nondividing liver cells. This system has no need for PMS, as the hepatocytes can activate precarcinogens. In addition, the nondividing cells have no background of thymidine incorporation (*4*).

Mammalian Mutagenicity Assays

Three assays of this type are in use. The first and most common one uses mammalian fibroblasts. In this assay mutants are recognized by the appearance of colonies resistant to the purine analogs, 6-thioguanine or 8-azaguanine (*5*) (Chart 6.1). These analogs are not cytotoxic but are activated to cytotoxic nucleotides by a "purine salvage" enzyme, hypoxanthine–guanine phosphoribosyltransferase (HGPRT), which is present in most cells. This enzyme reuses preformed purines for nucleic acid synthesis. However, it is not essential for cell survival because most cells are able to synthesize purines de novo. The mechanism of the HGPRT-catalyzed reaction is presented in Scheme 6.1.

Normal cells will not grow in cultures exposed to either 6-thioguanine or 8-azaguanine. However, in the presence of a mutagen, 6-thioguanine–8-azaguanine-resistant mutants that lack HGPRT arise and colonies are formed. Addition of PMS is necessary to activate precarcinogens. This assay is extremely sensitive because HGPRT is not an essential enzyme; thus, its deletion does not result in the formation of lethal mutants. The locus of HGPRT is on the X (sex) chromosome, which is highly mutable and has no duplicate.

A modification of this procedure uses freshly prepared hepatocytes as a

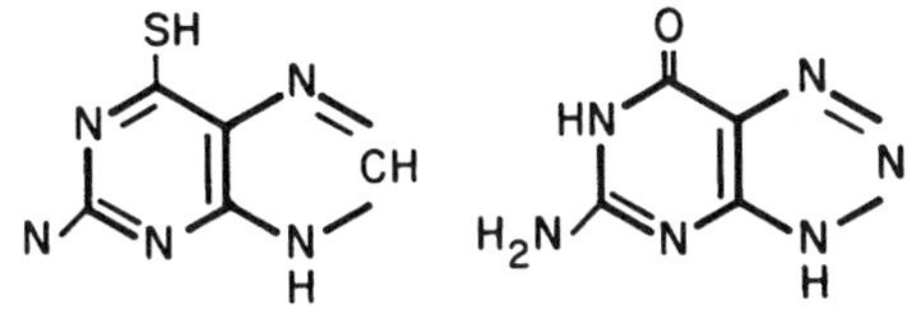

SH N N CH N N N H 1

O HN N N H2N N N H 2

Chart 6.1. 6-Thioguanine (1) and 8-azaguanine (2).

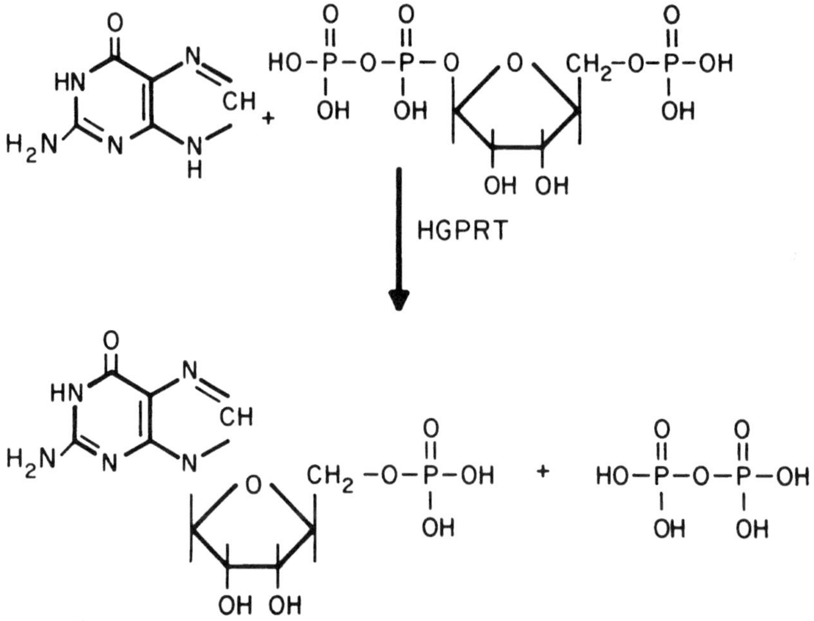

Scheme 6.1 Salvage pathway of purines, a mechanism of activation of purine analogs.

feeder layer. PMS is omitted in this assay because hepatocytes have xenobiotic-activating enzymes (*4*).

Another mammalian mutagenesis assay is based on mutation in the locus responsible for the synthesis of thymidine kinase, an enzyme required for activation of the antimetabolite iododeoxyuridine (*6*) (Scheme 6.2). Only mutated cells, which have lost kinase, form colonies in the presence of the antimetabolite. The kinase is not an essential enzyme, and thus no lethal mutants are produced.

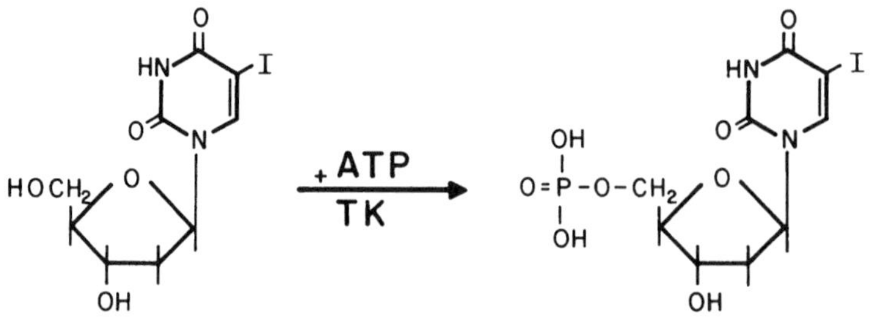

Scheme 6.2. Mechanism of activation of 5-iododeoxyuridine.

The third assay in this class scores for mutants resistant to the alkaloid ouabain (*7*). Ouabain resistance is derived from the mutation of the gene responsible for the synthesis of membrane ATPase, an enzyme involved in K^+–Na^+ transfer. Ouabain inhibits this enzyme noncompetitively and thus interferes with essential cell functions. This assay lacks sensitivity because the only mutants available for scoring are those that have the ATPase altered so that it retains its enzymatic activity, but does not bind ouabain. Mutants that have inactive ATPase are lethal and as such do not form colonies.

The assays discussed so far score for mutagens and, by inference, for genotoxic carcinogens. The two assay systems that follow are applicable to both genotoxic and epigenetic carcinogens.

Sister Chromatid Exchange Assay

This assay scores for exchange of loci between sister chromatids of chromosomes (*4*). The cells are grown in the presence of 5-bromodeoxyuridine (5BrdUR) for a period of time required for two rounds of DNA replication. 5BrdUR is incorporated in the newly synthesized DNA strand in place of thymidine. After the first replication, one DNA strand of one chromatid of the chromosome contains 5BrdUR; after the second replication, both strands of one chromatid and one strand of the second chromatid contain 5BrdUR. With fluorescent staining techniques, the two chromatids can be distinguished from each other. This procedure allows observation of the mutagen-induced exchanges of chromatid segments. This assay is very sensitive, but in most cases the chemical injuries responsible for these chromosomal lesions have not been identified.

Cell Transformation Assay

A cell transformation assay is the only test that directly scores for malignant transformation rather than for mutagenesis, as have the assays described so far. It is applicable to both genotoxic and epigenetic carcinogens (*8*).

Mammalian cells are grown on agar as a monolayer. When confluence is achieved, the growth of normal cells is arrested by contact inhibition. When a carcinogen is present in the culture, the cells that undergo malignant transformation continue to divide. Because there is no place to proliferate in the horizontal plain, the transformed cells pile on top of each other; thus, colonies are easy to score. PMS must be added to the culture to activate precarcinogens. Injection of these proliferating cells into animals produces tumors. This observation proves that the colonies indeed represent malignantly transformed cells.

Carcinogenicity Testing in Fish

Several test systems for carcinogens use fish. With these systems there is no need for elaborate cage sterilization and bedding changes. Thus, a much larger number of animals can be used at a lower cost. A comprehensive review of this subject has been published (*9*).

Biological Testing in Rodents

Bioassays (i.e., testing of chemicals in laboratory animals) give reliable information about carcinogenicity. In spite of species differences in susceptibility to carcinogens, every human cancer can be reproduced in animals, and most animals are subject to cancer. Because of the cost of bioassays (EPA estimates vary from $390,000 to $980,000 per assay) and because of the time involved (up to 30 months), it is not realistic to test all 50,000 compounds in common use. Therefore, a selection process for bioassay testing of chemicals has been instituted.

Currently two such testing programs are operating in the United States. An eight-member Interagency Testing Committee, representing different federal agencies and departments, recommends chemicals to the EPA Administrator for testing. The National Toxicology Program (NTP) Chemical Nomination and Selection Committee reports to the National Cancer Institute (NCI).

Selection for bioassay is based on the results of multiple in vitro tests and on consideration of chemical structures. Structure–activity relationship (SAR) studies have been done with many classes of compounds and, at least within some groups, fairly accurate predictions can be made as to the possible carcinogenicity of a compound.

The NCI has published *Guidelines for Carcinogenic Bioassay in Small Rodents*, which describes the minimum requirements for the design and execution of a bioassay (*10*).

The gist of these guidelines, in abridged form, follows.

1. Each chemical should be tested in at least two species and in both sexes (rats and mice are usually used).
2. Each bioassay should contain at least 50 animals in each experimental group.
3. Exposure to chemicals should start when the animals are 6 weeks old (or younger) and continue for most of their life span (for mice and rats, usually 24 months). The observation period should continue for 3–6 months after administration of the last dose.
4. One treatment group should receive the maximum tolerated dose (MTD), which is defined as the highest dose that can be given that would not alter the animals' normal life span from effects other than cancer. The other treatment group is treated with a fraction of the MTD.
5. The route by which a chemical is administered should be the same or as close as possible to the one by which human exposure occurs. The chemicals may be given by any of the following routes: orally (with food or water or by force-feeding), by inhalation, or topically (by application to the skin).

In some instances, two-generation bioassays are performed in which both generations are exposed to the potential carcinogen. The advantage of this procedure is that it exposes fetuses and very young animals, which are much more sensitive to chemical injury than adults. The animals that die during the study and the survivors that are killed at the completion of the study are examined for tumors. The results are evaluated statistically with a p value of 0.05, which means that the probability that the given results were obtained by chance is less than 5%.

The positive outcome of a bioassay indicates, but is not necessarily evidence, that an agent will be carcinogenic in humans. As of 1982, the International Agency for Research on Cancer (IARC) listed 142 substances experimentally shown to be carcinogens in animals. Of those, only 14 have been recognized as human carcinogens. A more in-depth treatment of this subject is available in reference 10.

Dose–Response Assessment

When extrapolating from bioassay-generated dose–response data to obtain a quantitative estimate of human risk, two parameters have to be considered: biological extrapolation and numerical extrapolation.

Biological Extrapolation

Metabolic differences separate humans and test animals, and laboratory animals are usually highly inbred whereas the human population is genetically highly heterogeneous. These contrasts generate a basic problem of how to adjust the dose measured in bioassays to the dose experienced in humans. Several approaches may be considered:

1. straight translation from animals to humans of milligrams per kilogram per day,
2. straight translation from animals to humans of milligrams per square meter per day,
3. straight translation from animals to humans of milligrams per kilogram per lifetime, and
4. in cases where the experimental dose is measured as parts per million (ppm) in food, water, or air, human exposure is expressed in the same units.

Table 6.1 shows, in relative terms, how the mode of translation affects the estimates of human risk. These data indicate that risk estimates may vary by as much as a factor of 40.

The National Research Council (NRC) study compared the incidence of

Table 6.1. Relative Human Risk Projected, Depending on How Dose Rate Is Scaled from Experimental Animals to Humans

Experimental Animal	*Base Unit (mg/kg/day)*	*Estimated Human Risk*		
		mg/m²/day	*mg/kg/lifetime*	*Food (ppm)*
Mouse	1	14	40	6
Rat	1	6	35	3

NOTE: Reproduced from reference 10.

site-specific chemical-induced tumors in experimental animals and humans. Five chemicals and cigarette smoking were evaluated; translation from bioassay to humans was based on milligrams per kilogram per lifetime.

For two of these chemicals (*N,N*-bis(2-chloroethyl)-2-naphthylamine and benzidine) and for cigarette smoking, the human incidence occurred as predicted from the animal study. However, for aflatoxin B_1, diethylstilbestrol, and vinyl chloride, human incidence was greatly overestimated (10, 50, and 500 times, respectively). Thus, the Consultative Panel on Health Hazards of Chemicals and Pesticides concluded that:

> Although there are major uncertainties in extrapolating the results of animal tests to man, this is usually the only available method. . . . Despite the uncertainties, enough is known to indicate what dependencies on dose and time may operate and to provide rough predictions of induced cancer rates in the human population.

Another problem is how to interpret the bioassay results if the response of the two animal species tested varies greatly or if only one responds positively. In spite of some controversy about how to handle such data, there is general agreement among U.S. federal agencies that the extrapolation should be based on results from the more sensitive species.

Numeric Extrapolation

To obtain meaningful results within bioassay limits, it is necessary to expose the test animals to relatively high doses of the potential carcinogen. A normal dose–response relationship can be demonstrated for high doses (Figure 6.1). However, humans are usually exposed to considerably lower doses of environmental carcinogens than those used in laboratory animals. The cancer incidence resulting from such low exposure is expected to be many orders of magnitude lower than that observed in bioassays.[1] The path of the dose–response curve for

[1]The following example illustrates this point. Let us consider a dose–response assessment for a suspected colorectal carcinogen. In the United States the frequency of colorectal cancer is on the average 1 in 2000 people (120,000 cases per year in the population of 245

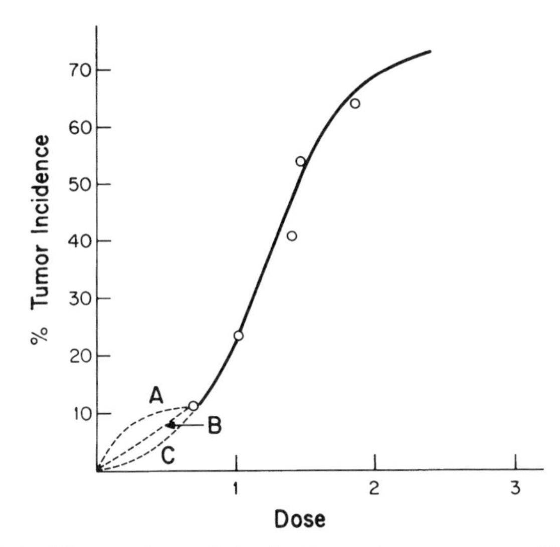

Figure 6.1. Possible ways of extrapolation of a bioassay dose–response curve. Key: A, superlinear; B, linear; C, infralinear.

exposures below the lowest observable bioassay exposure can be only guessed. Because of this uncertainty, the most frequently used extrapolation is a straight line from the lowest observed dose–effect point to the zero dose. However, other approaches have been suggested.

The infralinear extrapolation (Figure 6.1) can be obtained from models based on the best fit of observable data points into a mathematical equation. Unfortunately, in practice there are very few observable data points available (usually two or three). Thus, many models will fit the experimental dose–response curve equally well, which makes the extrapolated segment highly hypothetical.

Superlinear extrapolation can produce a series of hypothetical lines without providing a logical reason to put more faith in any particular one. The superlinear model concept is based on the observation that in some bioassays the lower doses were more effective in producing tumors than the higher ones. The problem with this approach is that the relatively low effectiveness of the higher doses was due to the agent's toxicity. At the higher doses many animals died before they developed tumors.

---

million). To demonstrate occurrence of one tumor, 2000 animals per each dose per gender would be needed. This would amount to 12,000 animals. Occurrence of one tumor in a group of animals can hardly be considered significant. Demonstrating the occurrence of a more significant number of tumors, for instance 10, would require 120,000 animals. Obviously, experiments on such a scale would not only be prohibitively expensive, but physically impossible to perform.

Dose–response assessment varies considerably according to the extrapolation model of the dose–response curve. Compared to straight-line extrapolation, the infralinear model underestimates and the superlinear model overestimates tumor incidence.

All these models are based on a generally accepted concept that there is no threshold dose below which there are no tumors. This point can be neither proven nor disproven; even if tumors cannot be demonstrated below a certain dose, perhaps if more animals were used some tumors would appear.

Negative Results

The fact that tumors were not detected in a test population of 100 animals does not indicate zero risk. According to statistical calculations, the absence of tumors indicates merely that there is a 95% likelihood that the actual incidence of tumors is no more than 0.45%. This estimate of tumor incidence that might escape detection represents the upper 95% confidence limit (*see* footnote 3 in Chapter 2) of an experiment with 100 animals.

It must be concluded that quantitative cancer risk assessment of environmental toxins is highly hypothetical.

Exposure Assessment

The factors to be considered in exposure assessment are:

1. Who and what is likely to be exposed to the compound in question?
2. How much exposure may be anticipated?
3. In which way, how long, and under what circumstances will the exposure occur?

To make calculations of the overall human exposure possible, certain standard values of human anatomy and physiology are set (*11*):

1. Mass (kg): man 70, woman 60, child 20.
2. Skin surface area (m^3): total (180 cm tall) 1.8, clothed with short sleeves 0.3, clothed with long sleeves 0.1.
3. Resting respiration rate (L/min): man 7.5, woman 6, child 4.8.
4. Respiration rate during light activity (L/min): man 20, woman 19, child 13.
5. Volume of air breathed (m^3/day): man 23, woman 21, child 15.
6. Fluid consumption (L/day): man 2, woman 1.4, child 1.4.
7. Food consumption (g/day): all humans 1,500.

The exposure assessment is not easy, not only because people move from place to place and engage in a variety of activities, but also because all possible routes of exposure have to be considered. Thus, for instance, to estimate the total human exposure to a carcinogen from contaminated groundwater, contributions of the following routes of exposure have to be calculated:

- direct exposure through drinking;
- exposure through inhalation from showering, bathing, and other uses of water;
- exposure through the skin by the body's contact with the contaminated water;
- exposure through ingestion of food that was in contact with the contaminated water.

Moreover, at each stage of this analysis the bioavailability for each route of entry and the metabolism of the carcinogen has to be considered.

A general criticism of the exposure assessment is that it is done for each carcinogen separately, whereas in a real-life situation, people may be exposed to several carcinogens at the same time. A cumulative exposure may have an additive, synergistic, or antagonistic effect. In addition, simultaneous exposure to inhibitors or inducers of xenobiotic-metabolizing enzymes may complicate the true picture even further.

Risk Characterization

The cancer risk may be expressed in several ways. The most common risk measure is the *individual lifetime risk.* This expresses the probability, such as 1 in 10,000 or 1 in 100,000 or so, that an individual will develop cancer during his or her lifetime because of the continuous exposure to a carcinogen. From the straight-line extrapolation of the dose–response curve to zero, the risk of cancer per unit of the carcinogen is calculated. This is called carcinogenic "potency" or "unit cancer risk". By multiplying potency by the exposure dose, the individual lifetime risk is obtained.

Population or *societal risk* is obtained by multiplying the individual risk by the number of people exposed. It expresses the number of cases that are due to one-year, or alternately, to lifetime exposure to a carcinogen. The time parameter has to be defined because the results will vary greatly, depending on whether the calculation is done for a year or for a lifetime.

The *relative risk* is expressed by dividing the risk (the incidence rate) in the exposed group by the risk in the unexposed group or in the general population.

The risk in the exposed group divided by the risk in the general population, corrected for factors such as age and time period, is called the "standardized mortality (or morbidity) ratio".

Finally there is also the "loss of life expectancy". This risk is calculated by multiplying the individual lifetime risk by the average remaining lifetime (assuming 72 years as an average life span).

Critique of Risk Assessment

Risk assessment as it is practiced in the present form was set in place in 1986 and was focusing specifically on carcinogenesis. Lately, risk assessment has been severely criticized by both the industries and some environmental groups. The industries were complaining that risk assessment, as it is conducted under the rigid rules of the EPA, frequently imposes unnecessary burdens on the industries for minimal benefits for protection of health and the environment. The environmentalists, on the other hand, maintain that risk assessment is inherently misleading. Their point is that science has no way of evaluating the effects of exposure to several chemicals simultaneously. Because everyone in the real world is exposed to multiple chemicals simultaneously, risk assessment is never describing the real world, yet almost always pretends to describe the real world. Risk assessment pretends to determine "safe" levels of exposure to poisons, but in fact it cannot do any such thing. Therefore, risk assessment provides false assurances of safety while allowing damage to occur (*12*). Besides, there are no agreed-upon ways of assessing health effects other than cancer, such as damage to the nervous system, immune system, or genes.

Ames and his co-workers (*13*) raised another criticism. They question the value of the bioassay for quantitative assessment of carcinogenicity in humans on the ground that the MTD is toxic enough to cause cells' death. This in turn allows neighboring cells to proliferate. In addition, the death of cells stimulates phagocytosis, and with it, release of oxygen radicals. Both cell proliferation and release of free radicals are important aspects of the carcinogenic process. In other words, use of sublethal doses (MTD) in itself potentiates the carcinogenicity of a compound.

Presently the EPA, recognizing that the conventional process of risk assessment is outdated, is revising the process to conform with new scientific information. Thus, more weight should be given to the structure–activity relationship, toxicity to genes, and mode of action. The revised process calls for addition of a narrative summary of the hazard characterization (*14*). Also, a new classification of hazardous substances with respect to their carcinogenic effect was proposed (*14*):

Category I. Carcinogenic risk to humans under any conditions.

Category II. Carcinogenic risk to humans, but only under limited conditions.

Category III. Although carcinogenic in animals, not likely to pose a carcinogenic hazard to humans.

Category IV. Either a demonstrable lack of carcinogenicity, or no evidence is available.

Despite all the revisions, risk assessment presents only an approximation of the real risk determination. Yet that is the best we have at present. Let us hope that as our scientific knowledge increases, new and more accurate ways of determining the risk to human health and to the environment will be forthcoming.

Ecological Risk Assessment

While human-health risk assessment, although far from perfect, has now been firmly in place for several years, a new concern has emerged regarding ecological risk assessment. Although ecological risk assessment is required by government agencies before implementation of new projects, there is a growing realization of the complexity of the problems. First of all, in contrast to human-health risk assessment, which concerns individuals of a single species, ecological risk assessment deals with populations of thousands of species. Second, because of the complexity and the interwoven nature of ecosystems, there is the problem of proper selection of an end point. An example of this complexity is the symbiotic relationship between freshwater mussels and fish. Larval mussels must attach to a particular fish species during development. Thus, the demise of a fish species will result in extinction of the mussel population. Further, there is a growing realization that degradation of an ecosystem can be due not only to chemical but also to biological and physical factors such as introduction of exotic species or land development. Even if chemicals alone would be considered, the multitude of chemical agents in the environment and their possible cumulative or synergistic effect make study of the adverse impact of a single chemical agent highly speculative.

In 1992 the EPA published *Framework for Ecological Risk Assessment.* However, an EPA advisory body called the Risk Assessment Forum is scrutinizing the field to lay the groundwork for new ecological risk assessment guidelines (*15*).

References

1. *Introduction to Biochemical Toxicology;* Hodgson, E.; Guthrie, F. E., Eds.; Elsevier: New York, 1980; Chapter 8, p 143.
2. Ames, B. N.; McCann, J.; Yamasaki, E. *Mutat. Res.* **1975,** *31,* 347.
3. Zeiger, E. *Cancer Res.* **1987,** *47,* 1287.
4. Williams, G. M.; Weisburger, J. H. In *Cassarett and Doull's Toxicology;* Klaasen, C. D.; Amdur, M. O.; Doull, J., Eds.; Macmillan: New York, 1986; Chapter 5, p 99.
5. Chu, E. H. Y.; Malling, H. V. *Proc. Natl. Acad. Sci. U.S.A.* **1968,** *61,* 1306.
6. Clive, D.; Flamm, W. G.; Machesco, M. R.; Bernheim, N. J. *Mutat. Res.* **1972,** *16,* 77.
7. Huberman, E.; Sachs, L. *Proc. Natl. Acad. Sci. U.S.A.* **1976,** *73,* 188.
8. Heidelberger, C.; Freeman, A. E.; Pienta, R. J.; Sivak, A.; Bertram, J. S.; Casto, B. C.; Dunkel, V. C.; Francis, M. W.; Kakungaj, T.; Little, J. B.; Schechtman, L. M. *Mutat. Res.* **1983,** *114,* 283.
9. *Use of Small Fish Species in Carcinogenicity Testing;* Hoover, K. L., Ed.; National Institutes

of Health: Bethesda, MD, 1984; National Cancer Institute Monograph 65; Proceedings of a symposium held at Lister Hill Center, Bethesda, MD.
10. Office of Technology Assessment. *Cancer Risk: Assessing and Reducing the Dangers to Our Society;* Westview: Boulder, CO, 1982.
11. Cohrsen, J. J.; Covello, V. T. *Risk Analysis: A Guide to Principles and Methods for Analyzing Health and Environmental Risk;* U.S. Council on Environmental Quality: Washington, DC, 1989.
12. *Rachel's Environment and Health Weekly;* No. 470, Nov. 30, 1995; Environmental Research Foundation, 105 Eastern Avenue, Suite 101, Annapolis, MD 21403–3300.
13. Ames, B. N.; Magaw, R.; Swirski Gold, L. *Science (Washington, D.C.)* **1987,** *236,* 271.
14. Hanson, D. J. *Chem. Eng. News* September 26, 1994, p 21.
15. Renner R. *Environ. Sci. Technol.* **1996,** *30*(4), 172A.

7 Occupational Toxicology

Threshold Limit Values and Biological Exposure Indices

Industrial workers make up the segment of the population that is most vulnerable to chemical injury. To protect them from occupation-related harm, the American Conference of Governmental and Industrial Hygienists publishes annually revised threshold limit values (TLVs) (*1*), guidelines for permissible chemical exposure at the work place.

TLV refers to concentrations of substances in parts per million or milligrams per cubic meter in the air to which most workers can be exposed on a daily basis without harm. These values apply to the work place only. They are not intended as guidelines for ambient air quality standards for the population at large.

Obviously, genetic variations and diverse lifestyles (such as smoking, alcohol use, medication, and drug use) must be considered. Hypersensitive individuals may be adversely affected by exposure to certain chemicals even within the limits of the TLV. Thus, TLVs should be treated as guidelines only and not as fixed standards. The recommended goal is to minimize chemical exposure in the work place as much as possible.

TLVs are expressed in three ways:

1. Time-weighted average (TLV–TWA) designates the average concentration of a chemical to which workers may safely be exposed for 8 h per day and 5 days per week.

2. Short-term exposure limit (TLV–STEL) designates permissible exposure for no more than 15 min, and no more than four times per day, with at least 60-min intervals between exposures.
3. Ceiling concentrations (TLV–C) are concentrations that should not be exceeded at any time.

How protective the TLVs are is being questioned. The 1990 report that analyzed the scientific underpinnings of the TLVs revealed that at the exposure at or below the TLV, only few cases showed no adverse effect (*2*). In some cases even 100% of those exposed were affected. On the other hand, there was a good correlation between the TLVs and the measured exposure occurring in the work place. Thus, it appears that the TLVs represent levels of contaminants that may be encountered in the work place, rather than protective thresholds.

Biological exposure indices (BEIs) provide another way of looking at exposure to chemicals. This method supplements air monitoring for compliance with TLV standards. BEIs are standards of permissible quantities of chemicals in blood, urine, or exhaled air of exposed workers.

These standards are useful in testing the efficacy of personal protective equipment and determining a chemical's potential for dermal or gastrointestinal absorption. Of course, BEI findings have to be interpreted carefully. The results may be affected by external factors such as lifestyle and exposure outside the work place.

Respiratory Toxicity

The morphology and physiology of the respiratory system and its role as an important route of entry for xenobiotics was discussed in Chapter 2. Respiration, the exchange of O_2 and CO_2 with blood, is only one of several functions of the lungs, albeit the most important of them. Other functions include excretion of gaseous metabolites and metabolism and regulation of circulating levels of vasoactive hormones such as angiotensin, biogenic amines, and prostaglandins (*3*).

Any damage to the lung tissue responsible for these regulatory functions will affect blood pressure and consequently the lungs' perfusion with blood. To maintain proper oxygenation of blood, a match is necessary between alveolar ventilation[1] (5250 mL of air min^{-1}) and the volume of blood perfusing the lungs (5000 mL min^{-1}). Any change in blood flow will perturb this ventilation–perfusion balance and result in dysfunction of the organism.

Toxins (gases, vapors, or aerosols) may injure respiratory tissue, or they may

---

[1]*Alveolar ventilation* is defined as the volume of gas available for exchange with blood during 1 min [alveolar ventilation = (tidal volume − residual volume) × (breaths per minute)]. *Residual volume* is the volume of gas remaining in the lungs after maximal exhalation. For a definition of tidal volume, *see* Chapter 2.

cause systemic toxicity by penetrating the tissue and entering the circulation. Injuries to the respiratory system vary in severity (depending on the agent and the degree of intoxication) from irritation to edema, fibrosis, or neoplasia. The site of toxicity depends on the water solubility of a gas or on the size of aerosol particles or droplets.

Irritation of Airways and Edema

Water-Soluble Gases

The upper respiratory system is susceptible to attack by water-soluble gases such as ammonia, chlorine, sulfur dioxide, and hydrogen fluoride. Before a gas can gain access to the tissue, it has to penetrate the mucous lining. This barrier imparts some protection against very small quantities of toxic gases, but it does not protect the tissue against large doses. Toxicity to the respiratory tissue in this region is most frequently manifested by irritation. However, edema may occur in more severe cases.

Edema results from damage to the cell membrane; this damage affects membrane permeability and causes release of cellular fluid. Swelling of the tissue, constriction of the airways, difficulty with breathing, and increased sensitivity to infection are manifestations of edema. The development of edema is a slow process. Because it may take many hours before it is fully developed, the affected individual may not be aware of the danger.

People with respiratory diseases, such as asthma or chronic bronchitis, are affected to a greater extent than healthy individuals. Although survivors may recover without permanent damage, very severe exposure to such water-soluble gases may be fatal.

Large Aerosol Particles

Aerosols of particles larger than 2 μm also cause damage to the upper respiratory system. Arsenic oxides, sulfides, and chlorides are used in a variety of industries, such as manufacturing of colored glass, ceramics, semiconductors, and fireworks and in hide processing. However, upper respiratory exposure to these compounds is most likely to occur in ore-smelting industries and in pesticide manufacturing.

In these cases, particles of arsenic compounds are usually too large to penetrate into the lung alveoli and are deposited in the nasopharyngeal region and in the upper bronchi. Their toxicity is manifested by irritation of the airways that results in a chronic cough, laryngitis, and bronchitis-like symptoms. Arsenic trioxide (As_2O_3) is a suspected human carcinogen; exposure to this compound should be kept to a minimum. Compounds considered to be carcinogens are listed and described by the National Toxicology Program in their annual report on carcinogens and in the monographs of the International Agency for Research on Cancer (IARC).

Chromium and its compounds are used in stainless steel manufacturing, chrome plating, pigment manufacturing, and hide processing. Hexavalent chromium compounds such as chromate (CrO_4^{2-}) and bichromate ($Cr_2O_7^{2-}$) cause nasal irritation, bronchitis-like symptoms, and (on chronic exposure) lung tumors and cancer.

Exposure to nickel and its monoxide (NiO) and subsulfide (Ni_2S_3) may occur during the processing of nickel ores. Because the ore dust particles are rather large, their toxicity is confined to the nasal mucosa and to the large bronchi. Nickel subsulfide in the form of dust or fumes is a confirmed human carcinogen of the nasal cavity.

Poorly Water Soluble Gas

Examples of poorly water soluble gases that penetrate deep into the lungs, causing damage to alveolar tissue, are ozone (O_3), nitrogen dioxide (NO_2), and phosgene ($COCl_2$). The mode of action of ozone and nitrogen dioxide is related to their oxidizing potential.

Peroxidation of cellular membranes causes edema. In addition, NO_2 reacts with alveolar fluid to form HNO_2 and HNO_3, corrosive acids that also damage the cells. Exposure to ozone may occur in a variety of industrial settings because ozone is used for bleaching waxes, textiles, and oils. Nitrogen dioxide is widely used in chemical industries and in the manufacture of explosives.

Some metals and their derivatives, such as cadmium oxide (CdO), nickel carbonyl [$Ni(CO)_4$], and beryllium also cause pulmonary edema. Cadmium oxide is used in the manufacture of semiconductors, silver alloys, glass, battery electrodes, and cadmium electroplating. The fumes of CdO consist of extremely fine particles that penetrate alveoli. Inhalation of such fumes leads to edema, pneumonitis, and proliferation of type I pneumocytes of the alveolar lining. Chronic exposure may result in emphysema. CdO is also listed by both the EPA and the International Agency for Research on Cancer (IARC) as a carcinogen that primarily induces prostatic cancer.

Nickel carbonyl is a highly volatile liquid used in nickel refining and nickel plating. Inhaled vapors cause pulmonary edema. In case of exposure, 48 h of surveillance is necessary.

Work-related exposure to beryllium dust may occur in the manufacture of ceramics and alloys and during the extraction of beryllium from its ore. The fine dust of beryllium enters alveoli and causes pulmonary edema. Chronic exposure leads to granulomatous pulmonary disease (referred to as berylliosis), which may progress to pulmonary fibrosis. Beryllium has been shown to be a carcinogen in animals and is also a suspected human carcinogen.

Phosgene, used in the preparation of many organic chemicals, is also manufactured as a war gas. It is highly toxic as it undergoes hydrolysis to CO_2 and HCl in the lungs. The liberated HCl causes damage to the alveolar cells and, in turn, severe edema. The onset of the edema may be delayed for as long as 48 h.

Paraquat

The herbicide paraquat (*see* Chapter 10) is highly toxic to the respiratory system. It causes pulmonary edema regardless of the route of entry into the system. Whether paraquat is inhaled or ingested, it enters the alveolar space and becomes concentrated in type II pneumocytes. Its toxicity probably results from generation of superoxide radicals ($\cdot O_2$) (*3*), which may cause peroxidation of cellular membranes. Paraquat is eliminated from the body by being actively secreted into the renal tubules. However, it also damages the tubules, and thus inhibits its own secretion. As a result, it accumulates in the blood and leads to pulmonary toxicity.

Diquat, a structural analog of paraquat, although equally toxic to cultured lung cells, does not exert pulmonary toxicity in vivo. This difference in activity probably occurs because diquat is not retained in the alveolar cells.

Pulmonary Fibrosis

Pulmonary fibrosis, also designated as pneumoconiosis, is another response of lungs to respiratory toxins. The initial injury to the cells is caused by physical rather than chemical action of minute solid particles or fibers. In the early stages of the disease, small (1–10 mm in diameter) islets of collagen are deposited in the pulmonary region. The islets grow progressively larger, eventually fusing into a network of fibers pervading the whole lung and leading to a loss of lung elasticity. In addition, blood vessels in the affected areas narrow, and alveolar walls are destroyed; the results are decompartmentalization of the alveoli and emphysema. The injury is assumed to be related to the activity of macrophages that engulf the injurious particles, which in turn damage lysosomal membranes and release lysozymes. The macrophages are digested by their own enzymes and release the engulfed particles; the process may then be repeated. Thus, a single particle is capable of destroying numerous macrophages.

Deposition of collagen probably results from the stimulation of fibroblasts by a factor, or factors, released from broken macrophages. Simultaneously, another factor, referred to as the lipid factor, is released and stimulates the generation of more macrophages (*4*). A cascade of events seems to lead to deposition of increasing amounts of collagen.

Silicosis

Silicosis results from chronic exposure to respirable particles of crystalline silica; amorphous forms do not cause this disease. Animal experiments indicate that inhalation of amorphous silica causes only minimal fibrosis. However, under such conditions only a small amount of silica was retained in the lungs. In contrast, when injected into the peritoneum or into the lungs, amorphous silica was more fibrogenic than crystalline quartz (*4*). Silicosis is frequently complicated by the onset of tuberculosis.

Black Lung Disease

Black lung disease, a common illness of coal miners, was for a long time thought to be caused by chronic exposure to coal dust because lungs of the deceased victims were blackened by coal. It appears now that the disease, which has all the characteristics of lung fibrosis, is most likely caused by silica dust produced in the process of coal mining.

Asbestosis

Asbestos is a group of hydrated fibrous silicates that are divided into two basic families: the curly, named "serpentine", and the rodlike, named "amphibole" (*5*). The types belonging to the amphibole family are the most pathogenic; their toxicity depends on the size of the fibers and perhaps on other physical properties. The most harmful fibers are 5 μm in length and 0.3 μm in diameter.

Asbestosis is encountered among workers employed in the mining of asbestos or in the construction or demolition of housing that contains asbestos. Cases of asbestosis have also been observed among janitors and plumbers working in schools and office buildings. In this case, the exposure comes from asbestos insulation of steam pipes and boilers.

In addition to fibrosis, the symptoms of asbestosis involve calcification of the lung and formation of mesothelial tumors. The latency period for mesothelial tumor development is unusually long. Up to 30 years may elapse between exposure and the clinical appearance of neoplasia. The widely publicized high incidence of asbestosis and related mesothelial and lung tumors that occurred during the 1970s was a result of asbestos exposure of shipyard workers employed by the U.S. Navy during World War II.

Asbestos fibers have the potential to migrate into the peritoneal cavity and cause tumors of the peritoneal mesothelium. Tobacco smoke potentiates the effect of asbestos and promotes lung tumor formation (*6*).

Pulmonary Neoplasia

One of the frequent causes of occupationally related pulmonary neoplasia is respiratory exposure to polycyclic aromatic hydrocarbons (PAHs). As discussed in earlier chapters, PAHs are carried into the lungs by minute particles of soot and fly ash. The risk of lung cancer from this source is greatest among coke oven and coal tar pitch workers. Tobacco smoke, which is the main cause of lung cancer overall, increases the risk of pulmonary neoplasia among the population exposed to PAHs at the work place.

The habit of smoking, which is rapidly decreasing among the more highly educated classes of society, is still very much ingrained among blue-collar workers. Unfortunately, the nature of their work makes this segment of the population most vulnerable to chemical injury.

The TLV–TWA values of the compounds and substances discussed in this section are presented in Table 7.1.

Allergic Responses

The Immune System

Functions of the Immune System

The immune system performs two essential roles: It provides resistance to infectious agents and surveillance against arising neoplastic cells. These functions are

Table 7.1. TLV–TWA Values of Some Compounds Affecting the Respiratory System

		TWA		
Substance	*Formula*	*ppm*	*mg/m³*	*Carcinogenicity*
Ammonia	NH_3	25	18	
Chlorine	Cl_2	0.5	1.5	
Sulfur dioxide	SO_2	2	5	
Hydrogen fluoride	HF	3[a]	2.5[a]	
Arsenic	As		0.2	
Arsenic trioxide	As_2O_3		2	Suspected
Chromate	CrO_4^{2-}		0.05	Established
Nickel	Ni		1	
Nickel subsulfide	Ni_3S_2		1	Established
Ozone	O_3	0.1	0.2	
Nitrogen dioxide	NO_2	3	6	
Phosgene	$COCl_2$	0.1	0.4	
Cadmium oxide	CdO		0.01[a]	Established
Nickel carbonyl	$Ni(CO)_4$	0.05	0.1	
Beryllium	Be		0.002	Suspected
Paraquat	—[b]		0.1	
Silicon dioxide	SiO_2		0.05–0.1[c]	
Asbestos	—[d]		0.2–2.0[c]	Established

[a]TLV–C (ceiling).
[b]*See* Chapter 6.
[c]Depends on the crystalline form.
[d]Hydrated calcium magnesium silicates of variable composition.
SOURCE: Adapted from reference 1.

performed through several highly specialized cells collectively referred to as leukocytes, or as they are commonly known, white blood cells. Leukocytes originate from the stem cells of the bone marrow. As they mature they differentiate as granulocytes, lymphocytes, and macrophages. The lymphocytes are further differentiated into T-lymphocytes, B-lymphocytes, and non-T-, non-B-lymphocytes.[2]

Mechanisms of Immune Responses

There are two mechanisms of immune responses:

- nonspecific or constitutive, and
- specific.

The nonspecific immune system does not require a prior contact with an inducing agent and lacks specificity for antigens. It constitutes the organism's primary defenses and involves two types of phagocytic cells: granulocytes (polymorphonuclear leukocytes, PMNs), and macrophages (mononuclear leukocytes, MOs); and two types of non-T, non-B lymphocytic killer cells: natural killer (NK) cells and antibody-dependent killer cells (antibody-dependent cellular cytotoxicity, ADCC) cells. The NK cells have a spontaneous cytolytic activity against many different tumor cells. The ADCC killer cells require antibody to lyse the target tumor cells (*see* the next section). These cells circulate in the blood, and their lifetime is 1 to 3 days.

The specific immune system requires activation by antigens. There are two types of specific immune responses: cell-mediated immunity, involving T-lymphocytes, and humoral immunity, involving B-lymphocytes.

T-lymphocytes act by developing into antigen-specific killer cells that lyse the foreign cells bearing that antigen on their plasma membrane. The development of these cytolytic T-cells (CTLs) requires cooperative interaction between the precursor CTLs, antigen-processing cells (usually macrophages), and other T-cells called T-helper cells. Humoral immunity involves B-lymphocytes which, after sensitization by an antigen, produce antibodies. Antibodies are proteins of a general structure consisting of two light and two heavy chains. The chains are connected with each other by S–S linkages (Figure 7.1). Each chain has a "variable region" and a "constant region". The variable region is responsible for the interaction with an antigen, whereas the constant regions of heavy chains are responsible for biological activation of ADCC killer cells, the granulocyte, and the macrophage.

The Antibodies' Mode of Action

The five general classes of antibodies are IgM, IgG, IgA, IgD, and IgE, where Ig stands for immunoglobulin. Their mode of action involves four pathways:

---

[2]The designations T and B indicate the primary lymphoid tissue where the maturation of lymphocytes occurs; T stands for thymus and B for bursa-equivalent.

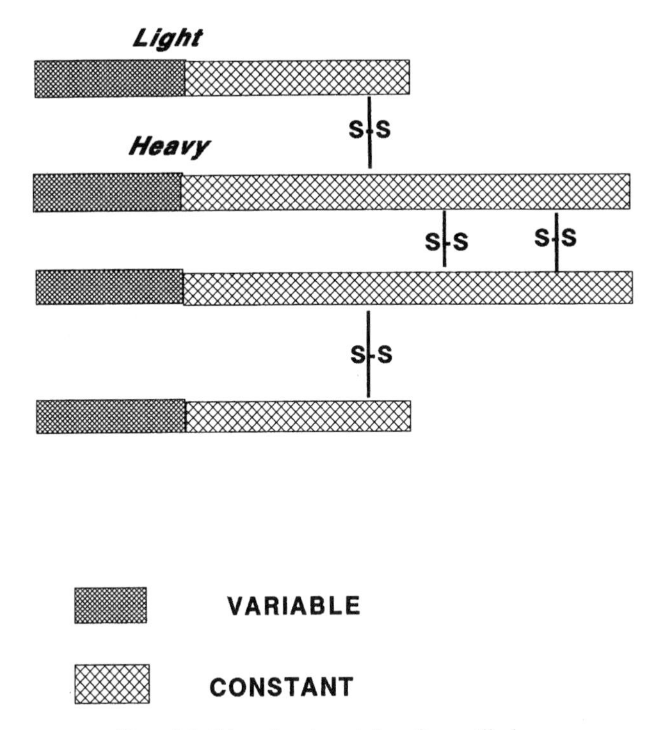

Figure 7.1. Schematic representation of an antibody.

- neutralization of viruses,
- opsonization, that is, inactivation of viruses and bacteria by coating,
- binding to antigens and linking them to ADCC killer cells, and
- complement fixation, that is, a cascade of events involving sequential binding to 20 serum proteins, resulting in generation of biological activities capable of cell lysing.

There are also other interactions and mutual reinforcements between the humoral and cell-mediated immune systems that are not discussed here. For more details on this subject, the reader is referred to reference 7.

Dysfunctions of the Immune System

An injury to the immune system may occur at doses of toxic agents much below those at which toxicity is apparent. Because immunocompetent cells require continued proliferation and differentiation, the immune system is very sensitive to agents that suppress cells' proliferation.

Assessment of an injury to the immune system may be based on any of the following symptoms: increased susceptibility to infections, changes in the peripheral leukocyte count and cell differential count, alteration in histology of lymphoid organs, and depressed cellularity of the lymphoid tissue.

Dysfunctions of the immune system may involve allergic reactions, immune suppression, uncontrolled proliferation, and autoimmunity. Allergic reactions occur when the immune system responds adversely to environmental agents. Examples of allergies are asthma and contact dermatitis. Examples of uncontrolled proliferation are leukemia and lymphoma.

Immune suppression may be a genetic phenomenon, but it may also be induced by drugs, infections, neoplasia (as in the case of leukemia), exposure to radiation, malnutrition, and environmental or occupational exposure to chemical agents.

Autoimmunity is the reactivity of the individual's system against its own tissue. It may have a genetic origin, or it may be due to exposure to environmental chemicals that bind to tissue or serum products. Consequently these modified "self-antigens" produce immune responses.

Allergies are a result of dysfunctions of the immune system. The immune system, which is designed to inactivate and eliminate foreign bodies, reacts abnormally in some individuals when challenged with specific substances. The most frequently encountered occupational allergies are asthma and contact dermatitis.

Common Agents

The agents that induce an allergic response vary greatly and can involve such things as organic chemicals, metals, dusts, and bacteria. Examples of some chemicals frequently responsible for occupation-related allergies are toluene diisocyanate, used in plastic and resin manufacturing; formaldehyde, widely used in manufacturing phenolic resins, in textile finishes, in the processing of hides, and in numerous other industrial processes; and hexachlorophene, used in manufacturing germicidal soaps and cosmetics. The chemical structures of these compounds are presented in Chart 7.1. Some metals such as beryllium, chromium, and nickel can cause contact dermatitis.

Allergies of Food Industries

A number of allergies affect workers employed in agricultural and food industries. Farmer's lung disease is a reaction to spores of thermophilic fungi, which grow in damp hay at temperatures of 40–60 °C. In sensitive individuals, the spores produce flulike symptoms, fever, malaise, chills, and aches.

A similar allergy, called bagassosis, occurs on exposure to the dust arising from bagasse, the dry sugar cane left after the extraction of sugar. The cause of the disease is probably not the dust itself, but rather microorganisms growing in the bagasse.

On the other hand, an allergy referred to as byssinosis, which occurs among

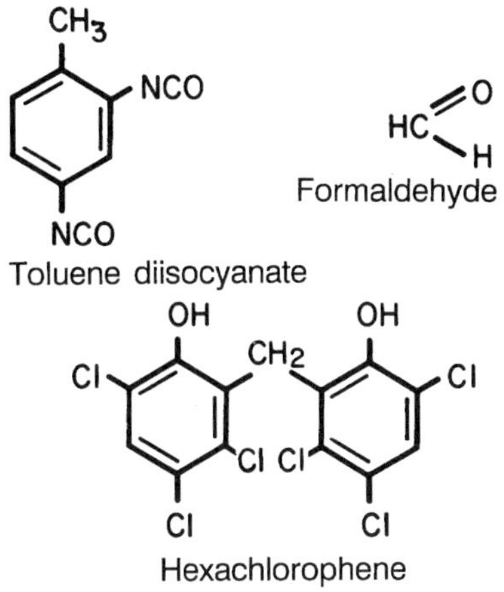

Toluene diisocyanate

Formaldehyde

Hexachlorophene

Chart 7.1. The most common industrial agents inducing allergic response.

both cotton pickers and cotton mill workers, seems to be caused by some agents present in the cotton fibers. Byssinosis is not limited to exposure to cotton; it also affects people exposed to flax and hemp dust.

Mushroom picker's lung, maple bark stripper's disease, and cheese washer's lung are other allergies affecting workers in the agricultural and food industries.

Nephrotoxins

Kidney Physiology

The physiological roles of the kidneys are excretion of waste and regulation of total body homeostasis. Each kidney contains about 1,000,000 basic functional units called *nephrons* (Figure 7.2). Nephrons perform three functions:

1. filtration of blood plasma in the glomerulus,
2. selective reabsorption by the tubules of reusable materials, and
3. secretion of waste products into the tubular lumen.

The glomerulus is a network of capillaries surrounded by a round, double-walled capsule, referred to as Bowman's capsule. The capsular space between the walls is continuous with the tubule. Plasma is filtered through the capillary walls of the glomerulus. While the filtrate enters the capsular space, the blood exits the glomerulus through the efferent arteriole, which then divides into multiple capillaries surrounding the tubule. At a blood flow rate of 1 L min^{-1}, the entire blood volume of the person passes through the kidneys in 4–5 min. The rate of filtration depends on the hydrostatic and oncotic pressures on both sides of the capil-

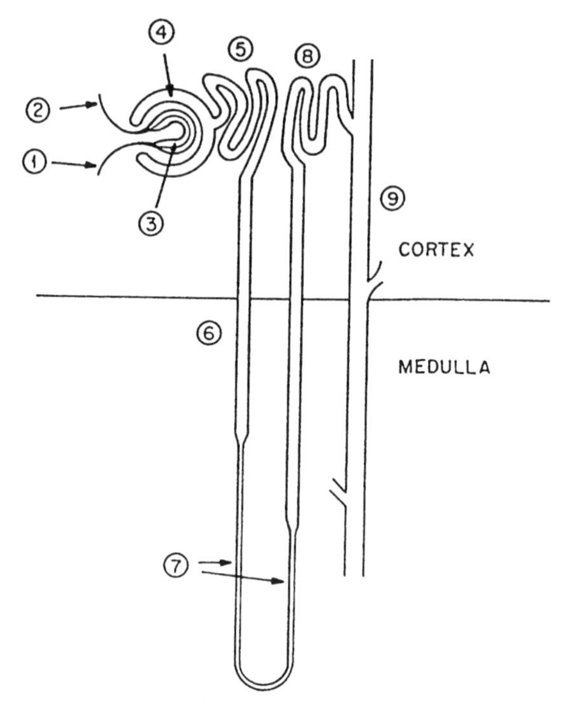

Figure 7.2. Schematic representation of a nephron. Key: 1, afferent arteriole; 2, efferent arteriole; 3, glomerulus; 4, Bowman's capsule; 5, proximal convoluted tubule; 6, pars recta of the proximal tubule; 7, loop of Henley; 8, distal convoluted tubule; and 9, collecting duct.

lary walls. (Oncotic pressure is the osmotic pressure plus the imbibition pressure of the hydrophobic colloids present in the system.) The rate of filtration is expressed by equation 7.1:

$$\text{SNGFR} = k \times a\,(P_c - P_s) - (p_c - p_s) \qquad (7.1)$$

where SNGFR is the single-nephron glomerular filtration rate, k is the permeability coefficient, a is filtration area, P is hydrostatic pressure, and p is oncotic pressure. Subscripts c and s refer to glomerular capillary and capsular space, respectively (*8*).

Composition of Fluids

In a normally functioning kidney, the composition of the filtrate is the same as that of the protein-free plasma; thus blood and its glomerular filtrate are initially isosmotic. Many of the substances in the filtrate (such as glucose, amino acids, electrolytes, and water) are reused by being selectively absorbed in the proximal convoluted tubule. The descending segment of the loop of Henle lacks a specialized, energy-dependent, absorption mechanism; it is permeable to water but not to solutes. Thus, the tubular fluid becomes concentrated (hyperosmotic) as water is removed by diffusion. This situation is reversed in the ascending segment of the loop, which is impermeable to water but permeable to NaCl. Here the fluid becomes hypoosmotic with respect to plasma.

While the reusable materials are absorbed from the tubular fluid, hydrogen and potassium ions and a variety of waste products (such as urea, uric acid, creatinine, and xenobiotics) are excreted into the tubular lumen. The final phase of urine production takes place in the collecting tubule where, depending upon the water–electrolyte balance in the body, urine is concentrated or diluted.

Autoregulation

Two regulatory systems provide for proper functioning of the kidneys. The first one, called autoregulation, concerns maintenance of a constant glomerular filtration rate, unaffected by blood pressure fluctuations. In response to certain stimuli, such as a decreased blood flow or decreased sodium concentration at the distal nephron, the "juxtaglomerular apparatus" located at the afferent arteriole releases the hormone renin. Renin reacts with a humoral factor produced by the liver, angiotensinogen, to form angiotensin I. This compound is then converted to a powerful vasoconstrictor, angiotensin II, by the converting enzyme located in the lungs. The result of this series of reactions is an increase in blood pressure and restoration of normal filtration rate.

Antidiuretic Hormone

The other regulatory system concerns body water. When the body begins to dehydrate, a sensor located in the anterior hypothalamic region of the brain triggers the release of antidiuretic hormone (ADH, also called vasopressin) from the pituitary gland. ADH acts via cyclic AMP (adenosine 5′-monophosphate) on receptors at the collecting tubule, making the tubule walls permeable to water. Thus, water is reabsorbed and urine is concentrated. A more detailed treatment of this subject is given in reference 8.

Chemical injuries to the kidney can be evaluated by urinalysis, blood analysis, or assessment of specific renal functions. The standard tests are urine specific gravity, pH, and concentration of electrolytes, protein, sugar, and blood urea nitrogen (BUN).

Renal Clearance

The concept of renal clearance was developed to express quantitatively the excretion of a substance by the kidneys. By definition, *renal clearance* represents "the volume of blood or plasma cleared of the amount of the substance found in 1 minute's excretion of urine" (*8*). The mathematical expression for renal clearance is presented in equation 7.2.

$$C = (U \times V)/P \quad (7.2)$$

where V is the rate of urine excretion in mL min^{-1}, U and P are the urinary and plasma concentrations of the test substance in mg dL^{-1}, respectively, and C (clearance) is expressed in mL min^{-1}.

The renal clearances of certain substances of known excretory behavior are useful in assessing specific renal functions, as shown in Table 7.2.

The reserve functional capacity of the kidneys is remarkable in that the removal of one kidney leads to prompt hypertrophy of the other one, without the slightest evidence of any functional impairment. Of the total kidney mass, 75% must be nonfunctional before any clinical signs appear.

Heavy Metals

Metallic mercury and its derivatives are widely used as catalysts and fungicides and for numerous industrial applications. The high volatility of mercury and its derivatives makes exposure especially dangerous, as they may enter the circulation easily via the respiratory route. Nephrotoxicity of mercurial species involves vasoconstriction and necrosis of the pars recta of the proximal tubule. The mechanism of cellular damage is not known, but it may be related to mercury's tendency to inactivate enzymes by reacting with sulfhydryl groups.

Cadmium was discussed earlier as a pulmonary toxin. About 15–30% of inhaled cadmium is absorbed into the circulation from the respiratory system (*9*). Cadmium injures the glomerulus and proximal tubules, as manifested by urinary excretion of proteins, amino acids, and glucose.

Table 7.2. Use of Renal Clearance for Assessment of Specific Renal Functions

		Excretory Behavior		
Renal Function	*Test Substance*[a]	*Filtered*	*Secreted*	*Reabsorbed*
Glomerular filtration rate	Inulin (120)	Yes	No	No
	Creatinine (95–105)	Yes	No	No
Renal plasma flow	Aminohippurate[b] (574)	Yes	Yes	No

[a]Numbers in parentheses indicate normal clearance values in milliliters per minute per 1.75 m^2 of body surface area.
[b]Aminohippurate is cleared completely from the blood during a single passage of blood through the kidneys.

Chromium, another respiratory toxin, is also a nephrotoxin. Hexavalent chromium, such as in chromate and bichromate, causes necrosis of the proximal tubule. At low doses the damage is limited to the convoluted part, but at high doses the whole proximal tubule is affected.

Lead is ubiquitous, and most public exposure comes from air, water, soil, or lead-based paint. Because lead has numerous industrial applications, industrial workers may be additionally exposed. Chronic exposure to lead initially causes damage to the proximal tubular cells. However, this damage may progress to irreversible interstitial fibrosis and vascular and glomerular sclerosis.

Halogenated Hydrocarbons

Some examples of organic nephrotoxins are carbon tetrachloride, chloroform, hexachlorobutadiene, and bromobenzene (Chart 7.2). Chloroform and carbon tetrachloride are widely used as solvents, especially for fats and waxes. In the past they were also used as fire extinguisher liquids. Their mode of action is not known, but chloroform is apparently converted by cytochrome P-450 to phosgene. Their site of toxicity is the proximal tubule.

Hexachlorobutadiene is an environmental pollutant and a specific nephrotoxin. Its mode of action is not known, but conjugation with glutathione may be the initial step in its conversion to a nephrotoxin (*10*). It acts on the pars recta of proximal tubules.

Bromobenzene is used as a solvent and as an additive to motor oil. It is speculated that it becomes a nephrotoxin upon activation by cytochrome P-450 to 2-bromoquinone (*10*).

The TWAs of these nephrotoxins are presented in Table 7.3.

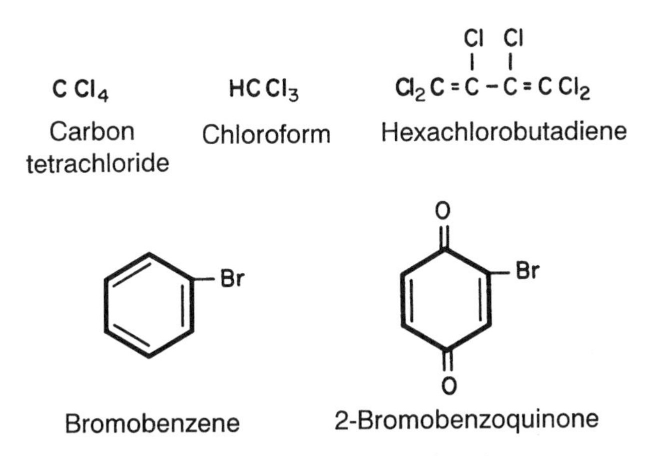

Chart 7.2. Examples of organic nephrotoxins.

Table 7.3. TLV–TWA Values of Some Nephrotoxins

Substance	TWA		Carcinogenicity
	ppm	mg/m^3	
Metallic mercury (vapor)		0.05	
Mercury (alkyl derivatives)		0.01	
Mercury (inorganic compounds)		0.1	
Lead and inorganic compounds (dust and vapors)		0.15	
Carbon tetrachloride	5	30	Suspected
Chloroform	10	50	Suspected
Hexachlorobutadiene	0.02	0.24	Suspected

SOURCE: Adapted from reference 1.

Liver Damage

Liver Physiology

Movement of Blood

The liver is the largest organ in all vertebrates, but it is absent in invertebrates. The structure of the liver is rather simple; it consists of a continuum of hepatic cells (called hepatocytes or parenchymal cells) perforated by a network of cylindrical tunnels. A mesh of specialized blood capillaries, called sinusoids, extends through these tunnels. The sinusoid walls are lined with phagocytic cells called Kupffer cells. Their role is to engulf and destroy unwanted matter (such as solid particles, bacteria, and worn-out blood cells) contained in the incoming blood.

Both venous and arterial blood enter the liver through a large indentation called the porta hepatis. The main blood supply comes to the liver from the intestinal capillaries. These capillaries join into larger vessels called mesenteric veins, which then merge with each other, as well as with veins from the spleen and stomach, to form the portal vein. Upon entering the liver, the portal vein bifurcates into right and left branches that further subdivide and eventually drain into the sinusoids. The blood perfuses the liver and exits by the hepatic veins, which merge into the inferior vena cava that returns the blood to the heart.

The hepatic artery, which branches from the aorta, supplies the liver with oxygenated blood. A constant supply is needed for the multitude of metabolic energy-requiring activities.

Waste Removal

Waste material is collected in bile-carrying canaliculi, which converge into progressively larger ducts. These ducts follow the portal vein branches, with the bile flowing in the direction opposite to that of the blood. The bile ducts eventually merge, in the porta hepatis, into the hepatic duct. From there the bile drains into the upper part of the small intestine, the duodenum. Most (90%) of the bile acid is reabsorbed from the small intestine and returned to liver. This is referred to as *enterohepatic circulation.* Outside of the porta hepatis a branch separates from the bile duct. This cystic duct ends in the gall bladder.

Nutrients

The nutrients and xenobiotics absorbed from the gastrointestinal tract are carried by the portal vein to the liver, where storage, metabolism, and biosynthetic activities take place. Glucose is converted by insulin to glycogen and stored. When needed for energy, it is degraded back to glucose by glucagon. Fat, fat-soluble vitamins, and other nutrients are also stored. Fatty acids are metabolized and converted to lipids, which are then conjugated with liver-synthesized proteins and released into the bloodstream as lipoproteins.

The liver also synthesizes a multitude of functional proteins, such as enzymes, antibodies, and blood-coagulating factors. As mentioned in Chapter 3, the liver is the principal (although not the only) site of xenobiotic metabolism. Mixed-function oxidases (cytochrome P-450), conjugating enzymes, glutathione conjugases, and epoxide hydrolase are all located in the liver.

Xenobiotics

The water-soluble metabolites of xenobiotics are released into the bloodstream to be processed by the kidneys for urinary excretion. Unused nutrients and some waste materials, such as degradation products of hemoglobin (bilirubin) and lipophilic xenobiotics that escape conversion to hydrophilic compounds, are excreted into the bile. This process returns them to the intestine, where they are either excreted with the feces or reabsorbed and brought back to the liver via enterohepatic circulation. Most xenobiotics that enter the body through gastrointestinal absorption are sent directly to the liver. Therefore, this organ is particularly sensitive to chemical injuries by ingested toxins.

Types of Liver Damage

Chemical injuries to the liver depend on the type of toxic agent, the severity of intoxication, and the type of exposure, whether acute or chronic. The six basic types of liver damage are fatty liver, necrosis, hepatobiliary dysfunctions, virallike hepatitis, and (on chronic exposure) cirrhosis and neoplasia. All these types of damage except for neoplasia (liver cancer) are discussed in this section; neoplasia is discussed in the section called **Hepatotoxins.**

Fatty Liver

Fatty liver refers to the abnormal accumulation of fat in hepatocytes. This condition is associated with a simultaneous decrease in plasma lipids and lipoproteins. The mechanism of fat accumulation is related to disturbances in either synthesis of lipoproteins or the mechanism of their secretion.

The onset of lipid accumulation in the liver is accompanied by changes in blood biochemistry; serum glutamic oxaloacetic transaminase (SGOT), serum glutamic pyruvic transaminase (SGPT),[3] alkaline phosphatase, and 5′-nucleotidase are elevated, whereas blood-clotting factors and cholesterol are lowered. Blood chemistry analysis is thus a useful diagnostic tool.

Necrosis

Liver necrosis refers to a degenerative process culminating in cell death. Necrosis can be limited to isolated foci of hepatocytes, or it may involve a whole lobe or both lobes. When entire lobes are involved it is referred to as massive necrosis. The mechanism of necrosis is unknown. The changes in blood chemistry resemble those encountered with fatty liver, except that they are quantitatively larger.

Hepatobiliary Dysfunctions

Hepatobiliary dysfunctions are manifested by the diminution or complete cessation of bile flow, referred to as cholestasis. Retention of bile salts and bilirubin occur as a result; retention of bilirubin leads to jaundice. The mechanism of cholestasis is not well elucidated, but changes in membrane permeability of either hepatocytes or biliary canaliculi, as well as canalicular plug formation, have been implicated (*11*).

The biochemical manifestations of cholestasis are slightly different from those of fatty liver and necrosis. SGOT and SGPT are elevated only slightly or not at all, but alkaline phosphatase, 5′-nucleotidase, and cholesterol are greatly elevated. These hepatobiliary dysfunctions are usually induced by drugs (such as anabolic and contraceptive steroids) but are not likely to be induced by occupational exposure.

Virallike Hepatitis

Virallike hepatitis is an inflammation of liver with massive necrosis caused by certain prescription drugs, such as chlorpromazine and isoniazid. The incidence of

---

[3]The alternate names for SGOT and SGPT are AST (aspartic transaminase) and ALT (alanine transaminase), respectively. AST and ALT are new names, but the old names are still in use.

this disease is very low and no dose–response relationship has been established (*11*).

Cirrhosis

Cirrhosis is characterized by deposition of collagen throughout the liver. In most cases cirrhosis results from chronic chemical injury, but it may also be caused by a single episode of massive destruction of liver cells. Deposition of fibrous matter causes severe distortion of blood vessels, therefore restricting blood flow. The poor blood perfusion disturbs the liver's normal metabolic and detoxification functions. Perturbation of the detoxification mechanism leads to accumulation of toxins, which cause further damage and may lead to eventual liver failure.

Hepatotoxins

A number of metals, organic chemicals, and drugs induce fatty liver and liver necrosis. In most cases, both conditions can be provoked by the same compound; this is true for chloroform, carbon tetrachloride, bromotrichloromethane, dimethylaminoazabenzene, and dimethylnitrosamine. However, certain compounds exert a specific action. Acetaminophen, allyl alcohol, bromobenzene, and beryllium produce necrosis but not fatty liver. On the other hand, allyl formate, ethanol, cycloheximide, and cesium produce fatty liver but not necrosis.

Occupationally, liver injury is most likely to occur following exposure to vapors of volatile halogenated hydrocarbons (such as chloroform, carbon tetrachloride, and bromobenzene), which may enter the bloodstream via the pulmonary route. However, hepatotoxins may enter the gastrointestinal tract, and hence the liver, in the form of fine particles. They are inhaled, then expelled from the bronchi or trachea into the oral cavity, and swallowed with saliva.

Animal experiments (*12*) have shown that cirrhosis can be induced by chronic exposure to carbon tetrachloride and to some carcinogens. Drugs such as methotrexate and isoniazid can also cause cirrhosis. However, the most frequent cause of cirrhosis in humans is chronic use of large quantities of alcohol (160 g per day for 5 years or more).

Although many naturally occurring and synthetic chemicals cause liver cancer in animals, the incidence of primary liver cancer in humans is rather low in the United States. Some of the naturally occurring liver carcinogens are aflatoxin (*see* Chapter 3), cycasin (a glycoside from the cycad nut), and safrol (occurring in sassafras and black pepper; Chart 7.3). Some of the synthetic compounds that cause liver cancer in animals are dialkylnitrosamines, organochlorine pesticides, some PCBs, dimethylbenzanthracene (Chart 7.3), aromatic amines (such as 2-naphthylamine and acetylaminofluorene), and vinyl chloride.

The most noted case of occupation-related liver cancer is the development of angiosarcoma, a rare malignancy of blood vessels, among workers exposed to vinyl chloride in polyvinyl plastic manufacturing plants.

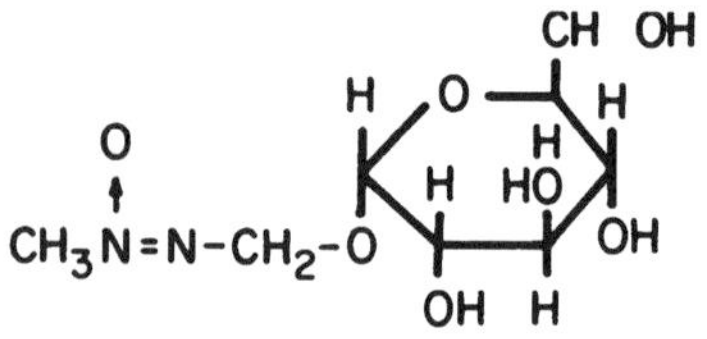

Cycasin

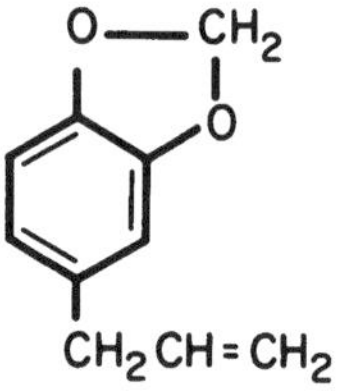

Safrol

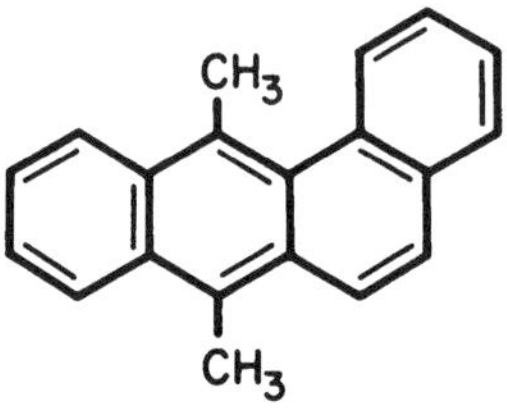

7,12-Dimethylbenz[*a*]anthracene

Chart 7.3. Examples of liver carcinogens.

Other Toxic Responses

The hematopoietic and nervous systems are frequently severely affected by industrial toxins.

Hematopoietic Toxins

Benzene, a component of motor fuel that is also widely used as an industrial solvent and as a starting material in organic synthesis, is a hematopoietic toxin. Chronic exposure to benzene vapors leads to pancytopenia, that is, decreased production of all types of blood cells (erythrocytes, leukocytes, and platelets). The long-term effect of benzene exposure is acute leukemia.

Lead is also a hematopoietic toxin. It interferes with the biosynthesis of porphyrin, an important component of hemoglobin. Severe anemia is one of the symptoms of lead poisoning. Lead is deposited in bones and teeth. Therefore, demineralization of bones, which occurs during pregnancy or as result of osteoporosis, causes release of lead into circulation and subsequently lead intoxication.

Neurotoxins

Metals such as lead, thallium, tellurium, mercury (especially its organic derivatives), and manganese are toxins of the nervous system. The nephrotoxicity of lead and the principal sources of lead exposure have been discussed. Lead and its compounds are also toxic to the central and peripheral nervous systems.

Chronic exposure to lead has different manifestations in adults than in children. In adults occupational exposure to lead fumes and dust causes a disease of the peripheral nervous system referred to as peripheral neuropathy. In children lead exposure is mostly from paint, water, and soil. It causes an alteration of brain structure, referred to as an encephalopathy.

The effects do not reflect the different routes of exposure. They vary because a child's blood–brain barrier is not as well developed as an adult's. This immaturity allows relatively easy access of the toxic metal to a child's brain, whereas the adult brain is protected. Some lead compounds are classified by the International Agency for Research on Cancer (IARC) as carcinogens.

These neurotoxic metals may also enter the system either by inhalation of vapors (mercury) or dust (tellurium, manganese), or by dermal absorption (thallium). The TLV–TWA values of these and other toxins are presented in Table 7.4.

Nonmetallic neurotoxins are frequently used in industry in the manufacture of chemicals and resins or as solvents. Some examples are hydrogen sulfide (which specifically paralyzes the nervous centers that control respiratory movement), carbon disulfide, *n*-hexane, methyl *n*-butyl ketone, and acrylamide. Exposure to all of these substances may occur through inhalation of vapors. In addition, carbon disulfide and acrylamide may enter the system by dermal absorption.

Table 7.4. TLV–TWA Values of Some Neurotoxins

		TWA	
Substance	*Formula*	*ppm*	*mg/m³*
Tellurium	Te		0.1
Thallium	Tl		0.1
Manganese	Mn		5 (dust)
Manganese	Mn		1 (fumes)
Acrylamide[a]	$CH_2{=}CH{-}CONH_2$		0.03
n-Hexane	$CH_3(CH_2)_4{-}CH_3$	50	180
Methyl *n*-butyl ketone	$CH_3CO(CH_2)_3{-}CH_3$	5	20

[a]Acrylamide is a suspected carcinogen.
SOURCE: TWA values are taken from reference 1.

n-Hexane and methyl *n*-butyl ketone are not toxic by themselves but are activated by cytochrome P-450 to the neurotoxic hexanedione ($CH_3COCH_2CH_2$-COCH) (*13*).

References

1. *Threshold Limit Values and Biological Exposure Indices for 1988–1989;* American Conference of Governmental Industrial Hygienists: Cincinnati, OH, 1988.
2. Roach, S. A.; Rappaport, S. M. *Am. J. Ind. Med.* **1990,** *17*(6), 727.
3. Menrel, D. B.; Amdur, M. O. In *Cassarett and Doull's Toxicology;* Klaassen, C. D.; Amdur, M. O.; Doull, J., Eds.; MacMillan: New York, 1986; Chapter 12, p 330.
4. Heppleston, A. G. *Br. Med. Bull.* **1969,** *25,* 282.
5. Mossman, B. T.; Bignon, J.; Corn, M.; Seaton, A.; Gee, J. B. L. *Science (Washington, D.C.)* **1990,** *247,* 294.
6. Office of Technology Assessment. *Cancer Risk. Assessing and Reducing Dangers in Our Society;* Westview: Boulder, CO, 1982.
7. Dean, J. H.; et al. In *Cassarett and Doull's Toxicology,* 3rd ed.; Klaassen, C. D.; Amdur, M. O.; Doull, J., Eds.; MacMillan: New York, 1986; Chapter 9, p 245.
8. Wallin, J. D. In *Review of Physiological Chemistry;* Harper, H. A.; Rodwell, V. W.; Mayers, P. A., Eds.; Lange Medical: Los Altos, CA, 1979; Chapter 39, p 626.
9. Goyer, R. G. In *Cassarett and Doull's Toxicology;* Klaassen, C. D.; Amdur, M. O.; Doull, J., Eds.; MacMillan: New York, 1986; Chapter 19, p 582.
10. Hook, J. B.; Hewitt, W. R. In *Cassarett and Doull's Toxicology;* Klaassen, C. D.; Amdur, M. O.; Doull, J., Eds.; MacMillan: New York, 1986; Chapter 11, p 310.
11. Plaa, G. L. In *Cassarett and Doull's Toxicology;* Klaassen, C. D.; Amdur, M. O.; Doull, J., Eds.; MacMillan: New York, 1986; Chapter 13, p 359.
12. Plaa, G. L. In *Cassarett and Doull's Toxicology;* Klaassen, C. D.; Amdur, M. O.; Doull, J., Eds.; MacMillan: New York, 1986; Chapter 10, p 286.
13. Norton, S. In *Cassarett and Doull's Toxicology;* Klaassen, C. D.; Amdur, M. O.; Doull, J., Eds.; MacMillan: New York, 1986; Chapter 13, p 359.

8

Air Pollution

Pollutant Cycles

It is somewhat artificial to consider air, water, and soil pollution separately because their effects are interchangeable. Chemicals emitted into the air eventually combine with rain or snow and settle down to become water and land pollutants. On the other hand, volatile chemicals from soil or those that enter lakes and rivers evaporate to become air pollutants. Pesticides sprayed on land are carried by the wind to become transient air pollutants that eventually settle somewhere on land or water. For discussion purposes, however, some systematic division appears to be advisable.

Although the problems of air pollution have been recognized for many decades, they were once considered to be only of local significance, restricted to industrial urban areas. With the current recognition of the destruction of stratospheric ozone, the greenhouse effect, worldwide forest destruction, and the acidification of lakes and coastal waters, air pollution assumes global significance.

Urban Pollutants: Their Sources and Biological Effects

The sources of urban air pollution are

1. power generation;
2. transportation;
3. industry, manufacturing, and processing;
4. residential heating; and
5. waste incineration.

Except for waste incineration, all of these pollution sources depend on fossil fuel and, to a lesser degree, on fuel from renewable resources such as plant material. Therefore, all of them produce essentially the same pollutants, although the quantity of each substance may vary from source to source.

The principal incineration-generated pollutants are carbon monoxide (CO), sulfur dioxide (SO_2), a mixture of nitrogen oxides (NO_x), a mixture of hydrocarbons, referred to as volatile organic compounds (VOCs), suspended particulate matter (SPM) of varying sizes, and metals, mostly bound to particles. Waste incineration, in addition, produces some chlorinated dioxins and furans that are formed on combustion of chlorine-containing organic substances.

Most of these air pollutants originate from geophysical, biological, and atmospheric sources. Their contribution to total air pollution is globally significant. This fact should not lead us into complacency about anthropogenic air pollution. In nature, a steady state has been established between emission and disposition of biogenic pollutants. Life on earth developed in harmony with these external influences. The steady state may be gradually changing, in the same way as the climate is changing, but these natural changes occur over a period of thousands or even millions of years.

In contrast, the present dramatic increase in the annual emission of pollutants generated by anthropogenic sources has occurred over a comparatively brief period of 200 years or so. Thus, it is not surprising that nature's steady state has been perturbed. The pH of water and soil is affected, crops and forests are damaged, and many species of plants and animals face extinction. In addition, the anthropogenic pollution sources are concentrated in certain (mostly populated) areas. Thus they have a greater health and environmental impact than most biogenic sources.

Figure 8.1 presents the 1985 emissions data of major urban air pollutants and the main contributors of each pollutant in the United States.

Carbon Monoxide (CO)

Most global emissions of this gas (60–90%) originate from natural sources, such as decomposition of organic matter and volcanic activities (*1*). The anthropogenic origin is primarily due to incomplete combustion of fossil fuel, particularly in internal combustion engines. Thus, motor vehicles are the main culprits (Figure 8.1). Carbon monoxide is a colorless, odorless, highly toxic gas. Its toxicity is due to its ability to displace hemoglobin-bound oxygen. The quantitative relationship between carboxyhemoglobin (HgbCO), oxyhemoglobin ($HgbO_2$), and the partial pressures of O_2 and CO is described by the Haldane equation:

$$HgbCO/HgbO_2 = K \times P_{CO}/P_{O_2} \qquad (8.1)$$

where K is a constant (245 for human blood at pH 7.4 and body temperature), and P_{CO} and P_{O_2} are the ambient partial pressures of CO and O_2, respectively.

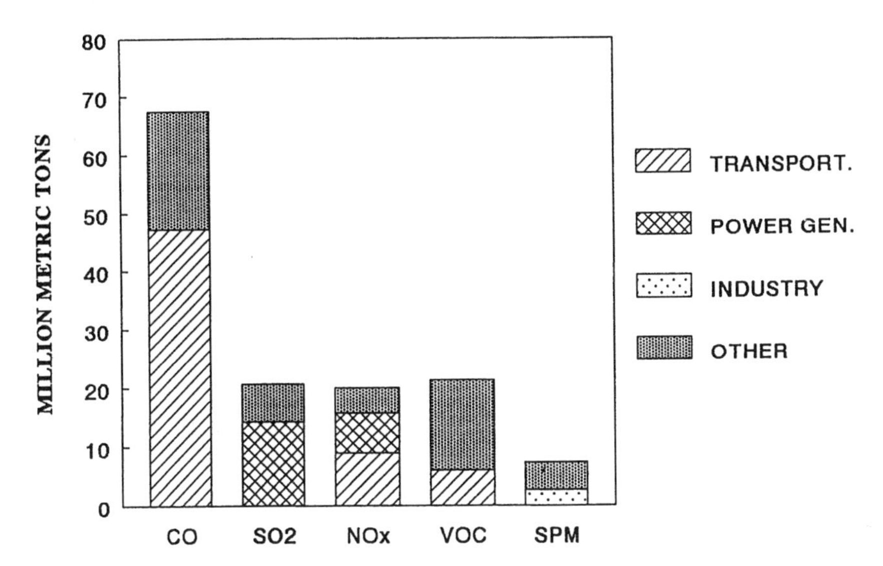

Figure 8.1. Emissions data of major urban air pollutants for 1985 and the main contributors of each pollutant in the United States. (Source: Adapted from reference 1.)

Equilibration of hemoglobin with the ambient carbon monoxide is a slow process, lasting several hours. The degree of intoxication depends on carbon monoxide concentration, the duration of exposure, and to a certain extent on the minute volume of respiration (*see* Chapter 2). Although timely removal of an intoxicated individual from the toxic environment fully restores physiological functions, the dissociation of carbon monoxide from hemoglobin takes considerable time. At one atmosphere pressure, removal of 50% of the gas takes 320 min.

No health effects are seen in humans at less than 2% carboxyhemoglobin content. However, at higher levels, an effect on the central nervous system has been noted in nonsmokers.[1] Cardiovascular changes have been observed at 5%. According to equation 8.1, 5% carboxyhemoglobin content will be achieved upon equilibration at 45 ppm ambient CO concentration. Thus, exposure to carbon monoxide is especially hazardous to people with heart conditions (*2*). More severe carbon monoxide intoxication involves headache, nausea, dizziness, and eventually death.

A lethal intoxication with CO can occur only in an enclosed space. In open spaces the effect of carbon monoxide is mitigated by dispersion. However, in heavy urban traffic carbon monoxide concentration may range from 10 to 40 ppm on the street and almost three times that inside the motor vehicles (*2*). Con-

[1]The content of carboxyhemoglobin in nonsmokers is 0.5–1%, whereas in smokers it may be as high as 5–10%.

centrations as high as 80 ppm have been encountered in tunnels and underground parking lots. Nothing is known about the health effects of chronic exposure to small doses of carbon monoxide. However, because exposure to CO in tobacco smoke is at least one factor contributing to coronary heart disease in smokers, one may speculate that continuous exposure to small quantities of CO may have a cumulative effect.

Although carbon monoxide has no direct impact on the environment, it has an indirect one on the greenhouse gases and on stratospheric ozone (*see* Chapter 9).

Sulfur Dioxide (SO_2)

Sulfur dioxide is a colorless gas of a strong suffocating odor, intensely irritating to eyes and to the upper respiratory tract. Globally, the natural and anthropogenic emissions of sulfur dioxide are more or less equal. Anthropogenic emissions, which predominate over land and in industrialized regions, are mainly produced by combustion of sulfur-containing coal and smelting of nonferrous ore. The natural sources of sulfur dioxide are volcanoes and decaying organic matter. In addition, dimethyl sulfide, which comes from the oceans, is converted in the atmosphere to sulfur dioxide.

The physiological effects of sulfur dioxide in experimental animals are manifested by a thickening of the mucous layer in the trachea and a slowing of the action of the mucociliary escalator (*2*). Sulfur dioxide, a water-soluble gas, is an irritant of the upper respiratory system and it does not penetrate significantly[2] into the lungs. At high concentrations most of it is normally detained in the upper part of the respiratory system and is eliminated by coughing and sneezing. However, some systemic absorption occurs through the whole respiratory system (*2*). Exposure to sulfur dioxide causes bronchial constriction and increases airflow resistance. Thus, it is particularly dangerous to people with respiratory problems. Sulfur dioxide also damages plants by causing bleaching of leaves.

Sulfur dioxide is readily adsorbed on tiny particles (byproducts of coal combustion, such as charcoal, ferric oxide, and metal salts). In the presence of moisture (i.e., in clouds or fog droplets) the particles catalyze oxidation of SO_2 to SO_3, which immediately combines with water to form sulfuric acid (H_2SO_4). When the moisture evaporates, the solid particles coated with sulfuric acid are left suspended in the air. About 80% of these particles are smaller than 2 μm in diameter (*3*). When inhaled, they penetrate into the tracheobronchial region and the alveolar spaces. SO_2 in the gas phase can also be converted to sulfuric acid, albeit at a slow rate, by reactions with free radicals. These reactions are more

[2]The fraction of SO_2 that penetrates the alveolar space is related to the concentration of gas in the inhaled air. At high concentration, 90% of it is removed in the upper respiratory system. At low concentration (1 ppm or less), 95% of the gas penetrates into the lungs.

pronounced in summer than in winter, because they require sunlight for generation of free radicals from the moisture in the air (*4*).

Animal studies (*3*) indicate that sulfuric acid's irritating effect on the respiratory system is 4–20 times stronger than that of sulfur dioxide. Sulfuric acid on the surface of particles is readily dissolved in pulmonary fluid. If present in a high enough concentration, it damages the respiratory tissue (*2*). The involvement of atmospheric sulfuric acid in acid deposition will be discussed in Chapter 10.

Nitrogen Oxides (NO_x)

Nitric oxide (NO) is formed by natural processes such as lightning and microbial digestion of organic matter. Microbial digestion first produces nitrous oxide (N_2O), which is then oxidized to NO. Anthropogenic formation of nitrogen oxides results from high-temperature combustion, whereby nitrogen in the air combines with oxygen. Nitric oxide is readily oxidized in the atmosphere to NO_2, and the mixture of both gases is referred to as NO_x. The total amount of NO_x formed during combustion and the ratio of NO to NO_2 depend on the fuel-to-air ratio and on the temperature of combustion.

Nitrogen dioxide is a reddish brown, irritating, and extremely toxic gas. When inhaled, it causes inflammation of the lungs, which after a delay of several days may develop into edema (swelling of the tissue, *see* Chapter 7). A short exposure to 100 ppm is dangerous and 200 ppm is lethal. At lower concentrations, such as 5 ppm, nitrogen dioxide may increase susceptibility to bronchoconstrictive agents (such as sulfur dioxide) in normal subjects, and at concentrations as low as 0.1 ppm (189 $\mu g/m^3$) in asthmatic subjects (*2*). Concentrations of 0.1 ppm or higher may occur in polluted urban air. In addition, data from animal experiments suggest that exposure to nitrogen dioxide increases susceptibility to respiratory infections by bacterial pneumonia and influenza virus (*2*). In general, emission of NO_x from stationary sources can be controlled better than that from motor vehicles. Also, pollution generated by motor vehicles occurs at the road level, whereas industrial pollutants are usually emitted through smokestacks and carried away by the wind. Although this high-altitude dispersion may reduce exposure of the urban population to NO_x, it probably has no effect on ozone and smog formation.

Photochemical Chain Reactions

The photochemical chain reactions that lead to tropospheric ozone and smog formation require both NO_2 and VOCs. NO_2 is split by sunlight into NO and a free-radical oxygen.

$$NO_2 + h\nu \rightarrow NO + O \tag{8.2}$$

where h is Planck's constant and ν is the light-wave frequency. The free radical reacts with molecular oxygen in a fast reaction to form ozone:

$$O + O_2 \rightarrow O_3 \tag{8.3}$$

However, ozone reacts with NO to regenerate both oxygen and NO_2:

$$O_3 + NO \rightarrow O_2 + NO_2 \tag{8.4}$$

Nitrogen dioxide is split again by sunlight, and the process is repeated over and over. Thus a steady state between NO_2 and NO, which is referred to as the *photostationary state* (*5*), determines the concentration of ozone. It is estimated that, in the absence of VOCs, the ratio of NO_2 to NO equals 1 at noon in North American latitudes. The resulting ozone concentration of about 20 ppb is far below the National Ambient Air Quality Standards (NAAQS) of 120 ppb (daily 1-h average) (*5*).

Because of a series of photochemical reactions involving hydroxyl radicals (·OH), VOCs in the air are converted to peroxy radicals that oxidize NO to NO_2.

$$ROO + NO \rightarrow RO + NO_2 \tag{8.5}$$

The depletion of NO shifts the NO_2/NO steady state in favor of ozone formation (equations 8.2 and 8.3). One of the substances occurring at high concentrations in polluted air is the peroxyacetyl radical. This radical, which oxidizes NO to NO_2, also reacts with nitrogen dioxide to form a lacrimator, peroxyacetyl nitrate ($CH_3C(O)O_2NO_2$) (PAN). The mixture of ozone, PAN, and other byproducts such as aldehydes and ketones creates a haze that is referred to as *photochemical smog.*

Photochemical Smog

Ozone is a respiratory toxin. Because it has low water solubility, it penetrates deep into bronchioles and alveoli. Acute exposure to ozone, which is mostly an occupational hazard, damages the respiratory tissue and causes edema, which may be fatal. Sublethal exposure increases sensitivity to bronchoconstrictive agents and to infections. Chronic exposure to ozone may lead to bronchitis and emphysema.[3] In addition, photochemical smog (i.e., ozone, PAN, and other byproducts) is an irritant of the mucous membranes, eyes, and skin.

The severity of photochemical smog depends, to a great extent, on climatic and topographic conditions. Persistent high-pressure systems tend to aggravate smog formation because they are characterized by intense sunlight and stable descending air that traps pollutants near the ground. In places surrounded by

[3]Emphysema is a condition characterized by decompartmentalization of alveoli. The surface area available for gas exchange is decreased, which causes difficulties in breathing.

mountains, the dispersing force of wind is diminished. Atmospheric temperature inversion also favors retention of photochemical smog near the ground. *Inversion* occurs when warm air aloft overlays colder air near the ground; thus the polluted air is prevented from rising above the inversion boundary.

Both ozone and PAN are toxic to plants. Whereas PAN affects mostly herbaceous[4] crops, ozone injures the tissues of all plants and inhibits photosynthesis. In addition, it increases the susceptibility of plants to drought and disease. With respect to plant damage, O_3, NO_2, and SO_2 act synergistically.

Photochemical oxidation and smog formation are the main known environmental and health hazards of NO_x emission. However, concern about the direct health effect of NO_x is growing. It appears that in significantly polluted urban areas, nitrogen oxides are responsible for a high frequency of respiratory diseases, such as bronchitis, pneumonia, and viral infections. There is also concern about their involvement in acid deposition; about one-third of the acid deposited is nitric acid.

Volatile Organic Compounds

VOCs originate from both anthropogenic and natural sources. The natural sources are vegetation, microbial decomposition, forest fires, and natural gas. According to an editorial published in *Science* (*6*), the natural emission of VOCs is estimated to be 30–60 million metric tons annually.

Anthropogenic emission results from incomplete combustion of fossil fuels and from evaporation of liquid fuels and solvents during storage, refining, and handling. The type of VOC emitted with flue gases or from the exhaust of motor vehicles varies with the type of fuel, the type of combustion (i.e., external or internal), and the presence or absence of pollution-abating devices.

Low-molecular-weight aliphatic, olefinic, and aromatic compounds, some of which are formed during combustion, are prevalent. At 500–800 °C, olefins and dienes tend to polymerize via free-radical formation to form polycyclic aromatic hydrocarbons (PAHs) (*7*).

Airborne PAHs are distributed between the gas phase and solid particles (byproducts of combustion, such as soot and fly ash). At least 26 airborne PAHs, some of them potential carcinogens and mutagens, have been identified and quantified (*7*). The most extensive study was done with benzo[*a*]pyrene (*see* Chapter 5). In general, the total concentration of PAHs in the air is about 10 times higher than that of benzo[*a*]pyrene, which has frequently been used as an indicator of the total concentration of PAHs in the atmosphere. Some reservations have been expressed as to the accuracy of this procedure (*7*). Contributions of various fuels and combustion techniques to the atmospheric emission of benzo[*a*]pyrene are presented in Table 8.1. According to these data, the greatest quantity of benzo[*a*]pyrene per BTU is produced by residential wood combus-

---

[4]Herbaceous plants do not have a woody stem and die entirely each year.

Table 8.1. Contributions of Fuels and Combustion Techniques to Atmospheric Emission of Benzo[*a*]pyrene

Fuel	*User*	*Benzo[a]pyrene (ng BTU^{-1})*
Coal	Utilities	0.056–0.07
Coal	Residences	0.12–61.0
Wood	Residences	27–6300
Oil	Residences	0.00026
Natural gas	Residences	0.02
Gasoline	Motor vehicles	0.6
Diesel fuel	Motor vehicles	2.3

SOURCE: Adapted from data in reference 10.

tion. Indeed, as shown in Figure 8.2, wood-burning in fireplaces and stoves contributed 85.5% to the total of 655 metric tons of PAHs emitted annually in the United States during the 1980s. The second-largest source was agricultural burning, and the third was forest fires (*8*).

Size of Particles

PAHs in the vapor phase do not present much of a health risk, but those bound to respirable particles do. The health effect of atmospheric carcinogenic PAHs is

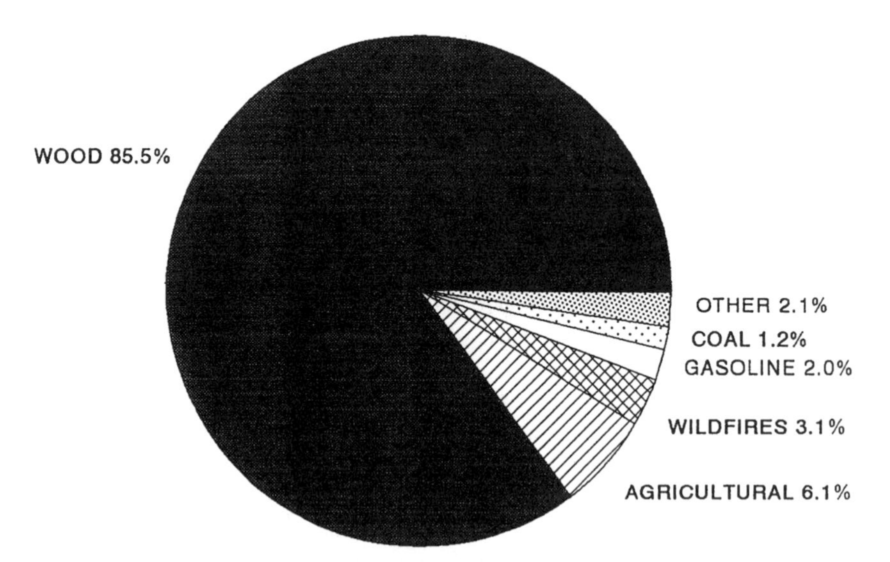

Figure 8.2. Emissions of PAHs in the United States during the 1980s. Annual emissions of PAHs total 655 metric tons. "Agricultural" refers to prescribed forest and agricultural burning; "wood" refers to wood-burning in fireplaces and stoves. (Source: Adapted from reference 8.)

related to the size of the particles with which they are associated, as only small particles penetrate the respiratory system. Particles having a diameter of 1 μm or less may penetrate the lungs. There the PAHs are desorbed and either activated to carcinogens by the pulmonary P-450 system or enter the circulation. The larger particles (2–5 μm) do not reach alveoli. These particles are expelled by the mucociliary escalator into the oral cavity, where they may be swallowed. In this case, the PAHs enter the circulation via the gastrointestinal route (*see* Chapter 2). According to some sources, the absorption of PAHs by the tissue and their carcinogenic potency may depend on the route of exposure (whether by respiration or ingestion with food) (*9*).

For benzo[*a*]pyrene, the allowable daily intake, defined as an intake associated with 1/100,000 increased lifetime risk of developing cancer for a human weighing 70 kg, is 48 ng per day. Human exposure (in nanograms per day) from various sources is as follows (*7*):

- air, 9.5–43.5;
- water, 1.1;
- food, 160–1600; and
- tobacco smoke, 400.

As can be seen, the relative cancer risk for the population at large from inspired benzo[*a*]pyrene is relatively low. Its concentration in the air is much below that in food and in tobacco smoke.

Exposure at Work and via the Food Chain

On the other hand, people in certain occupations, such as coke-oven workers and coal tar pitch workers, are at high risk. Their exposure may exceed that of the general population by a factor of 30,000 or more. In addition, urban-generated particles loaded with PAHs settle on land or water, and the carcinogens are likely to enter the food chain. Study of the sediment in the Charles River in Boston revealed a striking similarity between the composition of PAHs in the atmosphere and that in the river sediment (*11*). It appears that combustion of fossil fuels is the main source of water pollution with PAHs.

Benzene and Ethylene

Other hydrocarbons of interest are benzene and ethylene. Benzene is a human bone marrow poison and a carcinogen implicated as a cause of myelocytic and acute nonlymphocytic leukemia. Ethylene is one of the major products of automobile exhaust, but it may also be formed by other combustion processes. It contributes heavily to photochemical oxidants. Ethylene is a normal constituent of plants; it serves as a plant growth regulator and it induces *epinasty* (movement of a plant, such as folding and unfolding of a flower petal), *leaf abscission* (falling of leaves), and fruit ripening. Excessive external ethylene is therefore a plant toxin.

The involvement of hydrocarbons in photochemical smog formation was discussed earlier.

Airborne Particles

Particles are referred to as suspended particular matter (SPM). They may be divided into suspended solids and liquid droplets. Their effects on respiratory and systemic toxicity differ (*see* Chapter 2). The natural sources of airborne particles are dust, sea spray, forest fires, and volcanoes. Anthropogenic particles include solids ranging from 0.01 to 100 μm in diameter and minute droplets of sulfuric, sulfurous, and nitric acids. They are byproducts either of combustion (such as fly ash, soot, and numerous metals) or of industrial processes (such as milling and grinding).

In the atmosphere, continuous interaction takes place among various types of particles and between particles and the components of the gas phase. This interaction affects both the chemical composition and the size of the particles (*5*). Large particles (greater than 30 μm in diameter) may present a nuisance, but they do not have any serious health impact and they settle out rather quickly. In contrast, the atmospheric residence time of particles 1 to 10 μm in diameter is 6 h to 4 days; for particles smaller than 1 μm in diameter it is even longer.

Particles smaller than 5 μm in diameter enter the tracheobronchial and pulmonary region, where they irritate the respiratory system and aggravate existing respiratory problems. Their role as vehicles for transporting PAHs and sulfate and sulfite ions into the lungs has already been discussed.

SPM also has an environmental impact. Tiny sulfate particles, because of their light-scattering properties, are responsible for haze formation. This effect, which is amplified in the presence of high humidity, may persist for as long as a week. Soot particles, which have light-absorbing properties, also contribute to haze formation. SPM deposited on leaves inhibits absorption of carbon dioxide, plugs stomata (tiny orifices on the leaf surface for evaporation of water), and blocks sunlight necessary for photosynthesis.

Metal Pollutants

Among the metal pollutants, lead, mercury, and beryllium are of special interest because of their toxicity. With the gradual phasing out of leaded gasoline, the amount of airborne lead decreased considerably. Lead emissions in the United States declined from 144,000 tons in 1975 to 17, 900 tons in 1985 (*12*); 69% of it originated from combustion of leaded gasoline. At the same time, the contribution of municipal waste incinerators to lead pollution became more significant. Mercury and beryllium originate mainly from coal combustion. Regardless of their origin, both lead and mercury are essentially water and land pollutants. Their health and environmental impact will be discussed in the Chapter 10.

Atmospheric emission of beryllium has been estimated (*13*) to be 1134 metric tons annually. The major toxic effects of beryllium are pneumonitis (a disease characterized by lung inflammation) and berylliosis (a chronic pulmonary disease). Epidemiological studies suggest that it is also a carcinogen. It is not certain whether beryllium concentration in urban air is sufficient to create a health hazard for the population at large. In any case, beryllium represents an occupational hazard to workers involved in its production, processing, and use (*see* Chapter 7).

Nonmetal Pollutants

Fluorides and asbestos are nonmetal pollutants. Fluorine is a byproduct of coal combustion. It is released, entirely in the gas phase, in relatively large quantities. Being a reactive element, it combines readily with other atoms and molecules to form fluorides, which are respiratory irritants. They are also phytotoxins (*3*), and their main environmental impact is on plants. Fluorides cause leaf damage and eventual defoliation.

Airborne asbestos originates from industrial use and from the demolition of old buildings containing asbestos. Its health effects are mostly limited to asbestos workers and to workers who are incidentally exposed to asbestos while performing their duties. Therefore, exposure to asbestos is considered an occupational hazard. The health effects of this exposure are discussed in Chapter 7.

Trends and Present Status of Air Quality

Table 8.2 lists the U.S. National Ambient Air Quality Standards (NAAQS) and the World Health Organization (WHO) guidelines for the major urban air pollutants. Figures 8.3 and 8.4 show the trends in sulfur dioxide and suspended particle concentrations, respectively, in the air of selected cities in the United States and around the world during the last decade. The data indicate that in general, between 1976 and 1985 good progress toward abatement of sulfur dioxide pollution was achieved in industrialized countries. Houston is one exception where levels of sulfur dioxide increased considerably, probably because of the economic development during this period. Nevertheless, the levels remain well below the WHO guidelines and the U.S. NAAQS. It is important to note that the data presented in Figures 8.3 and 8.4 are the mean values of the residential, commercial, industrial, and suburban areas. Certain areas of a city evaluated by themselves may exceed the standards. For instance, in the residential area in the city center of New York, the mean daily concentrations were, for three monitoring periods, above the WHO guidelines (72 $\mu g/m^3$ in 1976–1978, 74 $\mu g/m^3$ in 1979–1981, and 65 $\mu g/m^3$ in 1982–85) (*1*). Among the cities of the industrialized world, Mi-

Table 8.2. NAAQS and WHO Guidelines

Pollutant	*Standard*
NAAQS	
Carbon monoxide	10,000 μg/m^3 or 9 ppm for 8 h
	40,000 μg/m^3 or 35 ppm for 1 h
Ozone	235 μg/m^3 or 0.12 ppm for 1 h
Nitrogen dioxide	100 μg/m^3 or 0.053 ppm per year
Sulfur dioxide	80 μg/m^3 or 0.03 ppm per year
	365 μg/m^3 or 0.13 ppm for 24 h
SPM	50 μg/m^3 per year
	150 μg/m^3 for 24 h
Lead	1.5 μg/m^3 per year
WHO guidelines	
Carbon monoxide	10,000 μg/m^3 for 8 h
Sulfur dioxide	40–60 μg/m^3 per year
	100–150 μg/m^3 for 98 percentiles[a]
SPM	60–90 μg/m^3 per year
	150–230 μg/m^3 for 98 percentiles[a]
Nitrogen oxides	150 μg/m^3 per day
	400 μg/m^3 per hour
Lead	0.5–1 μg/m^3 per year

[a]98% of daily averages must be below these values; no more than 7 days per year may exceed this value.
SOURCE: EPA communication and *Global Environment Monitoring System, Assessment of Urban Air Quality.*

lan stands out as exceptionally polluted; although the levels of sulfur dioxide decreased by nearly 45% between 1976 and 1985, they still considerably exceeded, by the last monitoring year, the WHO guidelines.

No progress in abatement of sulfur dioxide pollution has been achieved in cities of the developing nations. In some of them, as for example Teheran, Calcutta, and Beijing, pollution increased considerably during the monitoring period. This was probably a result of an attempt at industrialization with insufficient investment in modern technology.

Data in Figure 8.4 show the trends in mean daily concentrations of SPM in selected cities throughout the world. In North America progress in pollution abatement has been achieved, but in some cities the values are still in the marginal area of the WHO guidelines (between 60 and 90 μg/m^3). In Frankfurt, Brussels, and London the SPM pollution was already low at the beginning of the monitoring period. In addition, in two of these cities, Brussels and London, a further, albeit small, reduction in SPM levels has been achieved. On the other hand, in all five cities of the developing world listed here, SPM pollution highly exceeded the safe limits, and no real progress in its abatement could be achieved.

The situation is much less encouraging with respect to urban air pollution by

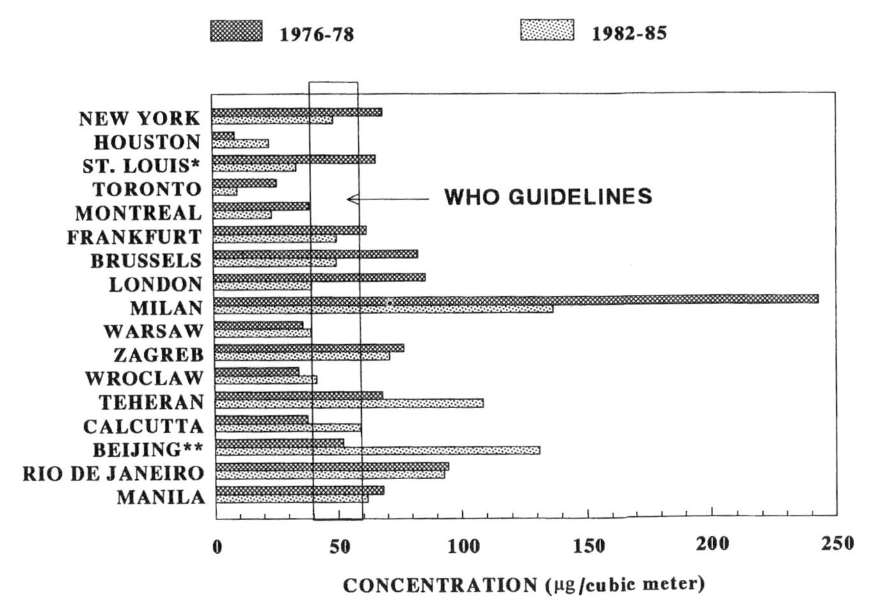

Figure 8.3. Trends in sulfur dioxide concentrations in the air of selected cities in the United States and around the world during the last decade. Reported values of each city are averages of commercial, residential, and industrial areas. The single asterisk indicates that the end period of the survey was 1979–1981; the double asterisk indicates that the initial period of the survey was 1979–1981. (Source: Adapted from reference 1.)

carbon monoxide and nitrogen oxides. In most countries during the period from 1973 to 1984, there was little change, or sometimes even an increase, in emissions of carbon monoxide and nitrogen oxides (*14*). Of 35 cities surveyed worldwide by the WHO and the United Nations Environment Programme (UNEP) for trends in ambient-air levels of nitrogen oxides, there was an annual decrease in 18 of them and an increase in 17 (*14*).

A summary of a WHO–UNEP air quality survey for the period from 1973 through 1985 in selected cities around the world is shown in Table 8.3. The WHO estimates that globally, out of 1.8 billion urban dwellers, nearly 1.2 and 1.4 billion live in areas with annual average levels of sulfur dioxide and SPM within the marginal limits or in excess of the WHO guidelines, respectively. One has to also be aware that compliance with the NAAQS or WHO guidelines does not necessarily assure lack of adverse health effects. It is emphasized by the WHO that "guidelines are only given for single pollutants; exposure to pollutant mixtures may lead to adverse effects at levels below the recommended guidelines for individual pollutants" (*15*). Thus, the goal should be to decrease air pollution as much as possible.

No air pollution data for individual cities are available beyond 1985. Howev-

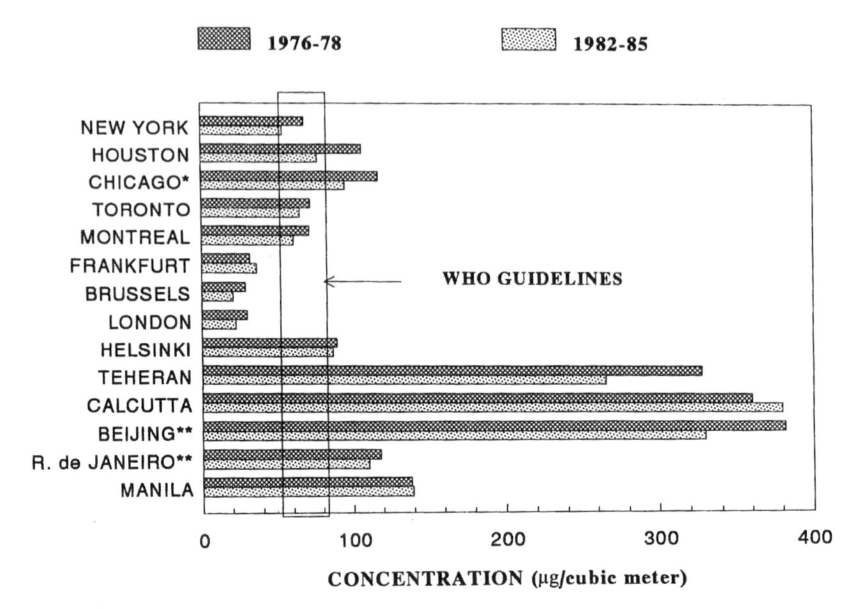

Figure 8.4. Trends in suspended particle concentrations in the air of selected cities in the United States and around the world during the last decade. Reported values of each city are averages of commercial, residential, and industrial areas. The single asterisk indicates that the end period of the survey was 1979–1981; the double asterisk indicates that the initial period of the survey was 1979–1981. (Source: Adapted from reference 1.)

Table 8.3. Percentage of Cities Exceeding the WHO Pollution Guidelines

Pollutant	*Number of Cities Surveyed*	*Percentage Exceeding WHO Guidelines*	
		Short-Term	*Long-Term*
Sulfur dioxide	54	45	30
SPM	41	55	60
Nitrogen oxides	28/42[a]	30	0
Carbon monoxide	15	55	NA[b]
Lead	23	NA[b]	20

[a]The first number refers to a short term, the second number to a long term.
[b]Data are not available.
SOURCE: Adapted from *Global Environment Monitoring System, Assessment of Urban Air Quality*; United Nations Environment Programme and World Health Organization: Geneva, Switzerland, 1988; Chapter 8, p 70. Reproduced with permission from reference 43. Copyright 1994 SFZ Publishing.

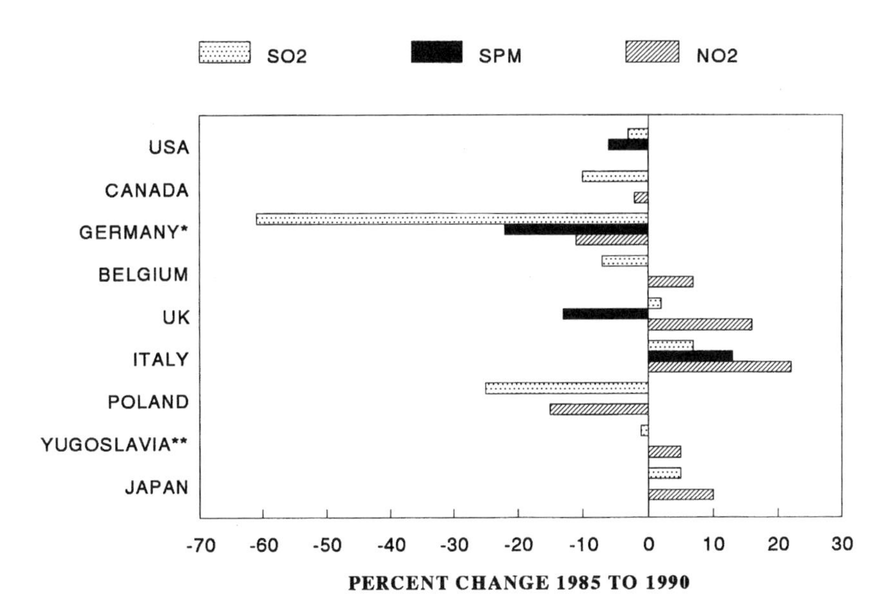

*Figure 8.5. Progress in abatement of air pollutant emissions between 1985 and 1990 in selected countries. Key: *, former West Germany only; **, former united Yugoslavia. (Source: Adapted from data in reference 16.)*

er, there are nationwide data on air pollutant emissions that can be used to compare the situation in 1985 with that in 1990 (*16*). As shown in Figure 8.5, except for Germany (former West Germany) and to a lesser extent Poland, the progress in abatement of air pollutant emissions was not spectacular in most of the countries listed in this figure; in some countries the situation has deteriorated.

Recent studies in American cities pointed to an association between air pollution with fine particulate matter, including sulfates, and excess mortality from lung cancer and cardiopulmonary diseases (*17*).

Pollution by Motor Vehicles

Gaseous and Vapor Pollution

The 1986–1987 American Lung Association survey (*18*) of nine U.S. cities (Baltimore, Chicago, Hartford, Houston, Los Angeles, Minneapolis/St. Paul, Philadelphia, St. Louis, and Washington, DC) revealed that all were in violation of the NAAQS for at least two pollutants, despite the fact that the deadline for compliance with the Environmental Protection Agency (EPA) standards was December 31, 1987. All but one (Minneapolis/St. Paul) exceeded the recommended ozone level, and all but two (Chicago and Houston) exceeded carbon monoxide levels. According to this survey, most of the noncompliance was due to pollution by motor vehicles.

An overview of air pollution caused by motor vehicles in selected cities around the world, compiled by the World Bank, is presented in Table 8.4.

In the United States, to conform with the Clean Air Act emission standards, all automobiles and trucks manufactured after 1975 are equipped with pollution-control devices. However, these devices perform satisfactorily only when properly maintained. Poor maintenance, tampering, and insufficient monitoring and inspection make the attainment of air quality standards problematic. In addition, the gains in air quality realized by installation of pollution-control devices are being offset by a steadily increasing number of motor vehicles on the road. In the United States the number of registered motor vehicles increased from about 108 million in 1970 to 180 million in 1986. If the present trend continues, this number may swell to 281 million, about one motor vehicle per person, by 2010.

Another problem is the escape of gasoline vapors into the air during refueling of motor vehicles. Devices for recovery of these vapors (stage II vapor recovery devices) are available, but their use is not enforced in most states.

The summation of the American Lung Association's survey states in part:

> The widespread failure of the metropolitan areas to meet the national air quality standards have substantial and direct public health effects. For the millions of Americans currently breathing unhealthy air, and especially for those people with existing lung disease or with a particular sensitivity or risk to air pollution, the findings of this study highlight a very real public health problem (*18*).

In another report, "Pollution on Wheels II", quoted in *Chemical & Engineering News* (*20*), the American Lung Association estimates that the annual health cost due to air pollution caused by motor vehicles is $4.5–$93 billion.

A 1989 survey (*21*) recorded the air quality inside 140 randomly chosen cars

Table 8.4. Contribution of Motor Vehicles to NO_x Emissions in Selected Cities

City	*Year*	*Percentage of Total Emissions of NO_x*
Mexico City	1987	64
Manila	1987	73
London	1978	65
Los Angeles	1976	71
Hong Kong	1987	75
Seoul	1983	60

SOURCE: Adapted from data in reference 19. Reproduced with permission from reference 43.

traveling the highways of southern California. The occupants of these cars were exposed to pollutant levels four times higher than those in the ambient air. Of the 16 pollutants measured, benzene levels were the highest.

Ozone pollution, generated mainly by motor vehicles and to a lesser extent by stationary sources, also affects agriculture. Concentrations of ozone drifting over some rural areas in the United States reach values as high as 50 to 60 ppb for an average period of 7 h/day. This level is sufficient to lower the yield of cotton and soybeans by 20% and that of peanuts by 15%. The yield of corn and wheat may be also affected, but to a much lesser extent (*22*).

Rubber and Asbestos

Tire wear is estimated as 360 mg/km per car (*23*); still, most of the pollution is restricted to the roadway and its vicinity. Rubber particles from tires contribute to air pollution and to water pollution as they are washed out with storm water into the watershed.

A study in the highly urbanized area of Los Angeles indicated that tire wear contributed 671 kg/day to aerosol organic carbon (2.4% of the total organic carbon in the air), whereas brake lining wear was estimated to be 1480 kg/day (*24*). A new study (*25*), which reported that urban air contains respirable black particles, probably originating from tires, appears to confirm the earlier findings. The major component of tires is natural latex. Proteins of natural latex are known to be antigens capable of eliciting hypersensitivity (*26*).

Sixty percent of the wear products of brakes are volatile materials such as CO, CO_2, and hydrocarbons; the other 40% are particulate matter. Only about 0.01% of this material is asbestos (*23*). These particulate wear products also present an urban air and water pollution problem.

The airborne respirable particles from tires and brakes may be, in part, responsible for the increasing incidence of asthma in the United States. According to a report from the National Center of Health Statistics, the prevalence of ever having asthma among 6- to 11-year-old children increased from 4.8% during 1971–1974 to 7.6% during 1976–1980 (*27*). The incidence was more prevalent among urban than rural children, thus providing additional indirect evidence that urban aerosols are the culprits.

Pollution by Industrial Chemicals

Toxic substances released into the air by industry have caused much concern. Although the Clean Air Act (Chapter 14) has a toxic substances provision, until recently only seven substances were regulated by the EPA: arsenic, asbestos, benzene, beryllium, mercury, radionuclides, and vinyl chloride. The Clean Air Act of 1990 increased the number of regulated toxic air pollutants to 189, but it will not be until 2003 that the law will be fully implemented (*see* Chapter 14).

Toxic Release Inventory and the Pollution Prevention Act

The Superfund Amendment and Reauthorization Act (SARA) of 1986 mandated that all industries producing, importing, or using more than 75,000 lb of a chemical (listed on the EPA index of toxic materials) annually have to report the toxic releases into the environment, and transfers of the toxic waste to other facilities. This is called the Toxic Release Inventory (TRI). In 1987 the reporting threshold was lowered to 50,000 lb per chemical, and in the following years to 25,000 lb. In the first year (1987) only 19,000 facilities (estimated 55–75% of all businesses required to file) complied with the regulation; by 1990 (the last year for which data were compiled) the number of reporting industries increased to 23,000. How complete was the latest compliance and how accurate were the data are not known. Figure 8.6 presents the TRI reports from 1987 to 1990, and Figure 8.7 shows the toxic releases for 1990 by industries. One can see that the largest quantity of toxins was released into the air and that the largest contributor was the chemical industry. Although more facilities complied with the TRI requirement in 1990 than in 1987, and despite the more stringent reporting requirements, the total amount of reported releases decreased from 6.7 billion pounds in 1987 to 4.8 billion pounds in 1990 (*28*). The figures for 1991 indicated a further 9% decline in the total amount of toxic releases. A breakdown of these latest figures according to the release category showed a 13% decline in toxic re-

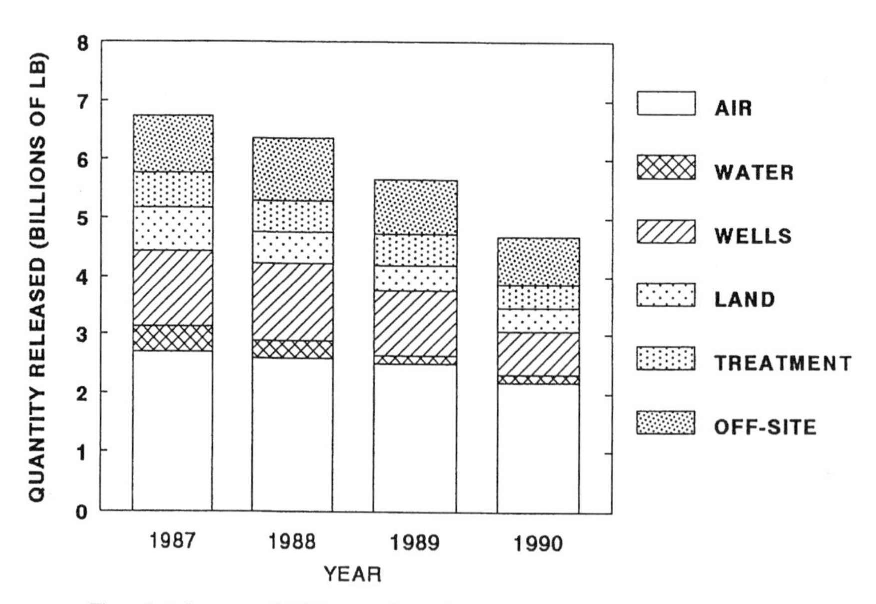

Figure 8.6. Summary of TRI reports from 1987 to 1990. Source: Adapted from data in reference 28. (Reprinted with permission from reference 43. Copyright 1994 S. F. Zakrzewski.)

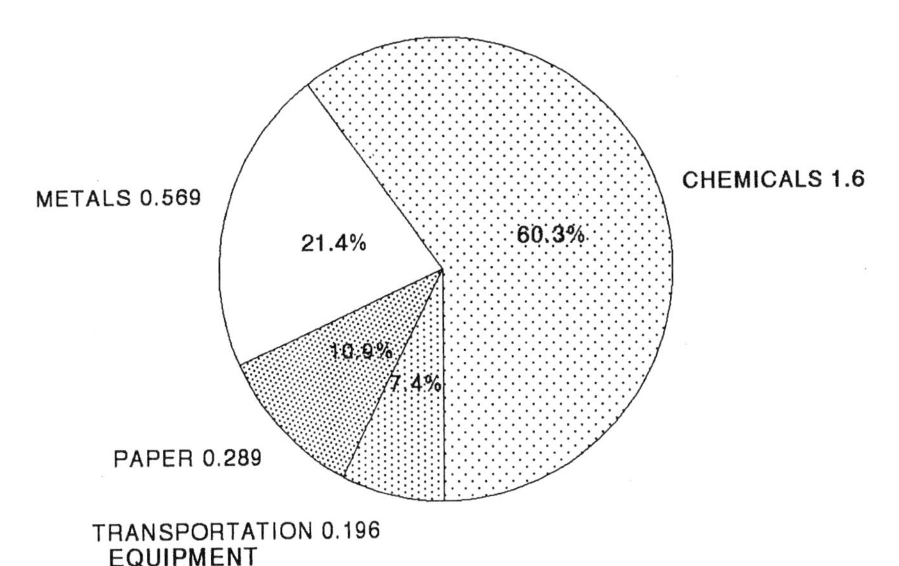

Figure 8.7. Toxic release inventory for 1990 by industries (in billions of pounds). Source: Adapted from data in reference 28. (Reprinted with permission from reference 43. Copyright 1994 S. F. Zakrzewski.)

leases into the air, 9% onto land, and 5% into deep wells. However, releases into water increased nearly 24% (*29*).

Although the decrease in toxic emissions during five years has not been spectacular, some progress in the right direction was achieved. It appears that the requirement for TRI reporting did motivate the industries to control their emissions. It made them aware of their contributions to the environmental blight and of the fact that their public image will suffer unless they clean up their act. In addition, the EPA is trying to enforce compliance with the law by conducting inspections and imposing stiff monetary penalties for noncompliance.

To motivate the industries to cut down pollution even further, the Pollution Prevention Act was enacted in 1990 and went into effect in 1992 (*30*). This act moved away from the past policies of regulations aimed at "end-of-pipe" pollution prevention, toward a voluntary program of pollution's source reduction. Within the frame of the Act, the EPA called for increased efficiency in the use of resources, such as raw materials, energy, and water; this can be achieved by investing in new technologies, by recycling instead of dumping, by personnel training, and by improving the manufacturing processes and management practices (*30*). Also, a new program called "33–50" was initiated. This program aimed at voluntary reduction in the releases and transfers of 17 toxic chemicals. It called for a 33% reduction of the 1988 releases of these compounds by 1992 and a

50% reduction by 1995. It is encouraging that this governmental initiative was generally well received by the industries. The latest TRI for 1994, which was published in the summer of 1996, revealed that the 50% reduction of releases was achieved in that year, i.e., one year ahead of the schedule.

The Chemical Manufacturers Association (CMA) responded with its own initiative called "CMA's Responsible Care", which encourages the affiliated industries to improve their waste-management practices. Unfortunately, there seem to be some inconsistencies and conflicts between the rhetoric of CMA's Responsible Care and CMA's lobbying effort.

Though these are positive turns of events, it is doubtful that the goal of zero discharges, advocated by some environmental organizations, can be ever achieved, at least not when population and demand for consumer goods keep growing.

Cancer Incidence

Unfortunately, a comprehensive epidemiological study of the impact of airborne toxic pollutants on human health is lacking. However, statistics indicate an increased cancer incidence in areas with a heavy concentration of chemical industries (Figure 8.8).

A case in point is the high incidence of a variety of cancers (including but

Figure 8.8. Cancer mortality among white males in the United States 1970–1980 (national average rate is 189 per 100,000). The black patches indicate areas of the highest (top 10%) mortality. (Source: Adapted from reference 32.)

not limited to cancers of the lung, brain, liver, and kidney, as well as miscarriages) reported by the press in the industrial corridor of Louisiana (*31*). This 85-mile corridor (popularly known as "cancer alley") begins in Baton Rouge and follows the Mississippi River to the southeastern outskirts of New Orleans. The corridor contains 7 oil refineries and 136 petrochemical plants, which produce 60% of the nation's vinyl chloride, 60% of all nitrogen fertilizers, and 26% of all chlorine. In that area alone approximately 400 million pounds of toxic chemicals are released annually into the air, including 500,000 pounds of vinyl chloride.

Statistics released from the National Cancer Institute indicate that 1970 cancer mortality in Louisiana exceeded the national average by 25%. In addition, it was reported (*31*) that cats and dogs in the industrial corridor were losing their hair and that Spanish moss began to disappear, as did crawfish from ponds and marshes.

Respiratory Problems

The National Cancer Institute's statistical review released in 1987 (*33*) recorded a 29.5% increase in the incidence rate of respiratory cancer (lung and bronchial) between 1973 and 1985. The number of smokers decreased between 1965 and 1985, from 43.3% to 30.8% of the population over the age of 20. Thus, it is unlikely that this increase may be attributed to smoking. Whether this trend is attributable to air pollution cannot easily be established, but airborne toxins should be considered as a contributory factor. Although the respiratory cancer is of great concern, it is not the only health problem caused by air polluted by toxic chemicals.

The effect of urban and industrial pollutants on human health in Eastern Europe has been documented. In the highly industrialized district of southwestern Poland, outdated industrial plants emit tons of sulfur dioxide, nitrogen oxides, chlorides, fluorides, vaporized solvents, and lead into the air. Bronchitis, tuberculosis, and pulmonary fibrosis (pneumoconiosis) (*see* Chapter 7) are more prevalent in this industrial district than anywhere else in the country. In one area 35% of the children and adolescents suffer from at least mild lead poisoning (*34*). In the highly polluted regions of the former Czechoslovakia, the frequency of respiratory diseases among preschool and school-age children was five times and three times higher, respectively, than it was among children from the less-polluted western region (*35*). Similarly, it has been noticed in Poland that the rates of chronic bronchitis were three times higher and asthma four times higher among army recruits from areas heavily polluted by sulfur dioxide than among recruits from the unpolluted areas (*35*). Overall life expectancy at birth, during the period 1985–1990, was 5% lower in Eastern than in Western Europe. On the other hand, infant mortality was nearly twice as high in the eastern countries (Poland, Czechoslovakia, and Hungary) than in West Germany (*35*).

Pollution by Incinerators

Another concern is the emission of airborne toxins by municipal and toxic waste incinerators. With the growing shortage of waste disposal sites and the increase in the cost of disposal, municipalities in the United States and around the industrialized world are tending to dispose of municipal waste by incineration and to use the heat produced for energy generation.

Facility Effectiveness

Incinerators built during the first half of this century are no longer in use because they do not meet present air quality standards. Although modern incinerators may meet the air quality standards for conventional pollutants, there is concern that incineration of chlorine-containing compounds, such as bleached paper and poly(vinyl chloride) plastics, produces toxic (and until recently, unregulated) dioxins and furans.

With the increasing use of disposable plastics and a variety of household chemicals that eventually end up in the waste stream, this concern seems to be justified. Epidemiological studies (*36*) point out the relatively high levels of dioxins in the milk of nursing mothers. This contamination may be attributable, at least in part, to waste incineration. In addition, waste incinerators contribute to air pollution by emitting toxic metals such as mercury, lead, zinc, cadmium, tin, and antimony.

Chemical Waste

Incineration of chemical waste presents a similar problem. According to Gross and Hesketh (*37*), the most modern controlled-air incinerators "are able to dispose of a wide variety of organic solid wastes." However, the efficiency of the destruction of these compounds is still open for debate. The law requires that the hazardous waste incinerators have a destruction-removal efficiency (DRE) of 99.99% for all hazardous waste and 99.9999% for "waste of special concern" such as PCBs and dioxins. However, according to the EPA scientists, none of the presently available incinerators can meet the governmental standards. Although some chemicals can be destroyed with 99.99% efficiency in test burning of a single compound, this does not mean that all compounds in a waste mixture will be destroyed with this efficacy, because the optimal destruction temperature may vary from compound to compound. For safety reasons, testing for combustion efficiency of highly toxic compounds, such as dioxins, is done with surrogate compounds that are supposed to be harder to destroy than the actual compound of interest. An assumption was made that if the 99.9999% DRE was achieved in the test, this DRE will also apply to dioxins or PCBs in the mixture of waste. Analysis of the results of actual burning revealed that if the test compound was present in a mixed waste at a concentration of less than 1000 ppm, its destruction was not nearly 99.9999% complete. Although the phenomenon is not well understood, the fact remains that the alleged completeness of destruction by incineration of highly toxic compounds in the waste stream may frequently be highly overestimated.

Several reports presented during the International Congress on Health Effects of Hazardous Waste, held in Atlanta, GA, in May 1993 indicated that people living downwind or in close vicinity to toxic waste incinerators had a greater prevalence of coughing, phlegm, wheezing, sore throat, eye irritation, emphysema, sinus trouble, and neurological diseases than those living upwind or some distance from the incinerators. Although none of these studies definitely proves the link between incinerators and a health hazard, they strongly suggest that such a link may exist (*38*).

Tall Stacks and Their Role in Transport of Pollutants

The Clean Air Act sets standards for local air quality. However, except for new stationary sources of pollution, which are required to install scrubbers for removal of sulfur dioxide from flue gases, it does not specify the means by which this air quality should be attained.

Thus, it was possible to make some smelters and coal-burning power plants conform to local air quality standards simply by increasing the height of their smoke stacks. Stacks over 200 feet high emit pollutants above the ambient air monitoring level. These pollutants are propelled with the wind for hundreds of miles. They settle, eventually, in a dry form or with rain or snow on land and water. This is known as *acid precipitation*; its effects will be discussed in Chapter 9.

Since 1970, 102 tall stacks (23 of them taller than 1000 feet) have been erected by utility companies in the United States (*39*). Legal action to outlaw tall stacks has been initiated by environmental organizations, and several bills concerning this issue have been proposed in Congress. In 1977 a "Tall Stacks" provision that prohibits the use of dispersion techniques as a means of conforming with NAAQS was added to the Clean Air Act (*see* Chapter 14). Despite this provision, the problem of airborne transport of pollutants still exists because of either loopholes in the law or lack of enforcement.

Indoor Air Pollution

Considering the indoor concentration of pollutants and the time spent indoors, the daily intake of some pollutants from indoor and outdoor air could be calculated (*40*) (Table 8.5). The EPA survey of air quality inside 10 buildings, conducted during the 1980s, identified 500 VOCs; the frequency of occurrence was in

Table 8.5. Comparison of the Daily Respiratory Intake of Pollutants from Outdoor and Indoor Air

	Intake (μg / day)	
Pollutant	*Indoor*	*Outdoor*
Formaldehyde	675	4.5
Toluene	1012	7.5
Respirable particles	1080	45
NO_2	270	7.5

SOURCE: Adapted from reference 40.

the following order: aliphatic hydrocarbons, aromatic hydrocarbons, and chlorinated hydrocarbons (*41*). A comparison of indoor and outdoor air quality in new hospitals, new office buildings, and new nursing homes is shown in Figure 8.9. A "sick building syndrome", which may cause a variety of illnesses, such as headaches, depression, fatigue, irritability, allergy-like symptoms, heart disease, and cancer, is a result of simultaneous exposure to a variety of chemicals. The most notorious causes of the sick building syndrome are xylenes and decane. They occur in some new buildings in concentrations 100 times higher than in the

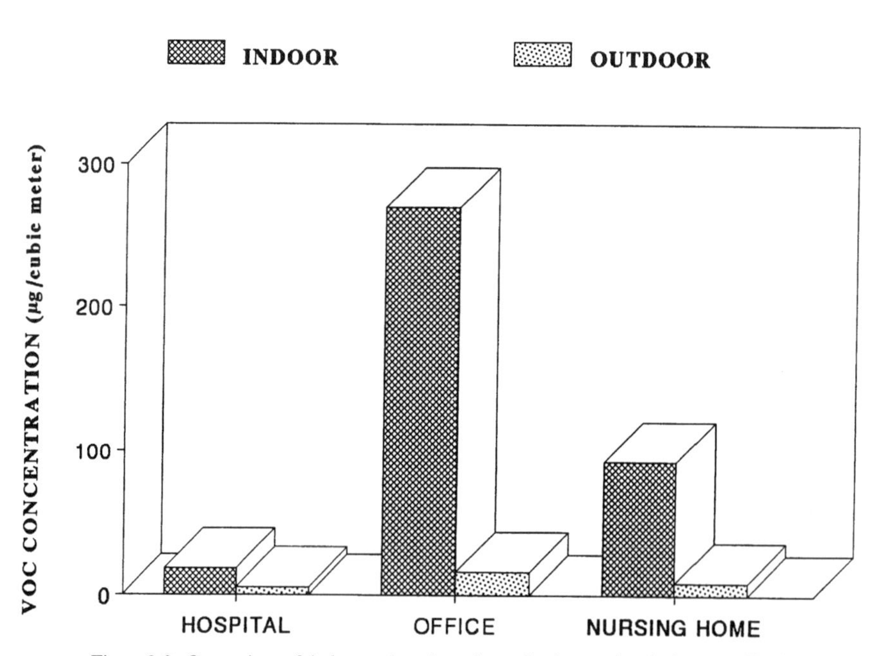

Figure 8.9. Comparison of indoor and outdoor air quality in new hospitals, new office buildings, and new nursing homes. (Source: Adapted from reference 40.)

outdoor air (*41*). As the buildings age, concentrations of chemical pollutants, in most cases, decrease substantially, making the buildings more livable.

Another problem is bacteria, viruses, fungi, and parasites originating from the forced-air heating systems, humidifiers, and air conditioners. These organisms may lead to allergic reactions or parasitic infections. It may be assumed that in old American buildings and in most European buildings, where central heating systems are based on steam or hot water circulating through radiators rather than forced-air circulation, the problem of bacteria, viruses, and fungi should be less critical.

In some parts of the world, including certain areas in the United States, the radioactive gas radon creates an indoor health hazard. Radon, a noble gas, is a product of disintegration of uranium, actinouranium, and thorium. Because these elements occur in soil and rocks, the building materials and soil under the buildings are the major sources of indoor radon. Water and natural gas are additional, albeit usually minor, sources of indoor radon. However, in certain locations, household water supply, especially from deep wells, may contain substantial quantities of this gas. Boiling water releases most of the radon, and that ingested by drinking cold water is quickly eliminated from the body without doing much harm. Thus, the main hazard of radon in household water is from breathing radon released into the bathroom air from showers or baths (*42*).

Despite the seriousness of the problem, so far the indoor air pollution in the United States remains, for the most part, unregulated. Indoor air pollution is aggravated in modern buildings because they are constructed with energy-saving in mind. Thus, the air exchange between inside and outside is restricted.

The main source of indoor air pollution in the developing countries is combustion of coal or biomass (wood, dung, agricultural waste, etc.) for heating and cooking in primitive, poorly vented stoves. The pollutants in that case are respirable particles coated with PAHs, nitrogen dioxide, sulfur dioxide, carbon monoxide, and a variety of VOCs. As mentioned earlier in this chapter, most of these pollutants are either irritants of tender tissues, respiratory and cardiovascular toxins, or both. In addition, some PAHs are carcinogens and mutagens mostly affecting the respiratory system. Sometimes, especially in rural houses, the concentration of certain indoor pollutants exceeds the WHO guidelines. Because women in the agricultural communities spend most of their time indoors performing household chores, exposure to fumes of biomass fuels might be the single most important health hazard for women (*40*).

References

1. World Resources Institute, International Institute for Environment and Development in collaboration with U.N. Environment Programme. *World Resources 1988–89, Atmosphere and Climate;* Basic Books: New York, 1988; Chapter 23, p 333.
2. Amdur, M. O. In *Cassarett and Doull's Toxicology;* Klaassen, C. D.; Amdur, M. O.; Doull, J., Eds.; MacMillan: New York, 1986; Chapter 25, p 801.

3. Waldbott, G. L. *Health Effects of Environmental Pollutants,* 2nd ed.; C. V. Mosby: St. Louis, MO, 1987.
4. MacKenzie, J. J.; El-Ashrey, M. T. *Ill Winds: Airborne Pollution's Toll on Trees and Crops;* World Resources Institute: Washington, DC, 1988; Chapter 5, p 33.
5. Seinfeld, J. H. *Science (Washington, D.C.)* **1989,** *243*(4892), 745.
6. Abelson, P. H. *Science (Washington, D.C.)* **1988,** *241*(4873), 1569.
7. Santodonato, J.; Howard, P.; Basu, D. *J. Environ. Pathol. Toxicol.* **1981,** *5*(1), 1.
8. Hileman, B. *Chem. Eng. News* February 8, 1988, p 22.
9. Menzie, C. A.; Potocki, B. B.; Santodonato, J. *Environ. Sci. Technol.* **1992,** *26*(7), 1278.
10. Harkov, R.; Greenberg, A. J. *Air Pollut. Control Assoc.* **1985,** *35,* 238.
11. Hites, R. A. In *Atmospheric Aerosol: Source/Air Quality Relationships;* Macias, E. S.; Hopke, P. K., Eds.; ACS Symposium Series 167; American Chemical Society: Washington, DC, 1981; p 187–196.
12. *Global Environment Monitoring System, Assessment of Urban Air Quality;* United Nations Environment Programme and World Health Organization: Geneva, Switzerland, 1988; Chapter 7, p 58.
13. Goyer, R. A. In *Cassarett and Doull's Toxicology;* Klaassen, C. D.; Amdur, M. O.; Doull, J., Eds.; MacMillan: New York, 1986; Chapter 19, p 582.
14. *Global Environment Monitoring System, Assessment of Urban Air Quality;* United Nations Environment Programme and World Health Organization: Geneva, Switzerland, 1988; Chapter 5, p 39 and Chapter 6, p 52.
15. *Global Environment Monitoring System, Assessment of Urban Air Quality;* United Nations Environment Programme and World Health Organization: Geneva, Switzerland, 1988; Chapter 8, p 70
16. World Resources Institute, International Institute for Environment and Development in collaboration with U.N. Environment Programme. *World Resources 1994–95, Atmosphere and Climate;* Oxford University: New York, 1994; Chapter 23, p 361.
17. Dockery, D. W.; Pope, C. A.; Xu, X. P.; Spengler, J. D.; Ware, J. H.; Fay, M. E.; Ferris, B. G. *New Engl. J. Med.* **1993,** *329*(24), 1753.
18. *Compliance of Selected Areas with Clean Air Act Requirements;* American Lung Association: New York, 1988.
19. World Resources Institute, International Institute for Environment and Development in collaboration with U.N. Environment Programme. *World Resources 1992–93, Atmosphere and Climate;* Oxford University: New York, 1992; Chapter 13, p 193.
20. Ember, L. *Chem. Eng. News* February 5, 1990, p 23.
21. *Chem. Eng. News* May 15, 1989, p 17.
22. MacKenzie, J. J.; El-Ashry, M. T. *Ill Winds: Airborne Pollution's Toll on Trees and Crops;* World Resources Institute: New York, 1988; Chapter 4, p 25.
23. Cadle, S. H.; Nebel, G. J. In *Introduction to Environmental Toxicology;* Guthrie, F. E.; Perry, J. J., Eds.; Elsevier Science: New York, 1980; Chapter 32, p 420.
24. Hildeman, L. M.; Markowski, G. R.; Cass, G. R. *Environ. Sci. Technol.* **1991,** *25*(4), 744.
25. Williams, B. P.; Buhr, M. P.; Weber, R. W.; Volz, M. A.; Koepke, J. W.; Selner, J. C. *J. Allergy Clin. Immunology* **1995,** *95*(1), 88.
26. Jeager, D.; Kleinhans, D.; Czuppon, A. B.; Baur, X. *J. Allergy Clin. Immunol.* **1992,** *89*(3), 759.
27. Gergen, P. J.; Mullally, D. I.; Evans, R., 3rd. *Pediatrics* **1988,** *81*(1), 1.
28. Hanson, D. *Chem. Eng. News* June 15, 1992, p 13.
29. Hanson, D. *Chem. Eng. News* May 31, 1993, p 6.
30. Thayer, A. M. *Chem. Eng. News* November 16, 1992, p 22.
31. Maraniss, D.; Weiskopf, M. *The Washington Post, National Weekly Edition* January 25–31, **1988,** *5*(14), p 8.

32. Pickle, L. W.; Mason, T. J.; Howard, N.; Hoover, R.; Fraumeni, J. F., Jr. *Atlas of U.S. Cancer Mortality Among Whites 1950–1980;* National Institutes of Health: Bethesda, MD, 1980; publication No. 87–2900.
33. National Cancer Institute. *1987 Annual Cancer Statistics Review Including Cancer Trends: 1950–1985;* National Institutes of Health: Bethesda, MD, 1987; publication No. 88–2789.
34. Lasota, J. P. *Sciences (N.Y.)* **1987,** *July/August,* 33.
35. *World Resources 1992–93, Central Europe;* Oxford University: New York, 1992; Chapter 5, p 57.
36. Pollock, C. In *State of the World 1987;* Brown, L. R., Ed.; W. W. Norton: New York, 1987; p 101.
37. Cross, F. L., Jr.; Hesketh, H. E. *Controlled Air Incineration;* Technomic: Lancaster, PA, 1985.
38. *Sci. News (Washington, D.C.)* **1993,** *143,* 334.
39. Thompson, R. In *Earth's Threatened Resources;* Gimlin, H., Ed.; Congressional Quarterly: Washington, DC, 1986; Editorial Research Reports.
40. *Global Environment Monitoring System, Assessment of Urban Air Quality;* United Nations Environment Programme and World Health Organization: Geneva, Switzerland, 1988; Appendix I, p 81.
41. Ember, L. R. *Chem. Eng. News* December 5, 1988, p 23.
42. *Radiation: Doses, Effects, Risks;* United Nations Environment Programme: Nairobi, Kenya, 1985.
43. Zakrzewski, S. F. *People, Health, and Environment;* SFZ Publishing: Amherst, NY, 1994; p 49.

9

Pollution of the Atmosphere

The Earth's Atmosphere

The earth's atmosphere consists of 78% (by volume) of N_2; 21% O_2; about 0.033% CO_2; trace amounts of noble gases, NO_x, and CH_3; and variable amounts of water vapor. At sea level, the amount of water vapor may vary from 0.5 g per kg of air in polar regions to more then 20 g per kg in the tropics.

The Standard Atmosphere

The standard atmosphere is a theoretical set of data that serves as a reference point for calculation of atmospheric changes due to the weather. The values are calculated for sea level conditions and correspond to a pressure of 760 mm of mercury (92.29 in., 1013.25 mbar), an air density of 1.22 kg/m^3, and a temperature of 15 °C (59 °F). The composition of the air within the troposphere, which is the lowest layer of the atmosphere, does not change with altitude; however, the pressure and temperature decrease with altitude. The relationship between altitude and pressure in the standard atmosphere is shown in Figure 9.1, and the relationship between altitude and temperature is shown in Figure 9.2. The rate of decrease of temperature with altitude (6.49 °C per km) is referred to as the "standard lapse rate". This rate is a strictly theoretical average value because the actual lapse rate varies depending on the weather. Because the air density is proportional to the pressure and inversely proportional to the temperature, it changes at the same rate as the pressure does.

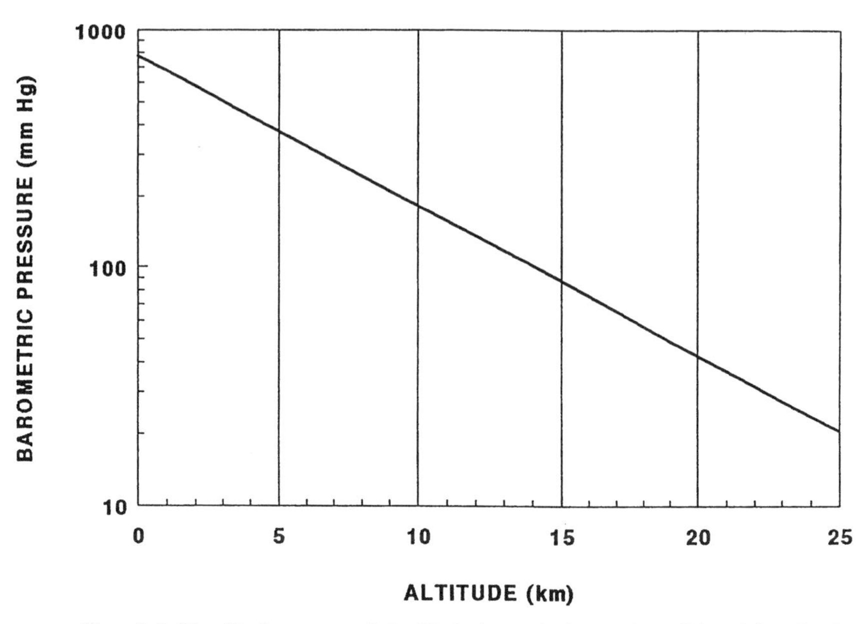

Figure 9.1. The altitude–pressure relationship in the standard atmosphere. (Adapted from data in reference 1.)

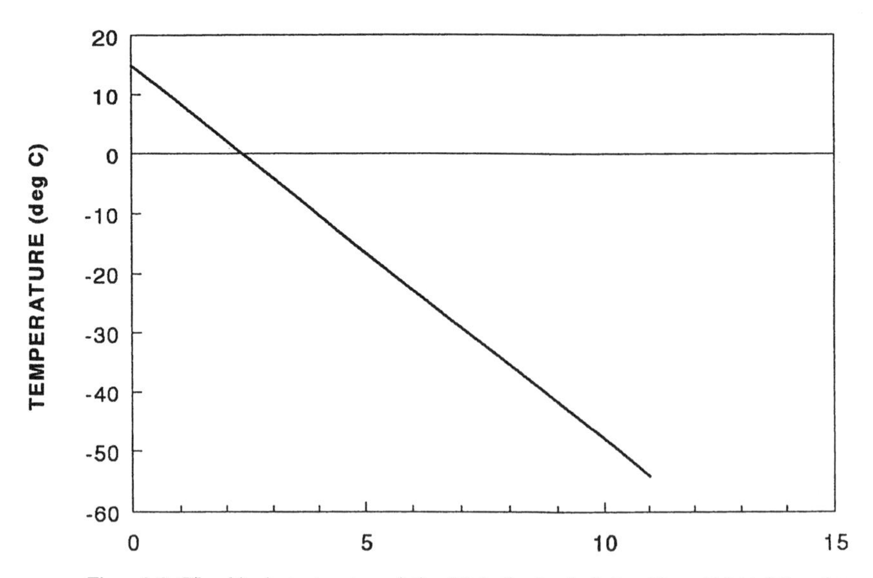

Figure 9.2. The altitude–temperature relationship in the standard atmosphere. (Adapted from data in reference 1.)

The Division of the Atmosphere

The atmosphere is divided into troposphere, stratosphere, mesosphere, and ionosphere (Figure 9.3). As shown in this figure, the division is based on temperature inversions that occur at the higher altitudes; the altitudes of these inversions vary with the season and with the geographic latitude. Although the general shape of the curves remains the same for all latitudes, the altitudes of the inversions are higher over the equator and lower over the poles; the curves presented in Figure 9.3 refer to middle latitudes. The boundary areas at each temperature inversion are called tropopause, stratopause, and mesopause, respectively.

Pollution of the atmosphere is generally the least appreciated of all environmental issues. The reasons are that it affects us neither directly nor immediately. Yet, next to overpopulation, this may be the most crucial issue affecting the survival of our civilization.

To appreciate the fragility of the earth's atmosphere, one has to consider its dimension in comparison to that of the globe. Let us imagine a globe 1 meter in

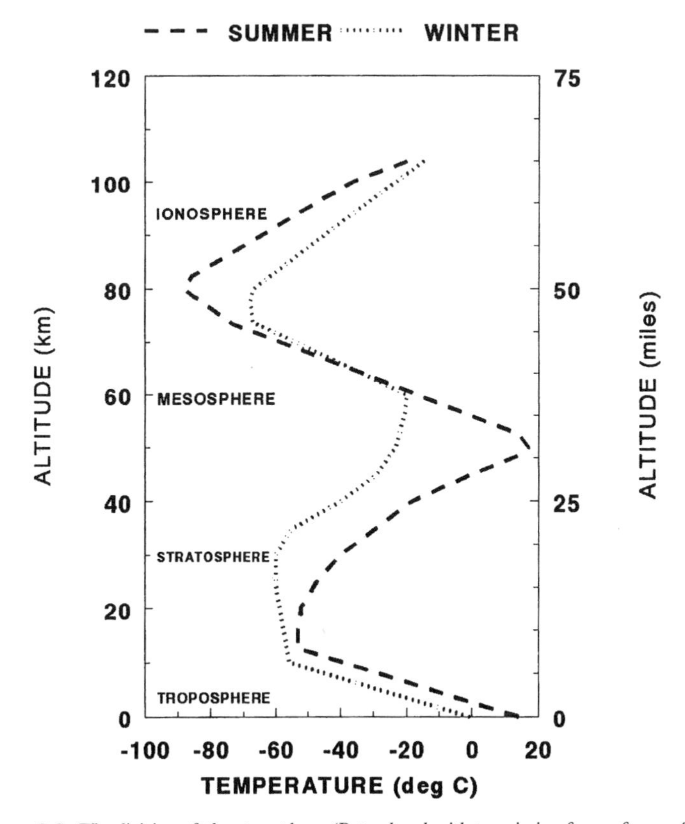

Figure 9.3. The division of the atmosphere. (Reproduced with permission from reference 46.)

diameter (the earth's equatorial diameter is 6378 km). The troposphere would then be 1.3 to 3.0 mm thick; the outer edges of the stratosphere would reach a height of 7.8 to 8.5 mm, and the outer edges of the mesosphere would be 12.5 to 14.0 mm above the surface of the globe.

Formation and Sustenance of Stratospheric Ozone

The solar radiation that penetrates the earth's upper, highly rarefied atmosphere strikes the oxygen molecules in the middle stratosphere, splitting them into single atoms. The highest concentration of the atomic oxygen occurs at an altitude of 30 to 40 km. The atomic oxygen is a very reactive species and interacts with the molecular oxygen forming ozone (O_3):

$$O_2 + h\nu \rightarrow O + O \qquad (9.1)$$

$$O_2 + O \rightarrow O_3 \qquad (9.2)$$

In the upper atmosphere the gases tend to separate according to their weight. The heavier gases settle down, whereas the lighter gases rise up. Because ozone is heavier than either oxygen or nitrogen, it tends to settle down. This movement is partially counteracted by the continuous stirring of the atmosphere. As the result of these competing forces, the highest concentration of ozone (a few parts per million) occurs at a level of 15 to 30 km. The increasing density of the atmosphere gradually attenuates the solar radiation, so that at altitudes below 25 km the photochemical ozone formation becomes extremely slow and eventually ceases completely (*2*). One would expect that a consistent bombardment of oxygen by the solar radiation will result in a continuous buildup of ozone. This does not happen, however, because as ozone is formed it is also destroyed by interactions with nitric oxide (equation 9.3) and hydroxy radicals (equation 9.4), and by the direct effect of solar radiation (equation 9.5) (*3*).

$$\begin{array}{c} O_3 + NO \rightarrow NO_2 + O_2 \\ NO_2 + O \rightarrow NO + O_2 \\ \hline O_3 + O \rightarrow 2O_2 \end{array} \qquad (9.3)$$

$$\begin{array}{c} O_3 + OH \rightarrow HO_2 + O_2 \\ O_3 + HO_2 \rightarrow HO + 2O_2 \\ \hline 2O_3 \rightarrow 3O_2 \end{array} \qquad (9.4)$$

$$O_3 + h\nu \rightarrow O_2 + O \qquad (9.5)$$

Thus, in an unpolluted atmosphere the stratospheric ozone concentration remains (within seasonal and latitudinal variations) relatively constant (*2*). Although the concentration of ozone in the stratosphere is only a few parts per million (ppm), it is sufficient to filter a part of the solar ultraviolet radiation, thus reducing the amount of radiation reaching the earth's surface.[1] The stratospheric ozone is popularly called a protective ozone layer. This name is somehow misleading because it reverses the cause–effect relationship. The name "protective layer" implies that the ozone is there for our and other species' protection. In fact, life on earth is what it is because it evolved according to the conditions imposed by the environment. If there were no ozone layer, it is likely that only aquatic life below the ocean surface, protected from the lethal radiation by a layer of water, could exist. Therefore, it may be expected that any perturbation of these conditions will have an effect on living matter.

Depletion of Stratospheric Ozone

Chlorofluorocarbons

In 1974, Molina and Rowland (*4*) first proposed that chlorine from a class of compounds designated as chlorofluorocarbons (CFCs) could cause stratospheric ozone depletion. CFCs were introduced in the 1930s and found numerous industrial applications as propellants for aerosols, plastic-foam-blowing agents, refrigeration and air conditioning fluids, cleaning fluids for electronic equipment, and fire extinguisher fluids. Their advantage is that they are chemically stable, nonflammable, and nontoxic. Ever since their introduction into commerce, the production and consumption of CFCs grew steadily until the 1970s. Then, because of the concern about their ozone-destructive potential, their use as aerosol propellants was banned in several industrialized countries, and their production declined. However, the production of CFCs increased again after 1982 because of the growing demand for foam insulation and for cleaning fluids in the electronic equipment and microchip industries.

The chemical stability of CFCs in the troposphere is a detriment to the environment. The two most damaging CFCs, $CFCl_3$ (CFC-11) and CF_2Cl_2 (CFC-12), have atmospheric lifetimes of 75 and 111 years, respectively. When released into the environment they rise slowly to high altitudes. In the lower stratosphere, they become exposed to intense ultraviolet radiation, which breaks them down,

[1]The ultraviolet radiation spectrum is divided into three regions according to wavelength: UV-A [below 280 nanometers (nm)], UV-B (280 to 315 nm), and UV-C (above 315 nm). The shortwave regions, UV-A and UV-B, are harmful to living organisms because they damage the deoxyribonucleic acid (DNA). The long-wave region, UV-C, is relatively harmless. Because UV-A is absorbed by the atmosphere and does not reach the earth's surface, our concern centers on UV-B. (1 nm equals one millionth part of a millimeter.)

causing release of chlorine radicals. Elemental chlorine destroys ozone as shown in equation 9.6:

$$Cl + O_3 \rightarrow ClO + O_2 \quad (9.6)$$

The ClO radical may further react with atomic oxygen to regenerate the chlorine radical according to equation 9.7:

$$ClO + O \rightarrow Cl + O_2 \quad (9.7)$$

Both ClO and Cl can be inactivated temporarily by reacting with nitrogen dioxide (equation 9.8) or methane (equation 9.9), respectively (*5*). Nitrogen dioxide is introduced into the stratosphere by oxidation of microbially produced nitrous oxide (N_2O). Methane originates from both natural sources and human activities.

$$ClO + NO_2 + \text{catalyst} \rightarrow ClNO_3 \quad (9.8)$$

$$Cl + CH_4 \rightarrow HCl + CH_3 \quad (9.9)$$

The Polar Vortex

During winter at the poles, a stream of air in the stratosphere (the *polar vortex*) encircles the polar regions. It isolates them from the warmer air of moderate zones. This polar vortex allows temperatures to drop to as low as –80 °C and –90 °C in the arctic and antarctic regions, respectively.

The polar stratospheric clouds (PSCs) that form at such low temperatures are the key to ozone destruction. There are two types of polar clouds: PSC I consists of nitric acid trihydride crystals ($HNO_3 \cdot 3H_2O$), and PSC II consists of ice. The process of PSC I formation involves a conversion of nitrogen oxides (NO, NO_2, and NO_3) into N_2O_5, and subsequent reaction of gaseous N_2O_5 with H_2O aerosol to form nitric acid (equation 9.10) (*6*).

$$N_2O_5\ (\text{gas}) + H_2O\ (\text{aerosol}) \rightarrow 2HNO_3\ (\text{gas}) \quad (9.10)$$

At temperatures of about 195 K (–78 °C), nitric acid freezes out as nitric acid trihydrate. The formation of chlorine nitrate ($ClNO_3$) depends on the availability of NO_2; thus, removal of free NO_2, referred to as denitrification, tends to decrease the content of $ClNO_3$ in the atmosphere. In addition, PSC I provides a catalytic surface for a heterogeneous reaction between $ClNO_3$ and HCl that leads to regeneration of active chlorine (equation 9.11).

$$ClNO_3 + HCl \rightarrow Cl_2 + HNO_3 \quad (9.11)$$

As the poles emerge from the polar night, the active chlorine species are converted by light to chlorine radicals (equation 9.12), which in turn react with ozone according to the following equations:

$$Cl_2 + h\nu \rightarrow 2Cl \tag{9.12}$$

$$2Cl + 2O_3 \rightarrow 2ClO + 2O_2 \tag{9.13}$$

ClO then enters the following chain reactions:

$$2ClO + \text{catalyst} \rightarrow Cl_2O_2 \tag{9.14}$$

$$Cl_2O_2 + h\nu \rightarrow Cl + ClOO \tag{9.15}$$

$$ClOO + \text{catalyst} \rightarrow Cl + O_2 \tag{9.16}$$

As the daylight period lengthens, the polar vortex dissipates and new air is brought to the region. This air carries with it nitrogen oxides, which inactivate the active chlorine (ClO) by forming $ClNO_3$. The process will then be repeated with the onset of polar winter. PSCs will catalyze the decomposition of chlorine nitrate and release active chlorine.

The depletion of ozone, which was first observed over Antarctica during austral spring, when it emerges from the winter darkness, appears to be spreading gradually to other latitudes. Ozone has been depleted by 5% or more since 1979 at all latitudes south of 60° S.

Health and Economic Implications

The depletion of stratospheric ozone affects not only the intensity of the UV-B radiation reaching the earth, but also the wavelength composition; it shifts more radiation toward the shorter, more damaging wavelengths. The risks to human health and to the survival of other species posed by UV-B radiation are estimated by a theoretical calculation of the potential for damaging the species' DNA (*7*). Using these criteria, the International Panel on Substances that Deplete the Ozone Layer estimated that since 1979, the annual DNA damage-dose increased 5% per decade at latitudes 30° N and 30° S, 10% per decade in the arctic region, 15% per decade at latitude 55° S, and 40% per decade at latitude 85° S. No significant increase was noted in the equatorial region (*8*). Because the intensity of UV-B radiation reaching the earth's surface is attenuated by cloud cover, suspended particles, and tropospheric ozone, the population in highly industrialized areas is, to a certain extent, protected from the harmful effects of UV radiation.

The increased intensity of UV-B radiation is likely to have a variety of effects on human health, such as an increased incidence of cataracts, skin cancer, and damage to the immune system. Indeed, animal experiments suggest that exposure to UV-B radiation will increase the frequency and severity of infectious diseases. It is now estimated that worldwide, for each 1% decrease in stratospheric ozone the cataract frequency will increase by 0.6 to 0.8%. A sustained ozone reduction of 10% will cause a 26% increase in nonmelanoma skin cancer and a lower, but still significant, increase in melanomas (*9*). In addition, a correlation has been found more recently between UV exposure and salivary gland cancer (*9*).

The future effect of ozone depletion on terrestrial plants is difficult to assess because many other factors, such as climatic changes associated with the greenhouse effect, may attenuate or aggravate the effects of increased intensity of the UV radiation. Plant species vary significantly in their responses to UV light. Among plants tested in the laboratory, many responded to UV radiation by exhibiting reduced growth, flowering, and photosynthetic activity (*10*). Some uncertainty exists about the magnitude of damage that may be inflicted on aquatic plants. It has been established that UV-B radiation greatly affects aquatic phytoplankton by damaging their mobility mechanism,[2] their DNA, and their photosynthetic apparatus (*11*). The decrease in marine plant growth could result in the demise of marine mammals, crustaceans, and fish species. Such changes could alter the whole marine ecosystem and further reduce the human food supply. Moreover, because marine phytoplankton account for more than half of the global carbon dioxide fixation, interference with this process may further augment the greenhouse effect (*11*).

Besides the biological impact, increased UV radiation will affect the durability of materials such as wood, paints, and plastics; the EPA estimates that a 10% depletion of the stratospheric ozone will cause $2 billion in damage to materials.

International Cooperation

The First International Conference on Substances that Deplete the Ozone Layer, spearheaded by the United Nations Environment Programme (UNEP), was convened in Montreal in October 1987. An agreement signed at the conference urged CFC producers to freeze production at the current level and to reduce it by 50% by 1998. By mid-1989, 36 countries had ratified this agreement.

Soon the measures approved by the Montreal convention were seen as highly inadequate. Even if CFC production were halted entirely, there are already enough CFCs in the stratosphere to carry on destruction of ozone for another 100 years. Although the results of the Montreal convention were meager, its significance should not be underrated. It marks the beginning of international cooperation in matters concerning protection of the global environment.

In May 1988 another conference, of a more local character, was convened in Colorado. This conference was jointly sponsored by the National Aeronautics and Space Administration (NASA), the National Oceanic and Atmospheric Ad-

---

[2]Some phytoplankton have the ability to adjust their position within the water column in response to changing light conditions (*11*).

ministration (NOAA), the National Science Foundation (NSF), the Chemical Manufacturers Association, the World Meteorological Association, and UNEP. The purpose of the conference was to discuss the results of the 1987 Airborne Antarctic Ozone Expedition, a joint project of Harvard University and NASA (*12*). The findings of this expedition were that "in 1987, the ozone hole was larger than ever. More than half of the ozone column was wiped out and essentially all ozone disappeared from some regions of the stratosphere. The hole also persisted longer than it ever did before, not filling until the end of November".

Because there were some indications of perturbed atmospheric chemistry in the north, NASA organized an airborne expedition to the arctic region during January and February 1989. Although the concentration of ClO in the arctic stratosphere was almost as high as that found over Antarctica (*13*), subsequent satellite observations indicated that there was no dramatic depletion of ozone. The explanation for this phenomenon was that in the arctic region the polar vortex disintegrated quickly after emergence from the polar night (*14*).

Phasing Out Fluorocarbons

In preparation for the Second International Conference on Substances that Deplete the Ozone Layer, scheduled for 1990 in London, diplomats, environmentalists, and CFC producers from the 36 nations that had signed the Montreal agreement gathered in Helsinki, Finland, in May 1989. They drafted the following proposal for possible total phasing out of CFCs and other ozone-destroying chemicals:

1. to phase out production and consumption of CFCs by the year 2000,
2. to phase out production and consumption of halons, carbon tetrachloride, and methyl chloroform as soon as feasible,
3. to commit themselves to speedy development of environmentally acceptable substitutes,
4. to make available to Third World countries all pertinent information, technologies, and training.

The provisions of this agreement allowed, in certain cases, production and use of the ozone-depleting substances after the year 2000, as long as this production did not exceed 15% of the 1986 production. This proposal was accepted and signed by the participating nations as amendments to the Montreal protocol during the 1990 London convention.

A significant development in this area was the announcement by DuPont, the world's largest producer of CFCs, that it will phase out production of these compounds by the year 2000. Substitutes for CFCs such as HCFC-141b ($C_2H_3Cl_2F$), HCFC-123 ($C_2HCl_2F_3$), HCFC-22 ($CHClF_2$), and HFC-134a (CH_2FCH) were developed to replace fully halogenated CFCs.

Fluorocarbons that carry hydrogen atoms (HCFCs) are decomposed significantly before reaching the stratospheric ozone layer (*15*). Their atmospheric lifetime and ozone-depletion potential are summarized in Table 9.1. The development of HCFCs is welcome news, but the EPA cautions that their real usefulness will depend on thorough assessment of their toxicity and the toxicity of their decomposition products. Although HCFCs are less damaging to stratospheric ozone than CFCs, nevertheless they carry some chlorine into the stratosphere. Therefore, according to EPA, they should be considered as transition substances, for use until better substitutes are developed (*16*). In addition, concern for the use of HCFCs and HFCs (that do not deplete ozone) centers on their properties as powerful greenhouse gases (GHGs).

Substitutes for foam-blowing processes that require neither CFCs, HCFCs, nor HFCs on also being developed. DuPont plans to use dimethyl esters to replace CFCs in the aerosol propellants (*17*) that are still used in Europe. The BASF Corporation introduced a foam-blowing process that eliminates use of CFCs entirely (*18*).

In 1992 a disturbing discovery of unusually severe ozone depletion in the northern hemisphere, over the populated areas of North America, northern Europe, and northern Asia, was reported. Because there are no PSCs at these latitudes, the atmospheric scientists speculated that the conversion of the inactive chlorine species to active chlorine may have been catalyzed by suspended sulfate particles (*19*). Although some sulfate particles normally occur in the stratosphere from natural sources, their unusually high concentration in 1992 was attributed to the eruption of Mount Pinatubo in the Philippines in June 1991. This eruption ejected 15–30 metric tons of sulfur dioxide into the atmosphere. Sulfur dioxide was promptly converted to sulfuric acid, which reacted in the stratosphere with metal salt particles, forming sulfate aerosol (*6*).

Because of this alarming news, the United States decided unilaterally to move the deadline for a complete elimination of ozone-depleting substances to the end of 1995. Shortly after, the signatories to the Montreal convention met again in Copenhagen. They followed the example of the United States and established 1996 as an international deadline for phaseout of ozone-depleting substances.

Table 9.1 Atmospheric Lifetime and Ozone-Depletion Potential of HCFCs

Compound	*Atmospheric Lifetime (years)*	*Ozone Depletion Potential*[a]
CH_2FCH_3	21	0
$CHCl_2CF_3$	1.9	0.016
CH_3CCl_2	8.9	0.081
$CHClF_2$	20	0.053

[a]CFC-11, used as a standard, is 1.00.

SOURCE: Reproduced in part from reference 15. Copyright 1989 American Chemical Society.

They also established restrictions on the use of HCFCs, requiring a freeze in their production by 1996 and complete elimination of their use by 2030.

At present, the depletion of stratospheric ozone is due mainly to the emissions of chlorine-containing compounds, such as CFCs, carbon tetrachloride, etc. However, a large-scale deployment of a supersonic transport may turn out to be still more destructive to the ozone layer than are CFCs. Fuel combustion is associated with formation of nitrogen oxides (NO_x). Although nitrogen dioxide (NO_2) protects ozone by binding the active chlorine molecules, nitric oxide (NO) has a high ozone-destroying potential, especially when emitted in the midst of the ozone layer where the supersonic airplanes fly (*20*).

The effect of restrictions imposed on the use of ozone-depleting chemicals by the Montreal protocol can already be perceived. The rate of increase in the atmospheric concentration of major CFCs is on the decline. However, one has to keep in mind that even immediate elimination of all ozone-depleting substances would leave enough chlorine radicals in the atmosphere to continue ozone destruction, albeit at a gradually decreasing rate, for another century.

Emission of CO_2 and Models of Climatic Changes

Life on earth depends upon a fixed supply of basic elements and substances, such as carbon, nitrogen, oxygen, and water. Because their supply is fixed, they must be continuously recycled. This process is referred to as biogeochemical cycling. At present, the biogeochemical cycling equilibria have been greatly perturbed by human activities.

The carbon cycle involves the exchange of carbon, mostly in the form of carbon dioxide, among the atmosphere, the biosphere (i.e., living plants and soil), and the oceans. The latter are the largest reservoirs of dissolved carbon dioxide. The biosphere and atmosphere hold about 2000 and 700 billion tons of carbon, respectively (*21*). The oceans hold about 14 times more than the biosphere and atmosphere combined. In addition, large amounts of carbon are stored in a nonexchangeable form as sediment in the oceans and in lesser amounts in the form of fossil fuels (i.e., oil, coal, and gas).

Atmospheric carbon dioxide is the mainstay of life support on earth; it is assimilated by green plants and subsequently converted to basic foods. In addition, O_2 is released during the assimilation process so that the reserves of oxygen in the atmosphere remain constant.

Temperature of the Earth

Carbon dioxide, together with water vapor, is responsible for maintaining the earth's temperature at a level that supports life as we know it. About half of the total solar energy that strikes the earth is absorbed by the earth. The rest is either reflected or absorbed by the atmosphere. About 50% of the absorbed thermal energy is consumed in evaporating water from the oceans, rivers, lakes, and soil, and about 10% by direct heating of the atmosphere. The remaining 40% is released as long-wave radiation. Carbon dioxide, water vapor, and small amounts of other gases in the atmosphere bounce 88% of this energy toward the ground, where it warms the surface of the earth; this is referred to as the natural greenhouse effect.[3] Thus, there is a correlation between the concentration of carbon dioxide in the atmosphere and the earth's temperature. The extent of the greenhouse effect is expressed by "radiative forcing", that is, the amount of heat (in watts) per square meter of the earth's surface area.

This correlation has been traced back for the past 160,000 years. Air bubbles trapped in the glacial ice core from Antarctica were analyzed for carbon dioxide, and the hydrogen-deuterium ratio was determined in ice of corresponding age. The deuterium content in rain and snow increases with increasing temperature. In areas where ice is permanent, the annual snowfall is packed into a distinct layer of ice. Thus the age of the ice samples being analyzed could be determined by counting the ice layers.

Figure 9.4 shows the plot of atmospheric carbon dioxide concentration and the antarctic air temperature, according to a study performed by a French–Soviet team at Vostok, Antarctica. According to these determinations, the highest temperature, 2.5 °C above the present one, occurred about 135,000 years ago when the concentration of atmospheric CO_2 reached its highest level of 300 ppm. The lowest temperature of the period, nearly 10 °C below the present temperature, occurred about 150,000 years ago and again 20,000 years ago at a CO_2 concentration of 185–195 ppm (*23*).

In 1990 a research team from the Freshwater Institute in Manitoba published (*24*) the results of 20 years of climatic, hydrologic, and ecological records in the Experimental Lakes Area of northwestern Ontario. According to this record, the air and lake temperatures in that area have increased by 2 °C and the average period of ice cover on the lakes has decreased by 3 weeks. Similarly, it has been noted that alpine glaciers are melting ten times faster than they did at the end of the last ice age (*25*), and that the ice cover on Mount Kenya decreased 40% from 1963 (*26*). The recent study of the Ok glacier in western Iceland revealed that the glacier shrank from 6 square miles in 1910 to its present size of 1

---

[3]Blackbody radiation (i.e., the maximum amount of energy that an object can radiate) increases with the blackbody's temperature. Measurements by satellites from above the earth's atmosphere of the total heat radiated by the earth indicate that the earth's surface temperature is −19 °C. This radiated energy is balanced by the solar heat absorbed by the earth. The actual average surface temperature of the earth is about 14 °C. This difference of 33 °C between the actual surface temperature and that observed from above the atmosphere is attributed to the greenhouse effect (*22*).

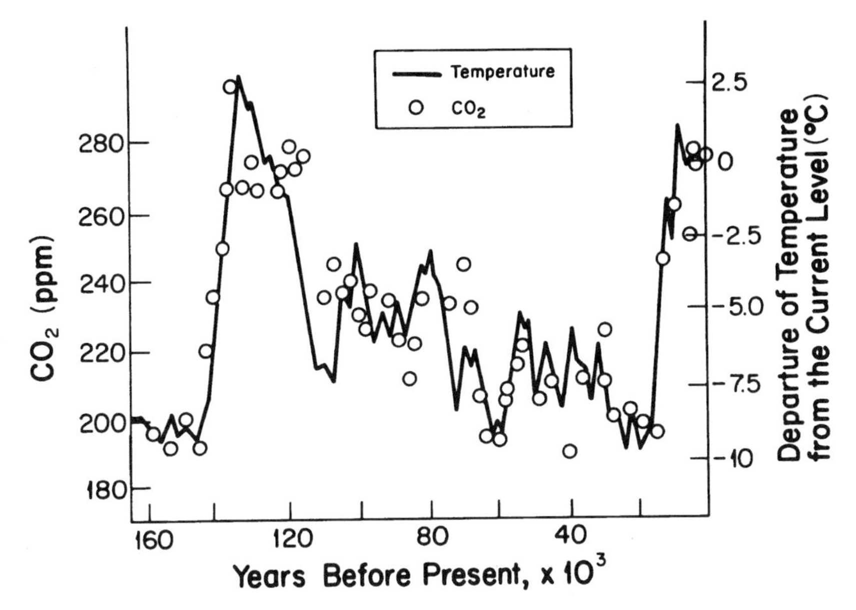

Figure 9.4. Plot of CO_2 content in the atmosphere and atmospheric temperature in the antarctic region. (Adapted from data presented in reference 23.)

square mile (*25*). In accord with these findings, actual measurements of the radiative forcing indicated an increase from 1 to 2.5 W/m^2 between the late 19th century and the present (*27*). Such change corresponds to a 1% increase in solar output. Although variations in solar output were observed during the past 100 years, they did not vary by more than 0.5%.

The annual emission of carbon from combustion of fossil fuels and wood increased from 93 million tons in 1860 to about 5 billion tons in 1987 (approximately 1 ton per person in 1987). Most of this increase in emission occurred during the last 30–40 years (*28*).

Factors Affecting Atmospheric Carbon Dioxide

Oceans

For years it had been thought that the oceans would remove excess carbon dioxide from the atmosphere. However, regular monitoring of atmospheric CO_2 since 1958 (*29*) has shown that the CO_2 concentration is rising at an average annual rate of 0.35%. The total increase since 1860 is 30%, with a present level of about 350 ppm. Studies on reconstruction of the earth's surface temperatures from records covering the last century were conducted at the Goddard Institute for Space Study and at the Climatic Research Unit (*30*). According to these studies, the earth's surface temperature has increased by 0.4–0.5 °C since 1880.[4]

Forests

Fossil fuel burning is not the only source of carbon dioxide emission. It is estimated that "slash and burn" forest clearing has released 90–180 billion tons of carbon since 1860. Presently, deforestation of tropical rain forests causes the release of 1.0–2.6 billion tons of carbon annually. This amount corresponds to between 20 and 50% of that released by fossil fuel combustion (*21*).

Clearing trees, even without burning, contributes to the greenhouse effect. Carbon contained in the stumps and wood left behind, as well as in the underlying soil, is either oxidized to CO_2 or digested by anaerobic microorganisms that release carbon in the form of methane (CH_4).

Methane and nitrous oxide (N_2O) are also GHGs. The estimated contributions of different gases to the greenhouse effect are as follows (*31*):

- CO_2, 49%,
- CH_4, 18%,
- CFC-11 and CFC-12, 14%,
- N_2O, 6%, and
- other, 13%.

Nitrous oxide occurs naturally as a product of metabolic activity of denitrifying bacteria. With the increasing use of nitrogen-containing fertilizers, the N_2O content in the atmosphere is rising.

Projections for the future indicate that by the year 2030 the methane contribution to the greenhouse effect may be 20–40% and the nitrous oxide contribution may be about 10–20% (*28*). Because trees assimilate atmospheric carbon dioxide, deforestation leads not only to increased emission of GHGs but also to decreased removal of carbon dioxide from the atmosphere.

Although carbon monoxide is not a GHG in its own right, it removes hydroxyl radicals that destroy GHGs. Thus, carbon monoxide emission contributes to the increase in concentration of GHGs (*32*).

---

[4]The data originally reported by the Goddard Institute indicated an increase in the earth's temperature of about 0.8 °C; those reported by the Climatic Research Unit indicated about an increase of about 0.6 °C. Subsequent research by T. Karl of the U.S. National Climate Data Center pointed out that the Goddard Institute values overestimated the temperature increase by about 0.38 °C, and those of the Climatic Research Unit overestimated the increase by about 0.15 °C.

Models of Climatic Change

The correlation between GHGs and atmospheric temperature is beyond doubt. However, how the increase in GHGs will affect the climate is open for discussion.

Computer-calculated projections of the effect of GHGs on the earth's temperature, modeled for three scenarios, are presented in Figure 9.5. The middle scenario reflects the present trend of emission of GHGs and assumes moderate climate sensitivity. (Climate sensitivity (Δt_{2x}) is defined as a degree of the earth's temperature change with a doubling of the CO_2 concentration. Estimates of Δt_{2x} vary from 1.5 to 4.5 °C.) According to this scenario, the average earth temperature should increase by 3.3 °C by 2100. The high scenario reflects accelerated GHG emissions and high climate sensitivity. The low scenario assumes radical curtailment of GHG emission and low climate sensitivity. The models also predict that the mean rate of evaporation and precipitation will increase by 2–3% for each degree of global warming.

Regional Patterns

How a 3.3 °C change in average global temperature may affect summer and winter temperatures and precipitation patterns at different latitudes is shown in Table 9.2. However, regional climate changes are difficult to predict, and different results may be obtained with a variety of models. The main problem is that the greenhouse effect can be modified by feedback mechanisms, of which we

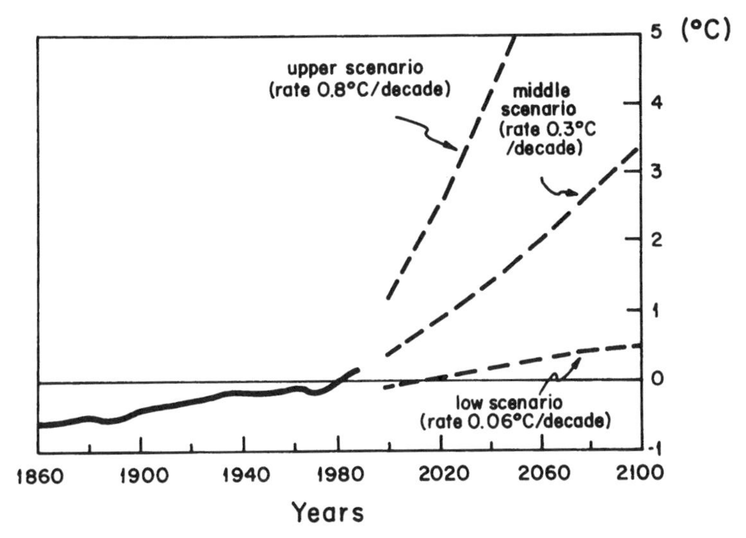

Figure 9.5. Computer projections of the anticipated change in mean global temperature resulting from emission of GHGs. The baseline refers to the present temperature. (Adapted from reference 33.)

Table 9.2. Predicted Regional Climate Changes

Latitude (deg)	*Change in Temperature (°C)*		*Change in Precipitation*
	Summer	*Winter*	
60–90	1.7–2.3	6.6–7.9	Increase in summer
30–60	2.6–3.3	4.0–4.6	Decrease in summer
0–30	2.3–3.0	2.3–3.0	Increase in places with heavy rain

SOURCE: Adapted from reference 33.

know little. The feedback mechanisms may aggravate the greenhouse effect (a positive feedback) or mitigate it (a negative feedback).

Effect on Vegetation

Regional warming and changes in precipitation patterns may cause a shift in agricultural areas, rendering some presently fertile regions unsuitable and opening new areas for agriculture. Similarly, the altered conditions will require a quick adaptation of tree species to the new climate. Failure to adapt within the short span of available time would result in eradication of some plant species. An increase of 3.3–4.0 °C in the average (winter–summer) temperature in the middle latitudes would require northward forest migration of 200–375 miles. Some trees, such as beech, can migrate only 12.5 miles per century. Spruce, the fastest migrant, can travel only 125 miles per century (*34*).

Although the earth has undergone periodic climatic changes during its existence, they usually occurred at the rate of a few degrees per tens of thousands of years. In contrast, the greenhouse effect is predicted to occur over one or two centuries. The shift of agricultural regions and eradication of tree species may be economically disastrous. It could result in desertification of some areas and consequently in food shortages and higher prices.

Effect on Oceans

Another concern regarding the warming of the earth is that higher ocean temperatures in the tropics will spur development of more frequent and more destructive hurricanes and typhoons.

Figure 9.6 depicts three scenarios for the change in ocean levels in response to the predicted increase in temperature. The rise in ocean levels projected by the middle and high scenarios reflects the anticipated melting of the polar ice cap and thermal expansion of water. The decrease of ocean levels anticipated by the low scenario is based on a prediction of increased snowfall, which would increase the mass of antarctic ice. This increased ice mass should result in loss of water

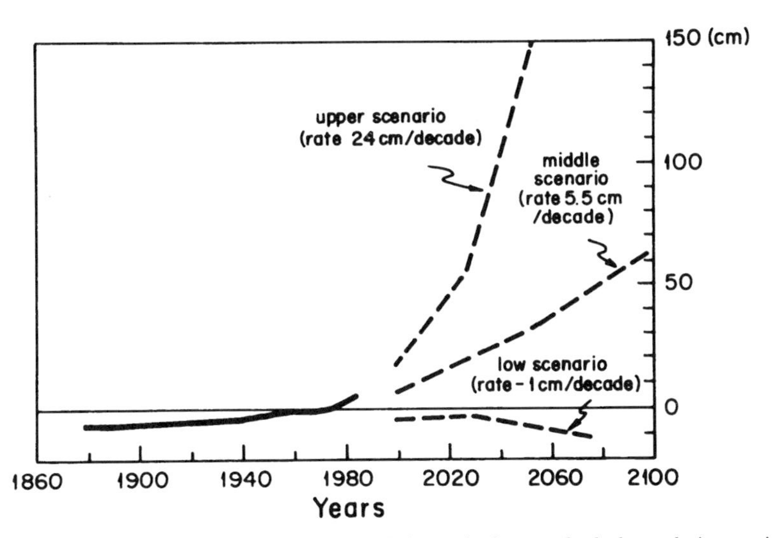

Figure 9.6. Computer projections of anticipated changes in the ocean levels due to the increase in global temperature. (Adapted from reference 33.)

from the oceans because of the cooling effect of the larger ice mass. In turn, this cooling would induce more ice formation.

A rise in ocean levels may have a disastrous environmental and economic impact. Inundation of low-lying coastal areas will decrease the availability of arable land, affect coastal infrastructure, and force abandonment of seashore residential areas. In addition, there will be adverse environmental effects such as salt intrusion into groundwater, rivers, wetlands, and soil. All these will also have an impact on agriculture. Some low-lying developing countries, such as Bangladesh or the Maldive Islands may be severely affected; for many of their inhabitants global warming means loss of their livelihood. Furthermore, increased frequency and destructiveness of hurricanes and typhoons, as mentioned earlier in this section, will cause additional damage and floods. The result would be considerable economic loss and food shortages.

Other Factors

Obviously, these forecasts contain a number of uncertainties. Much depends on future patterns of fossil fuel combustion, deforestation and reforestation, and other human activities. In addition, responses of global cycles cannot be predicted with certainty.

One unknown factor is the effect of clouds. An increase in temperature will certainly augment evaporation of water from oceans and inland waters. This evaporation may, or may not, result in increased cloud cover. High clouds may have a cooling effect by reflecting solar radiation, whereas low clouds may have a warming effect by trapping the earth's infrared radiation.

In addition to clouds, sulfate aerosols formed from SO_2 may mitigate the global warming because they reflect solar radiation. As pointed out earlier (*see* Chapter 8), about half of atmospheric SO_2 is of anthropogenic origin.

Another factor is the effect of global warming on ocean currents, which play an important role in moderating the climate of land areas. Our present understanding is that ocean currents are a consequence of water temperature, water salinity, and the earth's rotation. The two former factors affect water density. The denser water sinks, and a void created by the sinking water is replaced by the surface flow of water of lesser density. Furthermore, the earth's rotation creates a force, referred to as the Coriolis force, which deflects north–south and south–north movements of water into east–west and west–east directions. Melting of glaciers may locally affect water salinity, thus changing the existing currents; these changes may have a profound effect on the climate in different areas of the world (*35*).

There are at present two theories concerning the response of vegetation to global warming. One theory suggests that the increased concentration of carbon dioxide will stimulate plants' growth; this is called carbon dioxide fertilization. Accelerated plant growth and augmented photosynthesis will cause removal of carbon dioxide from the atmosphere, attenuating the greenhouse effect. The contrary theory speculates that the increased temperature will stimulate plants' respiratory activity and soil bacteria's metabolism, leading to increased production of carbon dioxide (*21*). The effect of increased concentration of carbon dioxide on plants has been recently scrutinized (*36*). It appears that some plants respond to "carbon dioxide fertilization" but some do not. Plants are classified into C3 and C4 plants according to the way in which they process the first product of CO_2 assimilation.[5] Most trees, and crops such as rice, wheat, potatoes, and beans belong to the C3 category. Grasses in tropical and subtropical areas, maize, sorghum, and sugercane belong to the C4 class. C3 plants, but not C4 plants, respond to CO_2 fertilization and grow bigger and produce more foliage, provided that more nutrients and water are also supplied. Considering all factors, it is doubtful that the increased plant growth could compensate for the increased release of carbon dioxide caused by augmented respiration and bacterial activity (*36*). In addition, the different responses of plants to increased tension of carbon dioxide may have a negative ecological impact because many plant species, unable to compete, may become extinct.

---

[5]The first step in photosynthesis involves incorporation of CO_2 into a five-carbon sugar, ribulose bisphosphate (RUBP). This reaction is catalyzed by an enzyme, RUBP carboxylase. The six-carbon sugar thus formed is unstable and decomposes rapidly into two three-carbon fragments in C3 plants, or into one four-carbon fragment and one two-carbon fragment in C4 plants. In addition, C4 plants have a pump that concentrates CO_2 near the active site of an enzyme crucial to photosynthesis (*36*).

Current Developments

During the International Convention of Atmospheric Scientists (also attended by representatives of the United Nations and political leaders of some nations), held in Toronto in June 1988, a warning was issued that the greenhouse effect had begun. Four of the previous 8 years (1980, 1981, 1983, and 1987) had been the warmest years since the recording of global temperatures began 134 years prior to the convention. The year of the convention, 1988, added to these statistics. The fact that the frozen earth beneath the arctic tundra in Alaska had warmed 2.2–3.9 °C over the last century (*37*) provides further support for the greenhouse theory.

It is impossible to prove whether the warm summers of the 1980s and, in particular, the heat and drought of 1988, were indeed the result of an oncoming greenhouse effect or whether they were part of periodic climatic fluctuations. Precise measurements of the earth's temperatures during 1979–1988 were recently obtained with microwave radiometry from satellites. Analysis of these data revealed considerable yearly variability in the global temperature during this period. However, no obvious trend (*38*) could be documented.

Ten years is too short a period to determine whether global warming is really taking place. The significance of this experiment is that it successfully demonstrated a new methodology that will allow precise monitoring of global temperatures in the future.

Preventive Action

Although there are some dissenting opinions, in general the scientific community considers greenhouse warming a serious threat requiring imposition of a system of global regulations aimed at limiting GHG emissions. Recently only 43 out of 300 scientists, mostly members of the American Meteorological Society, agreed to sign a statement opposing global warming initiatives (*39*). The sentiment against any formal restrictions on GHG emissions was also expressed by representatives of President Bush's administration on the grounds of an unproven assumption that any such moves may hurt the economy. The experts on economy are divided on this issue. An economist, William D. Nordhaus, asserted that agriculture, forestry, and fishery, the industries most likely to be affected by climate changes, represent only 10% of the U.S. economy; thus the net economic damage to the U.S. gross national product could be only 0.25% (*40*). The fallacy of this position is that food is the basic commodity for the survival of humanity. Shifting of the agricultural areas may have serious consequences on the global scale; it may create considerable economic chaos worldwide, especially in the developing countries. The United States would not be immune to the global economic and political upheaval resulting from widespread regional hunger. Moreover, a rising sea level will cause additional damage to agricultural land and infrastructure, resulting in hardship and expenditures.

Analysts in the U.S. Department of Energy estimate the cost of reducing carbon dioxide emissions by 20%, by the year 2000, at $90 billion annually. On the other hand, a private research group, the International Project for Sustainable Energy Paths, claims that large carbon dioxide cuts would benefit the economy if the carbon tax were put in investment credits or invested in energy efficiency. A similar view was expressed in an unpublished study prepared by the Environmental Protection Agency (*24*). The most recent analysis by academic researchers suggests "that a variety of energy efficiency and other measures that are now available could reduce U.S. emissions of greenhouse gases by roughly 10 to 40% of current levels at relatively low cost, perhaps at a net cost savings" (*41*).

The conflicting points of view of how to respond to the alleged threat of global warming are epitomized in two strategies: "No Regrets" and "Wait and See" (*42*). The "No Regrets" strategy stipulates that energy conservation and investment in new energy-efficiency technologies will promote economic development, increase employment, and improve the national balance of trade. Thus, even if the dismal consequences of a rapid climate change will not occur, the strategy will deliver demonstrable benefits. The "Wait and See" strategy, on the other hand, argues that, in view of the scientific uncertainties about regional climate changes, a shift from the present pattern of energy utilization will cause unnecessary hardship and stifle the economy. The problem with the "Wait and See" strategy, frequently embraced by the conservative policy makers, is that by the time the uncertainties about climate changes are resolved, it may be too late to intervene successfully. Thus, this strategy is a dangerous gamble with the welfare of future generations.

Since World War II it has been an accepted policy of the superpowers to withdraw resources from national economies to prepare for a war that may never happen. In comparison, a nonchalant attitude toward possible (albeit still uncertain) environmental disaster is difficult to comprehend.

Some increase in the earth's temperature is bound to occur, but the process of this warming can be slowed and eventually arrested. The more slowly the climatic changes occur, the easier and less painful will be the transition to new conditions. The way to slow this process is to decrease fossil fuel consumption (through more efficient automobiles and utilities, and more reliance on public transportation) and eventually to develop new nonpolluting energy sources. At the same time, the deforestation trend should be reversed by planting more trees and cutting fewer.

International Cooperation

It is encouraging that in 1988 the United Nations Environment Programme and the World Meteorological Organization sponsored the creation of the Intergovernmental Panel on Climate Change (IPCC). The IPCC was divided into three working groups: Great Britain was charged with responsibility for scientific matters, the Soviet Union with the study of the potential impact of climatic changes, and the United States with the development of policies. Several meetings of the IPCC have been held so far.

In November 1990, the Second World Climate Conference was convened in Geneva. The conferees confirmed IPCC findings that without reduction of GHGs, global warming will reach 2–5 °C over the next century. Furthermore, "If the increase of GHGs' concentration is not limited, the predicted climate changes would place stress on natural and social systems unprecedented in the past 1000 years" (*43*). It is regrettable, however, that the conference did not established any specific goals and deadlines for global limitation of GHG emission.

Global warming was again discussed during the United Nations Conference on Environment and Development in Rio de Janeiro during the week of June 3–14, 1992. A treaty on global warming was signed by the participants of the conference, but the postulates of the treaty were considerably watered down, and no targets or timetables for carbon dioxide emissions were set. As it was finally passed, the treaty set only nonbinding commitments for the industrialized nations to limit their GHG emissions.

As a follow-up of the United Nations Framework Convention on Climate Change, signed in Rio de Janeiro, representatives of 121 nations that ratified the Convention met in Berlin between March 27 and April 7 to discuss alterations to the treaty and ways to implement reduction of GHG emissions. Because of the disagreement between the industrialized and developing countries, little substantial progress was achieved. Eventually, the delegates declared that the treaty's current commitments are not adequate to protect the earth's climate. Furthermore, an agreement was achieved to begin negotiation, to be completed by 1997, setting specific reductions of GHGs after the year 2000 (*44*).

In the summer of 1995 the IPCC released a new report that states that the increase in average global temperatures of 0.3–0.6 °C observed during the past 100 years "is unlikely to be entirely due to natural causes" and that a "pattern of climate responses to human activities is identifiable in the climatological record" (*45*). On the basis of the latest global circulation models, which take into consideration the effect of sulfate aerosols, the IPCC projects an increase in the earth's average temperature over the next 100 years in the range of 1.0–3.5 °C, with the best estimate scenario being 2 °C. This warming is expected to raise the average sea level in the range of 15–95 cm, with the best estimate being 50 cm.

References

1. *Encyclopedia Britannica;* Chicago, IL, 1969; Vol. 15, p 285.
2. *Scientific Assessment of Stratospheric Ozone: 1989*; U.N. Environment Programme: Nairobi, Kenya, 1990; Vol. 1, Chapter 1, p 1.
3. Zurer, P. S. *Chem. Eng. News* May 28, 1993, p 8.
4. Molina, M. J.; Rowland, F. S. *Nature (London)* **1974,** *249,* 810.
5. McElroy, M. B.; Salawitch, R. J. *Science (Washington, D.C.)* **1989,** *243,* 763.
6. Brasseur, G.; Granier, C. *Science (Washington, D.C.)* **1992,** *257,* 1239.

7. Setlow, R. B. *Proc. Natl. Acad. Sci. U.S.A.* **1974,** *71,* 3363.
8. Madronich, S.; Bjorn, L. O.; Ilyas, M.; Caldwell, M. M. In *Environmental Effects of Ozone Depletion: 1991 Update;* U.N. Environment Programme: Nairobi, Kenya, 1991; p 1.
9. Longstreth, J. D.; de Gruijl, F. R.; Takizawa, Y.; van Leun, J. C. In *Environmental Effects of Ozone Depletion: 1991 Update;* U.N. Environment Programme: Nairobi, Kenya, 1991; p 15.
10. Teramura, A. H.; Tevini, M.; Borman, J. F.; Caldwell, M. M.; Kulandaivelu, G.; Bjorn, L. O. In *Environmental Effects of Ozone Depletion: 1991 Update;* U.N. Environment Programme: Nairobi, Kenya, 1991; p 25.
11. Haeder, D. P.; Worrest, R. C.; Kaumar, H. D. In *Environmental Effects of Ozone Depletion: 1991 Update;* U.N. Environment Programme: Nairobi, Kenya, 1991; p 33.
12. Zurer, P. S. *Chem. Eng. News* May 30, 1988, p 16.
13. Zurer, P. *Chem. Eng. News* March 6, 1989, p 29.
14. *Chem. Eng. News* March 13, 1989, p 19.
15. Zurer, P. *Chem. Eng. News* October 9, 1989, p 4.
16. Zurer, P. *Chem. Eng. News* July 16, 1990, p 5.
17. *Chem. Eng. News* May 29, 1989, p 12.
18. Stinson, S. *Chem. Eng. News* June 5, 1989, p 5.
19. Zurer, P. *Chem. Eng. News* May 18, 1992, p 27
20. Zurer, P. *Chem. Eng. News* June 24, 1991, p 23.
21. Postel, S.; Heise, L. In *State of the World 1988;* Brown, L. R., Ed.; W. W. Norton: New York, 1988; p 83.
22. Schneider, S. H. *Are We Entering the Greenhouse Century? Global Warming;* Sierra Club Books: San Francisco, CA, 1989.
23. Barnola, J. M.; Raynard, D.; Korotkevich, Y. S.; Lorius, C. *Nature (London)* **1987,** *329,* 408.
24. Schindler, D. W.; Beaty, K. G.; Fee, E. J.; Cruikshank, D. R.; DeBruyn, E. R.; Findley, D. L.; Linsey, G. A.; Shearer, J. A.; Stainton, M. P.; Turner, M. A. *Science (Washington, D.C.)* **1990,** *250,* 967.
25. Denniston, D. *World Watch* **1993,** *6*(1), 34.
26. Hileman, B. *Chem. Eng. News* April 27, 1992, p 7.
27. Whigly, T. M. L.; Pearman, G. I.; Kelly, P. M. *Indices and Indicators of Climate Change in Confronting Climate Change;* Mintzer, M., Ed.; Cambridge University: Cambridge, England, 1992.
28. Postel, S. In *State of the World 1987;* Brown, L. R., Ed.; W. W. Norton: New York, 1987; p 157.
29. Kou, C.; Lindberg, C.; Thomson, D. J. *Nature (London)* **1990,** *343,* 709.
30. Schneider, S. H. *Science (Washington, D.C.)* **1989,** *243*(4892), 771.
31. Zurer, P. *Chem. Eng. News* October 2, 1989, 15.
32. World Resources Institute, International Institute for Environmental Development in collaboration with U.N. Environment Programme. *World Resources 1988–89, Atmosphere and Climate;* Basic Books: New York, 1988; Chapter 23, p 333.
33. Jeager, J. *Developing Policies for Responding to Climatic Changes;* a summary of the discussions and recommendations of the workshop held in Villach (September28 –October 2, 1987) and Bellagio (November 9–13, 1987) under the auspices of Beijer Institute, Stockholm; World Meteorological Organization and U.N. Environment Programme; 1988; WC 1 P–1, WM O/TD, No. 225.
34. Roberts, L. *Science (Washington, D.C.)* **1989,** *243*(4892), 735.
35. Hileman, B. *Chem. Eng. News* March 13, 1989, p 25.
36. Bazzaz, F. A.; Fajer, E. D. *Scientific American* **1992,** p 68.
37. Brown, L. R.; Postel, S. In *State of the World 1987;* Brown, L. R., Ed.; W. W. Norton: New York, 1987; p 3.

38. Spencer, R. W.; Christy, J. R. *Science (Washington, D.C.)* **1990,** *247,* 1558.
39. *Chem. Eng. News* March 9, 1992, p 14.
40. Zurer, P. S. *Chem. Eng. News* April 1, 1991, p 7.
41. Parry, L. M.; Swaminathan, M. S. In *Confronting Climate Change;* Mintzer, I. M., Ed.; Cambridge University: New York, 1992.
42. Mintzer, I. M. In *Confronting Climate Change;* Mintzer, I. M., Ed.; Cambridge University: New York, 1992.
43. O'Sullivan, D.; Zurer, P. *Chem. Eng. News* November 19, 1990, p 4.
44. Zurer, P. *Chem. Eng. News* April 17, 1995, p 7.
45. *IPCC Climate Change 1995: The IPCC Second Assessment Report;* Cambridge University: Cambridge, England, 1996.
46. Zakrzewski, S. F. *People, Health and Environment;* SFZ Publishing: Amherst, NY, 1994.

10

Water and Land Pollution

Freshwater Reserves

Water covers 70% of the earth's surface. Only 3% of this is freshwater, which is indispensable in sustaining plant and animal life. The amount of freshwater is maintained constant by the *hydrological cycle.* This cycle involves evaporation from oceans and inland waters, transpiration from plants, precipitation, infiltration into the soil, and runoff of surface water into lakes and rivers. The infiltrated water is used for plant growth and recharges groundwater reserves.

Although the global supply of available freshwater is sufficient to maintain life, the worldwide distribution of freshwater is not even. In some areas the supply is limited because of climatic conditions or cannot meet the demands of high population density. In other places, although there is no shortage of freshwater, the water supply is contaminated with industrial chemicals and is thus unfit for human use. Moreover, fish and other aquatic species living in chemically contaminated water become unfit for human consumption. Thus, water pollution deprives us and other species of two essential ingredients for survival: water and food.

An example of hydrologic changes caused by urbanization is given in Figure 10.1. Conditions before and after urbanization were measured in Ontario, Canada, by the Organization for Economic Cooperation and Development (*1*).

In the urban setting, pervious areas are replaced with impervious ones (such as streets, parking lots, and shopping centers). Groundwater replenishment is greatly reduced and runoff is considerably increased by these changes. Thus, urbanization not only contributes to water pollution; it also increases the possibility of floods.

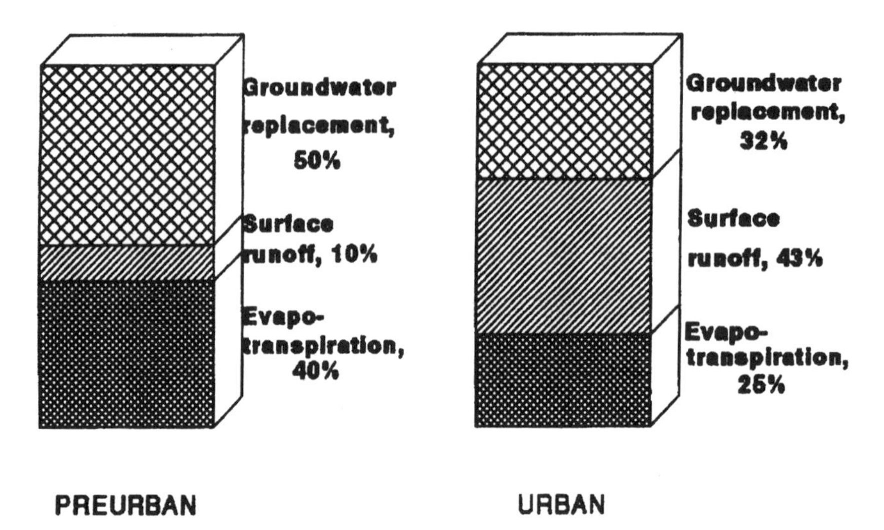

Figure 10.1. Effect of urbanization on disposition of rainwater. The study was conducted in Ontario, Canada, by the Organization for Economic Cooperation and Development. (Adapted from data in reference 1.)

Sources of Water Pollution

In industrialized countries, the sources of water pollution may be divided into point and nonpoint sources. *Point sources* have a well-defined origin, such as the outlet from a plant or from a municipal sewer line. *Nonpoint sources* lack any well-defined point of origin.

Although both types of pollution source present a serious problem, point sources can be controlled, at least in principle. Nonpoint sources, however, are difficult to control. Sources and types of nonpoint pollution in impacted rivers and lakes in the United States are shown in Tables 10.1 and 10.2. *Impacted* waters are those that are moderately or severely polluted, so as to interfere with their designated use (*1*).

Urban Pollutants

The sources of urban pollutants are municipal sewage, runoff from city streets and landfills, and industrial effluents.

Municipal Sewage

Municipal sewage consists mainly of human and animal waste; thus it is rich in nitrogen-containing organic nutrients. In addition, it contains grit, suspended soil,

Table 10.1. Sources of Nonpoint Pollution of Rivers and Lakes

Source	*Rivers*	*Lakes*
Agriculture	64	57
Land disposal	1	5
Construction	2	4
Hydromodification	4	13
Urban runoff	5	12
Silviculture	6	1
Resource extraction	9	1
Other	9	7

NOTE: All values are percentages.
SOURCE: Adapted from data in reference 1.

detergents, phosphates, metals, and numerous chemicals. Raw sewage entering streams and lakes stimulates excessive growth of aquatic bacteria, algae, and other plants. This excessive growth is referred to as *eutrophication*. Growth of microorganisms and bacterial digestion of the decaying plants consume the oxygen dissolved in the water. Because aquatic species require 5–6 ppm of dissolved oxygen, excessive growth causes oxygen depletion and thus kills fish by suffocation.

Table 10.2. Relative Contribution of Nonpoint Pollutants

Pollutant	*Rivers*	*Lakes*
Sediment	47	22
Nutrients[a]	13	59
Toxins	6	3
Pesticides	3	1
BOD[b]	4	3
Salinity	2	3
Acidity	7	4
Other	18	5

NOTE: All values are percentages.
[a]Phosphates and nitrates. Freshwater and seawater contain most of the nutrients required for growth of aquatic plants. The main exceptions are phosphates and nitrates, which are in limited supply. Thus release of nitrates and phosphates into lakes, rivers, and estuaries leads to their eutrophication.
[b]Biological oxygen demand, defined in the section **Metabolizable Organic Matter**.
SOURCE: Adapted from data in reference 1.

Metabolizable Organic Matter

The degree of pollution with metabolizable organic matter can be determined by a test called biological oxygen demand (BOD). This test measures the amount of oxygen needed by aquatic microorganisms to decompose organic matter during a 5-day period. Hence, metabolizable organic pollutants are referred to as BOD pollutants.

The removal of BOD contaminants, grit, soil, detergents,[1] and metals can be achieved relatively easily with a well-functioning wastewater-purification plant. However, the removal of phosphates and nitrates requires advanced treatment, and many plants are not equipped with an advanced treatment stage. Such plants may represent a considerable source of water pollution with nutrients (*see* the section **Nutrients and Pesticides** in this chapter). Thanks to the passage of the Clean Water Act in 1972, the percentage of the U.S. population served with wastewater treatment facilities increased from 40% in 1970 to 72% in 1985 (*2*). The remaining 28% of people not connected to sewage-treatment facilities represents rural and suburban populations that utilize septic tanks for disposal of their waste. Although septic tanks do not present much danger to surface water, they are frequently a source of groundwater contamination.

Synthetic Organic Chemicals

The removal of some synthetic organic chemicals from wastewater may present a problem. The synthetic chemicals found in municipal wastewater originate from both household use and industry.

Ordinary households in an industrialized society use substantial amounts of organic chemicals such as cleaning fluids, pharmaceuticals, cosmetics, and paints. Residual quantities of these substances may end up in the sewage. Hospitals, universities, dry cleaning establishments, garages, and other small commercial shops are not permitted to dispose of their chemical wastes through the sewers. Obviously, illegal dumping may occur; it is therefore the responsibility of municipal authorities, in charge of wastewater treatment, to watch for improper disposal.

The problem may occur when industrial plants contract with the city to dispose of their liquid waste through the municipal sewer system. Although the Clean Water Act (CWA) of 1972 requires that industrial plants prepurify their effluent before discharging it into municipal sewers, there is always potential for contamination with toxic compounds that are not well-identified. The removal of such chemicals from wastewater may be difficult and expensive. Furthermore,

---

[1]Detergents consist of hydrocarbon chains terminating with a hydrophilic ionizing group, such as phosphate or sulfate. Use of phosphate detergents is discouraged, as the phosphate contributes to the eutrophication of streams and lakes. Detergents are biodegradable in principle, but those with branched chains are degraded slowly. Thus some detergents may escape bacterial digestion in the course of the sewage-purification process.

most municipal sewage-purification plants are not equipped for this challenge. Toxic chemicals in sewage create potential hazards to aquatic life and inhibit the biological process of degradation of contaminants. In addition, they potentiate the toxicity of sewage sludge that must be disposed of in landfills.

Storm Water Runoff

Storm water runoff from cities and villages presents another problem. This runoff contains salts from road deicing, street refuse, animal waste, food litter, residue from atmospheric deposition of sulfuric and nitric acid, metals, asbestos from automobile brakes, rubber from tires, hydrocarbons from motor vehicle exhaust condensates, oil and grease, soil and inorganic nutrients from construction sites, and a variety of other chemicals.

Some localities possess a combined sanitary–storm sewer system. In such cases, the storm sewage undergoes purification. However, a severe downpour may exceed the capacity of the wastewater-purification plant. Rough sewage may then drain into the receiving waters.

In the absence of a combined system, storm runoff is a nonpoint source of pollution. As such, it is difficult to control. After a heavy downpour, the runoff from city streets and construction sites and leachates from landfills may bring a considerable quantity of pollutants into streams and lakes. Research (*1*) shows a heavy impact of urban nonpoint pollution on freshwater quality. In highly urbanized areas it may even surpass the impact of rural pollution (*1*).

Lead Pollution

Although lead pollution is essentially an urban problem, agricultural land, lakes, and rivers are also frequently affected. Lead has many toxic effects, including inhibition of red blood cell formation, kidney damage, and damage to the nervous system (*see* Chapter 7).

Sources

The sources of lead pollution are leaded gasoline, lead-based paint, and waste disposal. Except for some rural vehicles, the use of leaded gasoline has been practically eliminated in the United States, and thus the concentration of airborne lead is insignificant. However, large quantities of lead have accumulated in the soil as a result of decades of burning leaded gasoline. According to Environmental Protection Agency (EPA) estimates, the lead level in the soil along heavily traveled roads can reach 10,000 ppm or more (*3*). Use of leaded gasoline by farm vehicles, which is still allowed by law, is responsible for pollution of agricultural land. Runoff and seepage from lead-polluted soil leads to contamination of surface and groundwater.

For years lead-based paint was considered a problem only in old dilapidated tenements, where small children ingested crumbling wall paint. Although this concept was modified by subsequent research, the theory still prevailed, until very recently, that the primary exposure route is inhalation and incidental ingestion of household dust originating from lead-based paint (*3*). Accordingly, most prevention methods focused on removal of lead-based paint from older houses. More recently, studies of lead content in urban and suburban soil were performed in several large cities (Baltimore, MD, Minneapolis–Saint Paul, MN, and New Orleans, LA) and numerous smaller cities in Louisiana and Minnesota (*4, 5*). The results indicated a correlation between soil contamination and geographical site of the city. The contamination of the soil was found to be highest in the city-center, decreasing exponentially toward the outskirts regardless of the age of buildings (*6*). Also, contamination was substantially higher in the large cities than in the smaller ones. This pattern of contamination indicates that the most significant sources of lead in the soil are industrial processes, waste incineration, and decades of burning leaded gasoline (*6*). The fact that lead-loading in the soil between streets and buildings is on the order of 10^6 times greater than that in indoor dust lends additional support to the view that the lead-based paint is not the major factor in lead intoxication.

Lately there has been great concern about lead in drinking water. Even if the water source is not contaminated, lead may leach into drinking water from lead pipes or pipe solders. High amounts of lead have been found in old water fountains in offices and schools. Although this source of lead exposure is not confined to children alone, exposure of children is of particular concern. In view of the hazard of chronic lead exposure, the EPA has revised its limits for lead in drinking water from 50 to 10 ppb.

Toxic Symptoms in Children

Children are particularly susceptible to low-level lead intoxication, which creates a type of encephalopathy referred to as subclinical toxicity. No clinical symptoms of intoxication are observable. The brain damage is manifested by the child's neurophysiological behavior such as hyperactivity, unruliness, and a low IQ score. A 1986 EPA report cited 10–15 mg/dL in blood as enough to cause neurological deficiency.

Soil Erosion

Soil erosion is a natural phenomenon caused by water and wind; the rate of erosion depends on the degree of terrain coverage with trees or grasses, on the intensity and seasonal distribution of rainfall, and on the slope of the terrain. Agricultural practices that strip the plants' coverage from the soil accelerate this natural event. At present, soil erosion has become one of the most destructive aspects of agriculture; it causes silting of lakes and rivers, it causes pollution of surface water with nutrients and pesticides, and it affects the fertility of the land. In the United States, the sediment makes up 47% of all nonpoint river pollutants and 22% of all lake pollutants (*see* Table 10.2). Although most of the sediment originates from rural areas, dirt from urban centers may also contribute significantly in certain cases.

Sediment left by the runoff of topsoil from fields and dirt from urban centers represents a major ecological and economic problem. It creates water turbidity that reduces light penetration, decreasing plants' growth and diversity, it stifles the habitat by reducing the survival of eggs and young, and it helps to transport nutrients and toxic pollutants. It also accelerates the demise of lakes, streams, reservoirs, harbors, and irrigation canals by filling them with silt (*2*). Soil erosion may assume alarming proportions with poor farm management, cultivation of land unsuitable for agriculture, overgrazing, and deforestation. The problem is growing worse, especially in developing countries, as forests are cut and more land is cleared for agriculture. In Central America, for example, deforestation reduced forest cover from 60 to 40% between 1960 and 1980. Soil erosion there has become such a problem that siltation has clogged hydroelectric reservoirs, irrigation canals, and coastal harbors (*4*). Similarly, in the Philippines deforestation of 1.4 million hectares of an upland watershed and unsuitable agricultural practices between 1967 and 1980 were paralleled by a 121% and a 105% increase in the annual sedimentation rate in two major reservoirs, respectively (*7*).

Binding of Pollutants

The capability of soil to bind and transport pollutants depends on the nature of the soil as well as on the chemical and physical properties of the pollutant. Soil consists of inorganic components and of organic substances originating from plant and animal material. Inorganic components of the soil are classified as follows: sand, 0.02–2 mm; silt, 0.002–0.02 mm; and clay, <0.002 mm in diameter. Organic substances are referred to as humic substances (if completely decomposed and chemically rearranged) or as nonhumic substances (if only incompletely decomposed). Nonhumic substances constitute only 10–15% of the soil organic matter. Although the total organic matter comprises 0.1–7.0% of the soil, it may coat the inorganic components and block their adsorptive functions (*8*).

Soil organic matter is responsible for binding nonionic and hydrophobic compounds. The inorganic matter interacts with ionic and polar compounds; it also has cation-exchange capacity. The size of soil particles is important. The large surface area associated with very small particles provides a greater number of binding sites than the surface area of large particles. Water solubility of a pollutant is another property that affects its interaction with the soil. Water solubility, in turn, is affected by factors such as salt concentration, pH, the presence of other organic compounds, and temperature.

Cropland Fertility

Soil erosion is an important issue because it contributes to water pollution and affects cropland fertility. This problem becomes critical when combined with overgrazing and cultivation of agriculturally marginal land. The rate of soil erosion in the United States is 18 metric tons per hectare per year (1982 estimate). In some developing countries it is much higher. For instance, in Ethiopia it reaches 42 (1986 estimate) and in Kenya 72–138 (1980 estimate) metric tons per hectare per year (*9*). The effect of such a rapid loss of topsoil on the future ability of developing countries to produce enough food for their ever-growing populations may be catastrophic.

Although the predominant effect of erosion is loss of the topsoil, in some extreme cases soil erosion leads to terrain deformation by creation of gullies; reclamation of a land so distorted is almost impossible. Other causes of land degradation are depletion of nutrients, compaction of the soil by cattle or heavy machinery, waterlogging, salinization, and acidification. The other causes notwithstanding, soil erosion is still the major cause of soil degradation. It is responsible for 84% of the loss of the agricultural land in the world as a whole (*10*).

Salinization

Excessive salt accumulated in the upper layers or on the surface of the soil inhibits plant growth, and consequently the fertility of the land declines. In extreme cases certain areas may become sterile. Salinization may happen as result of salt loading or salt concentration.

Salt loading occurs when fields containing excessive salt are irrigated and properly drained; the salt may be washed into the streams that serve as the sources of irrigation water. Thus, each successive farmer downstream uses water of higher salinity than the upstream neighbor. Eventually, the stream becomes polluted by high salt loads. This situation is detrimental to aquatic life, as well as to the land.

Salt concentration, on the other hand, occurs with waterlogging in areas where large amounts of water are lost through evaporation. Waterlogging may occur when field drainage is impaired or when the groundwater table is too close to the surface. Standing water elutes salt from the ground. As the water evaporates, the salt concentration builds up near or on the surface of the land.

Nutrients and Pesticides

Runoff from farms causes pollution by nutrients such as nitrates and phosphates from fertilizers and by animal waste originating from feedlots. Both nutrients and animal waste contribute to eutrophication of lakes and streams.

Nutrients

Nitrates are of special concern because of their potential toxicity. With their high water solubility, they leach easily from the soil and contaminate surface as well as groundwater. In the soil (and in the oral cavity; *see* Chapter 3) they may undergo reduction to nitrites. When ingested via drinking water, nitrites may cause methemoglobinemia[2] and hypertension in children. The chemical reaction between nitrites and some pesticides may lead to formation of nitrosamines, which are known carcinogens and mutagens. Exposure to nitrites may cause gastrointestinal cancer, and prenatal exposure may lead to fetal malformations.

In contrast to nitrates, phosphates move primarily with the eroding soil. Even when applied to the field as a soluble orthophosphate, it soon reverts to an insoluble form that is readily adsorbed to soil particles. As a result, phosphate builds up in the sediment. However, under some circumstances there may be exceptions to this behavior. A study of pollution of U.S. coastal estuaries suggests that most of the phosphate, at least in the brackish waters of an estuary, exists in solution rather than being bound to the sediment (*11*).

Manure is a good fertilizer if used in moderate quantities on the fields. Large quantities of manure that accumulate in cattle feedlots produce leachate rich in organic nutrients as well as in phosphates, nitrates, and ammonia, creating a hazard of groundwater and surface water pollution. An example of a major ecological threat caused by excessive manure accumulation is the pork industry in the Netherlands. The 14 million animals in the southern part of the country have produced more manure than the country can use for its agriculture. As a result, in many areas water is highly polluted and surface layers of the soil are saturated with phosphates and nitrates (*12*). The accumulated manure also contributes to air pollution by releasing nitrous oxide, which is formed in the soil from ammonia by oxidizing bacteria. N_2O is converted in the air to nitric acid and as such, it is responsible for about 20% of acid deposition in the Netherlands (*13*).

Pesticides

Although pesticides constitute a small percentage of total water pollutants, one should not be lured into complacency about their use. Pesticides (whether insecticides, herbicides, or fungicides) by their very nature and purpose are poisons. Even if their amount is minimal in comparison to that of silt, their impact on the environment may be considerable. Since 1962, the use of pesticides in the United States has increased more than twofold. It now endangers groundwater quality in most of the states.

---

[2]A condition characterized by methemoglobin accumulation in the blood. Methemoglobin is a form of hemoglobin in which iron is oxidized to the trivalent state. As such, it is unable to carry oxygen.

The EPA has issued policies for groundwater protection from pesticides. These policies mandate restrictions on the use of pesticides in areas where their concentration in drinking water approaches the maximum amount allowable under the Safe Drinking Water Act. If the contamination is severe, the use of pesticides is outlawed (*14*).

Persistence in the Environment

Concern with pesticides centers on their properties, such as selective toxicity, persistence in the environment, bioaccumulation potential, and mobility. Persistence in the environment is perhaps the most crucial factor in their acceptability. Accordingly, they are divided into three groups: persistent, which decompose by 75–100% within 2–5 years; moderately persistent, which decompose within 1–18 months; and nonpersistent, which decompose in 1–12 weeks.

Decomposition of pesticides may occur by bacterial digestion as well as by photochemical and chemical reactions. It is frequently catalyzed by metals, soil components, or organic compounds. The reactions involve oxidations, reductions, hydrolyses, interactions with free radicals, and nucleophilic substitutions involving water. The fact that a pesticide "decomposes" (i.e., loses the activity for which it was designed) does not necessarily mean that it becomes a harmless substance.

Food Chain

Bioaccumulation is a function of the lipid–water partition coefficient of a substance and its refractivity to degradation and biotransformation. Bioaccumulation potential increases with increasing lipid solubility. In general, bioaccumulation is higher in aquatic than in terrestrial organisms. Pesticides accumulated in a terrestrial or aquatic organism may be biomagnified in the food chain; the degree of biomagnification is dependent on the length of the food chain.

Pesticides adsorbed onto soil particles may end up in the sediment at the bottom of lakes or rivers. They may enter phytoplankton, which are then consumed by higher organisms. These higher organisms are in turn consumed by still higher organisms, and so on. At each successive step of consumption, concentration of the substance increases. As an example, bioaccumulation of polychlorinated biphenyls (PCBs) in the food chain is presented in Table 10.3. Although PCB is not a pesticide, it has many physicochemical characteristics in common with chlorinated hydrocarbon pesticides.

Another problem with pesticides is their lack of specificity. Pesticides are designed to be more toxic for insects than for birds or mammals, but they usually do not distinguish between different species of insects. Thus they kill not only the pest against which they were applied but also other insects that might be natural predators of the pest, or which may serve as food for fish and birds. In addition, pesticides entering a watershed in high concentration may be harmful to fish. Rachel Carson (*16*) described spectacular fish kills caused either by aerial spray-

Table 10.3. Biomagnification of PCBs in the Food Chain

Species	*Concentration (ppm)*	*Degree of Magnification*
Phytoplankton	0.0025	1
Zooplankton	0.123	49.2
Rainbow smelt	1.04	416
Lake trout	4.83	1,932
Herring gull eggs	124	49,600

SOURCE: Adapted from data in reference 15.

ing or by release of insecticides into waterways. In the summer of 1950 the coniferous forests of New Brunswick (Canada) were sprayed with DDT (dichlorodiphenyltrichloroethane) to combat infestation by spruce budworm. The pesticide killed not only the pest against which it was intended, but also insects that served as food for young salmon and trout. The levels of DDT in brooks and rivers intersecting the forest reached toxic concentrations. Through the combined effect of food deprivation and toxicity, a large fish population was exterminated. In another case, which occurred in 1961 near Austin, Texas, large quantities of toxaphene and chlordane were dumped into a storm sewer by a pesticide manufacturing plant. The chemicals were then flushed into the Colorado River (Texas); they killed fish as far as 200 miles downstream from the release point.

Another problem with continuous use of chemical pesticides, especially when the same crop is planted on the same field over and over again, is a gradual selection of pests that either no longer respond to a pesticide, or require a larger application. In the end, while the use of pesticides keeps increasing, their effectiveness is declining. Moreover, expanded use of pesticides increases the cost of farming and inflicts more ecological damage.

The main classes of pesticides, their use, their solubility in water, and the mode of their transport in the soil are presented in Table 10.4. The chemical structures of some of these pesticides are shown in Charts 10.1–10.4. An in-depth treatment of this subject is presented in references 8 and 17.

Restrictions

Some of the most persistent pesticides (such as DDT, dieldrin, chlordane, and toxaphene) have been banned from use in the United States since 1978, and the use of others has been restricted. Despite the ban, residues of these pesticides still persist in the environment. Partially they are vestiges of prior use, and partially they are being transported by air. Since the ban against their use does not preclude manufacturing and export, it is likely that they are transported from Mexico or from South or Central America, where they are still in common use.

Table 10.4. Main Classes of Pesticides and Their Characteristics

Class	*Use*	*Persistence*	*Solubility in Water*	*Transport in Soil*
Chlorinated hydrocarbons	Insecticides	High	Extremely poor to insoluble	Soil erosion
Cationic heterocyclics	Herbicides	High	Good	Soil erosion
Triazines	Herbicides	Moderate	pH dependent	Soil erosion
Phenylureas	Herbicides	Moderate	Variable	Leaching (if highly soluble)
Dinitroanilines	Herbicides	Moderate	Poor	Soil erosion
Phenoxyacetic acid derivatives	Herbicides	Short	Good	Soil erosion
Phenylcarbamate derivatives	Herbicides	Short	Good	Soil erosion
Ethylenebis-(dithiocarbamate) metal derivatives	Fungicides	Short	Moderate	Unknown
Pyrethroids	Insecticides	Short	Extremely poor	Soil erosion
Organophosphorus	Insecticides	Short	Good	Leaching
Carbamates	Insecticides	Short	Good	Leaching

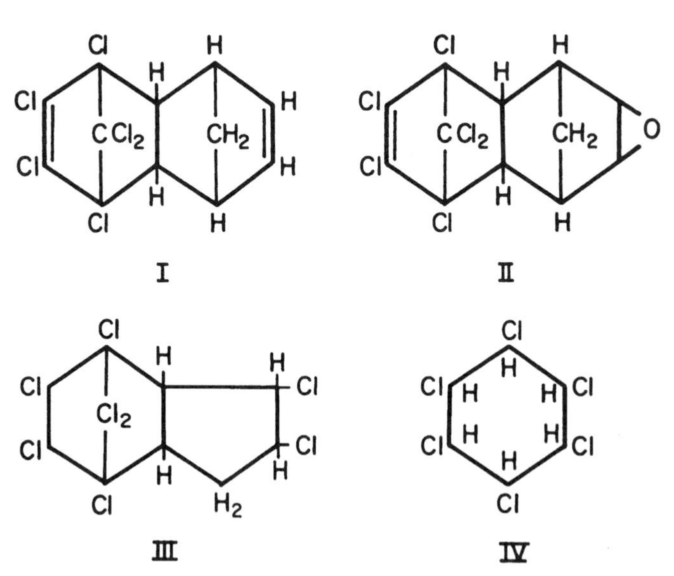

Chart 10.1. Chlorinated hydrocarbon insecticides: I, aldrin; II, dieldrin; III, chlordane; IV, lindane. DDT (see *Chart 3.10 in Chapter 3) belongs in this group.*

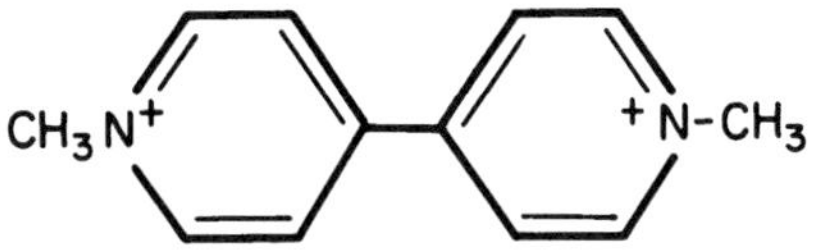

I

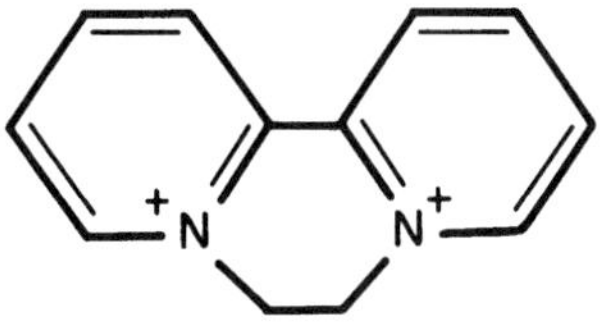

II

Chart 10.2. Ionic heterocyclic herbicides: I, paraquat; II, diquat.

A provision to ban or at least severely restrict export of pesticides that are not approved for use in the United States was introduced by the U.S. House and Senate into a 1990 farm bill. This provision was killed by the Bush administration. The so-called Circle of Poison Prevention Act was again introduced by Sen. Patrick J. Leahy of Vermont in 1991. This legislation would "ban the export of pesticides that cannot be used on food domestically or cannot be present on food consumed in the United States of America" (*18*). Although the bill has never been acted upon, the proposed legislation was strongly opposed by the National

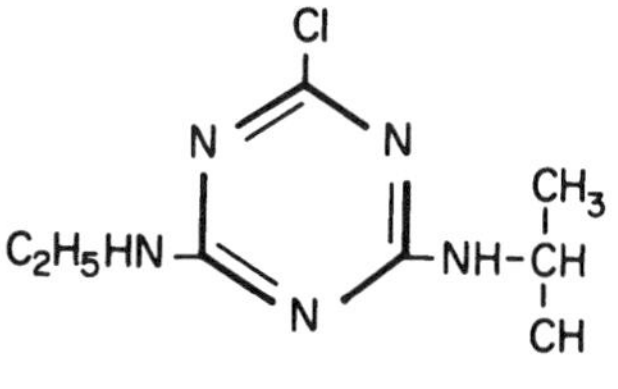

I

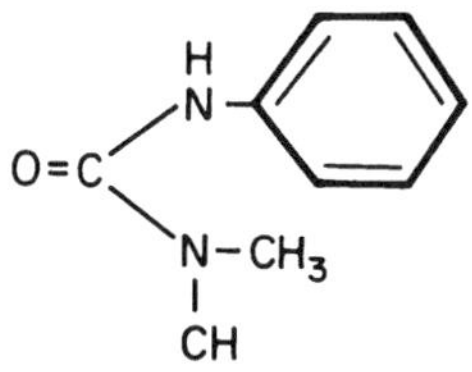

II

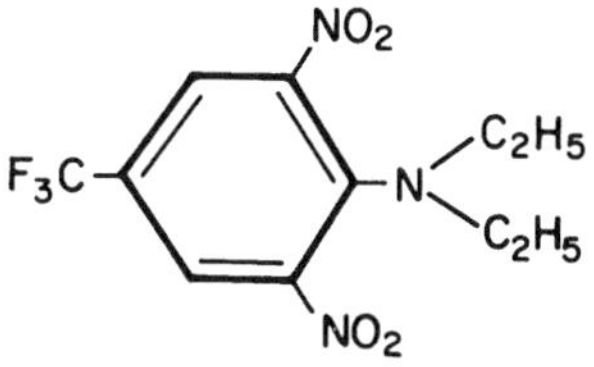

III

Chart 10.3. Miscellaneous moderately persistent herbicides: I, atrazine, a triazine derivative; II, monuron, a phenylurea derivative; III, benefin, a dinitroaniline derivative.

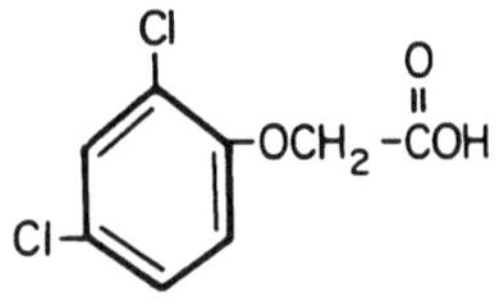

Chart 10.4. Miscellaneous nonpersistent pesticides: I, 2,4-D, a derivative of phenoxyacetic acid; II, 2,4,5-T, a derivative of phenoxyacetic acid; III, chloropropham, a phenylcarbamate derivative; IV, maneb, an ethylenebis(dithiocarbamate) metal derivative; V, carbaryl, a carbamate derivative. Examples of organophosphorus and pyrethroid derivatives are presented in Scheme 4.1 and Structure 4.1 in Chapter 4, respectively.

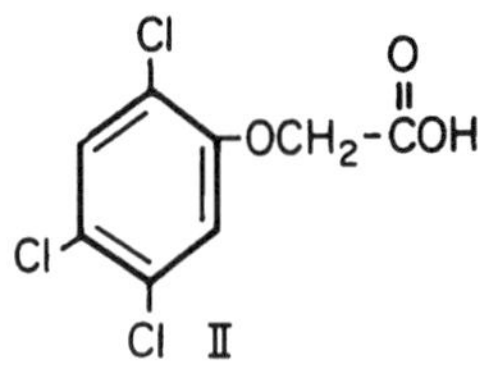

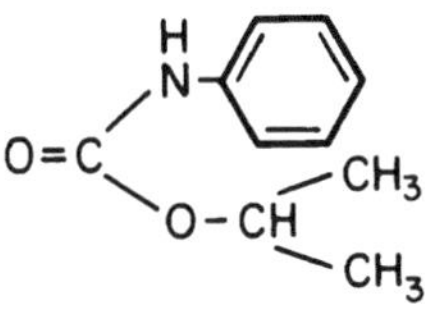

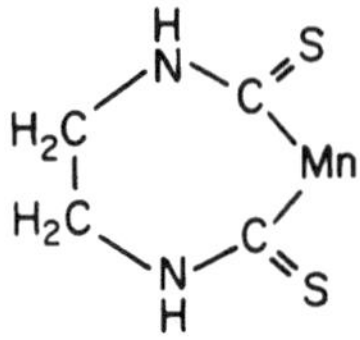

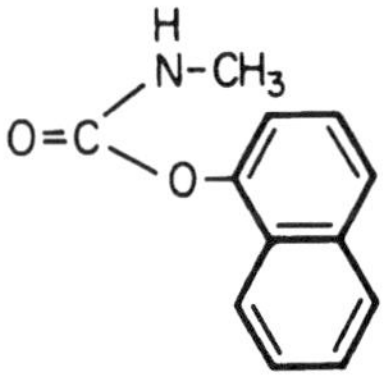

Agricultural Chemicals Association. The Association claimed that passage of the Circle of Poison Prevention Act will cost U.S. industry $750 million and will stifle agricultural chemical research and development (*19*). The ban on export of pesticides not registered in the United States was now incorporated into the Clinton administration's bill for pesticide–food safety reform (*20*).

Health and Environmental Effects

Concern about the health effects of chlorinated hydrocarbon pesticides stems from the observation that many of them, such as DDT, aldrin, and chlordane, were shown to produce liver cancer in rodents. Another type of potentially carcinogenic pesticide is represented by ethylenebis(dithiocarbamate) metal derivatives, of which the main representatives are maneb (manganese derivative) and zineb (zinc derivative). Although they are not carcinogenic in their own right, they are degraded and metabolized to a known carcinogen, ethylene thiourea, which may contaminate vegetables grown on soil treated with ethylenebis(dithiocarbamate) (*17*).

Recently, concern about effects of pesticides on human health and on the ecosystem began to move beyond cancer. It appears that some chlorinated hydrocarbon pesticides exert a multitude of toxic effects. These pesticides are neurotoxic, mutagenic, and teratogenic, they exert toxic effects on the reproductive system, and they suppress the immune system. It has been suggested that these compounds act by mimicking or inhibiting estrogen receptors (*21*). Xenoestrogens, as they are called (*see* Chapter 5), not only affect women's health, but are also believed to be responsible for a decrease in sperm count and a rise in testicular cancer in humans, as well as abnormal sexual development in some wildlife species (*22*). In some cases, as for instance in the case of the chlorinated derivatives of phenoxyacetic acid, 2,4-D and 2,4,5-T (the defoliant "Agent Orange"), the toxicity, especially the teratogenic activity, may be due in part to the always-present byproducts of their synthesis, the extremely toxic 2,3,7,8-tetrachlorodioxin.

The direct health impact of pesticides on the human population is difficult to establish. Limited epidemiological studies showed elevated frequency of some types of cancers among workers involved in manufacturing (*23*) or application (*24*) of pesticides; however, the effect of pesticides on the population at large has been explored only marginally. A new study suggests some correlation between levels of organochlorine pesticides in blood and breast cancer (*25*). Public concern is centered on a possible health hazard arising from traces of pesticides, as potential carcinogens, on fruits and vegetables. How valid is this concern is a subject of controversy within the scientific community. Some scientists claim that the carcinogenic hazard from residues of pesticides is insignificant compared with that of the background level of natural carcinogens (*26*); others disagree.

So far no link has been established between consumption of fruits and vegetables contaminated with traces of pesticides and any adverse health effect. However, one case of people becoming sick after eating watermelons contaminated with a pesticide, aldicarb, has been recorded. Ever since this incident, which was blamed on improper application of the pesticide, use of aldicarb on watermelons has been banned by the EPA.

Recently two reports appeared concerning pesticides in children's diets; one was published by the National Research Council and the other by a private organization called the Environmental Working Group. Both reports urge the EPA to develop special pesticide standards for children, stricter than those applicable to the adult population. They also recommend the study of children's diets to get a better idea of the actual intake of pesticides by the children (*27*).

According to some scientists, the controversy about pesticides on fruits and vegetables draws attention away from the more real potential problem: the health hazard caused by exposure to pesticides in the air, originating from such activities as control of mosquitoes or weeds along roads, from spraying of golf courses and suburban lawns, and from aerial spraying of fields and forests (*28*). Although the health hazard due to these activities is difficult to determine, it cannot be disregarded. A new epidemiological study indicates that dogs of home owners who spray their lawns with the herbicide 2,4-D, or who have their lawns commercially treated, were more likely to develop canine malignant lymphoma than dogs of home owners who do not spray their lawns (*29*). This study suggests that the extensive application of lawn herbicides may have human health implications. Also, a correlation between childhood brain cancer and exposure to insecticides has been reported (*30, 31*).

Because of the environmental problems caused by persistent pesticides, there is now a tendency to use, whenever possible, the nonpersistent ones that by definition decompose in 1–12 weeks. Unless a pesticide is mineralized (i.e., decomposes to CO_2 and water), we know nothing about the environmental effects and health hazard of the decomposition products. Moreover, the EPA's inspector general recently expressed concern about "inert" ingredients, which in fact constitute the bulk of commercial preparations of pesticides; there are 1400 "inerts", some of them known as hazardous substances and the rest of unknown toxicity (*32*).

The trends in the use of pesticides during the last 15 years fluctuated greatly from country to country. For instance, whereas in the United States there was a drop of 19%, in Canada the use of pesticides more than doubled. A slight increase also occurred in most of the European countries.

Alternative Agriculture

Public concern over the presence of pesticide residues in fruits and vegetables and water pollution problems caused by conventional agricultural practices have led to a new trend in food production, *alternative agriculture*. The aim of alternative agriculture is to limit dependence on fertilizers and pesticides and to prevent soil erosion. The techniques involve crop rotation, diversification of crops and livestock, use of nitrogen-fixing legumes, use of biological pest control, new tillage procedures, and planting cover crops after the harvest to prevent soil erosion (*33*). Although alternative agriculture is at present in an experimental stage, it may eventually offer a means to sustainable and nonpolluting food production.

Wetlands and Estuaries

Wetlands and estuaries represent an important ecological and economic resource. There are two types of wetlands: freshwater wetlands, and tidal marshes associated with the estuaries at the seashore. An estuary forms when an inlet of a river valley is invaded by the sea tide and seawater spills over the tributary valleys, forming an intricate network of little bays and inlets. Estuaries are normally bordered by tidal marshes that are formed by freshwater, but during the tide they become inundated by salty seawater; as result their water is brackish. A tropical counterpart of tidal marshes is mangroves. Mangroves are dense thickets of shrubs and trees characterized by arched roots emerging from the mud and joining the trunks above the water surface.

Both freshwater and tidal marshes have rich vegetation that abounds with grasses that supply winter food for ducks and geese. They act as giant water-purifying filters, attenuate floods, and provide a variety of food necessary to maintain species diversity and ecological balance. Mangroves, besides supplying food for aquatic species, also prevent coastal erosion.

The Loss of Wetlands

Estuaries are extremely rich in both land and ocean nutrients. The coastal rivers carry fertile silt that supports vegetation, which in turn provides for a chain of life. Estuaries are breeding shelters for many species of fish and shellfish. The most commercially valuable ocean fish (other than tuna, lobster, and haddock) depend on estuaries for food and propagation (*34*).

The greatest danger to wetlands comes from land development. Because about two-thirds of the world population lives along the coastlines and most rivers drain into coastal waters, the integrity of the tidal marshes and estuaries is threatened. In the United States, the population living within 50 miles of the shoreline doubled between 1940 and 1980 (*35*). Nearshore construction, landfilling, and dredging pollute coastal waters. Many coastal and inland wetlands are being drained or filled for residential or commercial construction, road building, farmland, and other uses. It is estimated that between 1956 and 1986, 11 million acres of wetlands were drained in the United States (*34*).

Coastal development in the tropics may also endanger the integrity of coral reefs. Coral plays an important role in the preservation of marine ecological balance because it serves as a shelter and a breeding place for many fishes. It also protects shores from erosion. Coral can thrive only in symbiotic relationship with photosynthesizing organisms called zooxanthellae (zooxanthellae provide coral with nutrients). When the coastal waters become turbid because of soil runoff from construction sites, light penetration is reduced, zooxanthellae die, and so does coral.

Another danger is pollution carried by rivers. Rivers carry urban, industrial, and agricultural pollutants that empty into the estuaries. The buildup of pollutants at the coast threatens the marine life. Poorly treated sewage and agricultural runoff introduce nutrients and BOD pollutants that stimulate growth of algae, depleting water of oxygen. Some algae, especially those having a red or brown color, known as red or brown tide, are toxic and kill fish and aquatic mammals feeding on fish.

In the United States, under the Public Trust Doctrine (*36*), the state or federal government may restrict development of land designated as wetlands. In 1991, President Bush, under pressure from land developers, land owners, and the oil industry, proposed to reclassify the definition of wetlands. Because not all wetlands are under water all year round, the definition of what is and what is not a wetland is somewhat arbitrary.[3] However, changing the existing definition to allow more development is a dangerous precedent. When all wetlands newly opened for development will be gone, there may be renewed pressure by the developers to change the definition again; this may lead to a gradual disappearance of all wetlands with catastrophic ecological consequences.

The Case of Chesapeake Bay

The Chesapeake Bay is one of the largest estuaries in the world; it is 195 miles long and its width varies between 3.4 and 35 miles. The drainage area of the bay covers 64,000 square miles, which includes six major rivers (Susquehanna, Patuxent, Potomac, Rappahannock, York, and James) that supply almost 90% of the freshwater input into the bay. Water quality in the Chesapeake Bay has been deteriorating gradually since the industrial revolution, but this decline accelerated rapidly since the late 1950s. Presently about 50% of the bay area is moderately to heavily polluted with nutrients and BOD contaminants, and perhaps to a lesser extent with heavy metals and pesticides. The ecological damage to the Chesapeake Bay is of major concern because the bay is a valuable source of seafood, a habitat for waterfowl, a stopover for migratory birds, and a unique recreational resource.

The extensive study of the causes of the bay pollution conducted by the EPA revealed that 78% of the nitrogen and 70% of the phosphates entering the bay were carried by three major rivers from upstream sources. Most of the phosphates originated from point sources, mainly wastewater treatment plants, whereas most of the nitrates were derived from agriculture (*2*). Some pollutants originated from as far away as Pennsylvania and New York. Another study conducted by the Environmental Defense Fund pointed out that the airborne transfer of nitric acid and ammonia has contributed considerably to the nitrogen loading of the bay (*37*). Table 10.5 shows the sources of the bay nitrogen inputs.

Realizing the ecological and economic consequences of progressive deterioration of water quality in the estuary, the Chesapeake Bay Program was initiated. The program called for a 40% reduction of phosphates and nitrogen input into the bay by the year 2000. This goal was to be achieved by providing governmental subsidies for improved agricultural practices and reduction of urban point and nonpoint pollution, and by withdrawal of subsidies for crops planted on the erodible soil. Although since 1987 the phosphate and nitrogen loading of the bay decreased by 7%, at this rate of progress, and considering the anticipated population growth and development in the watershed area, it is unlikely that the 40%

[3]The official definition (established in 1989) of a wetland is as follows: "A wetland is any depression where water accumulates for seven consecutive days during the growing season, where certain water-loving plants are found, and where the soil is saturated enough with water that anaerobic bacterial activity can take place."

Table 10.5. Contribution of Various Sources to the Nitrogen Loading of the Chesapeake Bay

Source	*Amount (million kg/year)*	*Percent of Total*
Airborne nitrate	143	23
Airborne ammonia[a]	79	13
Animal waste	195	32
Fertilizers	158	25
Point sources	42	7
Total	616	100

[a]Originates by evaporation of ammonia-containing fertilizers.
SOURCE: Adapted from data in reference 37.

goal will be met by the year 2000 (*2*). Thus, unless the continuous population growth and development will be arrested, and pollution prevention measures will be introduced and strictly enforced, the rehabilitation of the bay's water might be unattainable.

Industrial Pollutants

Industrial waste consists of a variety of pollutants, including sludges from the steel industry; toxic chemicals from chemical, mining, and paper industries; BOD contaminants from food processing plants; heat from power plants (conventional and nuclear) and from steel mills; and pH changes from the mining industry.

According to the Toxic Release Inventory in 1991, 190 million pounds of reportable hazardous waste were released into water, 360 million pounds on land, and 690 million pounds into deep wells. An additional 1.12 billion pounds were transferred to other facilities for treatment or disposal (*38*).

Definition of the Problem

The problem of toxic pollutants is difficult to handle because of the great variety of chemicals involved. They represent a hazard not only to aquatic life, but also to human health, either through direct exposure or indirectly through consumption of contaminated fish or waterfowl. The degree of hazard depends on the pollutants' toxicity, rate of discharge, persistence and distribution in the aquatic system, and bioaccumulation potential. Persistence is a function of the toxins' biodegradability in water and of their vapor pressure. Some highly volatile compounds, when discharged into water, evaporate and become air pollutants.

The health risk cannot be well defined because little or no information is available on the toxicity of most commercial chemicals (*39*). According to the data published in 1984 by the National Research Council (*40*), very little is known about the toxicity of approximately 79% of commercial chemicals. Fewer than 10% were examined for carcinogenicity, mutagenicity, and reproductive toxicity (*40*). Obviously, nothing is known about pollutants that are byproducts of industrial processes and were never intended for commercial use.

Mercury

One of the ubiquitous water pollutants is mercury. In humans, toxicity of mercury involves severe neurological disturbances manifested (in order of severity) by loss of sensation in the extremities, an unsteady gait, slurred speech, tunnel vision, loss of hearing, convulsions, madness, and death. In the past, most mercury contamination resulted from the dumping of inorganic mercury into lakes, streams, and seas. Although inorganic mercury is toxic, it is not easily assimilated by biological organisms. However, under anaerobic conditions it is converted into extremely toxic methyl- and dimethylmercury. These compounds penetrate biological membranes readily and subsequently undergo bioaccumulation.

The most dramatic case of mass mercury poisoning attributed to consumption of fish and other seafood contaminated with methylmercury occurred in 1956 in Japan. A mercury catalyst used in a chemical plant was discarded as waste sludge into Minamata Bay. The mercury was converted by aquatic biota to methylmercury, and eventually toxic amounts of it accumulated in fish and shellfish. The disease and its causes were not identified until 1963 (*41*).

During the 1960s and early 1970s, a great deal of mercury was dumped by industrial plants into the Great Lakes. As a result, fishing in Lake Erie, Lake St. Clair, the Detroit River, and the St. Clair River was stopped by both U.S. and Canadian authorities (*42*).

After the dumping of mercury ceased, it was generally believed that the problem of pollution by mercury was solved. However, in the late 1980s the problem resurged, though not in quite as severe a form as before. Although mercury in the environment originates from natural sources such as volcanoes and geologic deposits, it turned out that the anthropogenic sources contribute 75% to the global atmospheric load; the main sources are coal combustion (65%) and solid waste incineration (25%) (*43*). Research showed that the total concentration of mercury in the atmosphere has doubled since the 19th century.

Atmospheric mercury is carried to the earth with rain; it settles on land and it is carried into the lakes, ponds, and rivers with the runoff from fields. The concern for water pollution with mercury is dwarfed in comparison to that for pollution with chlorinated organic compounds. Nevertheless, the problem is serious. This is best illustrated by the fact that 12 states (Massachusetts, New York, Florida, Ohio, Connecticut, Michigan, Virginia, Tennessee, Minnesota, Wisconsin, California, and Oklahoma) enacted a fish-advisory for mercury. The Food and Drug Administration set the upper limit for mercury in fish at 1 ppm (1 μg/kg or

1000 ng/kg); fish exceeding this content of mercury may be banned from interstate commerce.

On the basis of the investigation of the Minamata Bay incident, the World Health Organization (WHO) established the human toxic dose of mercury in fish at 4300 ng/kg/day. To be on the safe side, WHO recommended that human uptake of mercury should not exceed 430 ng/kg/day. Because small children and fetuses are more sensitive than adults, they should not be exposed even to such small doses.

Other Heavy Metals

Many heavy metals are toxic and can be taken up from soil by the plants. Their toxicity is discussed in Chapter 7. In contrast to most organic pollutants, metals do not decompose in nature, and they remain in the environment until they are physically removed. An example of pollution with heavy metals is the contamination of the Hudson–Raritan Estuary with copper, mercury, lead, nickel, and zinc. The main contributors of these pollutants are the industrial plants in New York and New Jersey that discharge their effluents through municipal sewage (*44*). It appears that the legally mandated pretreatment of industrial effluents is not working satisfactorily.

Other cases of significant industrial pollution of U.S. rivers and coastal waters have been reported. In some areas of Galveston Bay, Texas, heavy metals exceed the EPA water quality standards. This pollution is attributed partially to industrial effluents and partially to waste disposal (*45*).

Polychlorinated Biphenyls

General Electric has two plants along the upper Hudson River that manufacture capacitors. During the 1950s and 1960s (i.e., before discharge permits were required; *see* Chapter 14), the plants kept discharging about 30 lb of PCBs per day into the river. Most of this discharge was adsorbed onto soil particles and settled to the bottom of the river with the sediment, which was retained in place by a dam located downstream from the plants. When the obsolete dam was removed in 1973, the disturbed sediment was swept down with the current, and 40 miles of the Hudson River downstream from the plants became heavily contaminated (*46*). Levels of PCBs in most edible fish exceeded the 5-ppm safety limit set by the Food and Drug Administration. As a result of high levels of contamination with PCBs, sport fishing in the Hudson River was completely wiped out and commercial fishing was curtailed to 40% of the precontamination level.

The legal battle that issued between the New York Department of Environmental Conservation and the General Electric Company, in response to the Hudson River pollution, was settled in September 1976. The provisions of the settlement obligated the General Electric Company to reduce emissions of PCBs as of the settlement date and cease using them completely by July 1977. However, the problem of how to handle the contaminated sediment remained unsolved. In 1988 General Electric researchers presented evidence that PCBs are biodegradable under the conditions that occur in the Hudson River sediment (*47*). Thus the company proposed to do nothing and let nature take its own course. The officials of New York State, on the other hand, being skeptical about the efficiency of the self-cleaning, were inclined to have the most contaminated stretch of the river dredged. The cost considerations notwithstanding, the problem remains of how to dispose of the dredged sediment safely.

Although the use of PCBs is now banned in the United States, considerable quantities of this toxin have accumulated in the environment and are still present in old electrical equipment, which is in use or discarded.

Being highly lipid-soluble, PCBs have a great bioaccumulation potential (*see* Table 10.3). Though their acute toxicity in animals is low (Aroclor 1254 was reported to have an oral LD_{50} in rats between 250 and 1300 mg kg^{-1}), chronic exposure is very harmful. PCBs have immunosuppressive activity, are tumor promoters, and interfere with calcium utilization; thus they can affect eggshell formation in birds (*8*, *46*). They are classified as carcinogens by both EPA and the International Agency for Research on Cancer. The induction of xenobiotic-metabolizing enzymes by PCBs was discussed in Chapter 3.

Dioxins

Since the late 1980s there has been concern about water pollution by dioxins associated with the paper-manufacturing industry. Dioxins comprise a group of 75 compounds with the same ring structure but varying degrees of chlorination. The most toxic compound in this group is 2,3,7,8-tetrachlorodibenzo-*p*-dioxin, referred to here as TCDD (for the structure *see* Chart 3.10 in Chapter 3).

Health and Ecological Effects

The LD_{50} of TCDD in rats is 0.022 and 0.045 mg kg^{-1} in males and females, respectively. It is teratogenic and carcinogenic in rodents. The toxicities established in humans include chloracne, porphyria,[4] liver damage, and polyneuropathies[5] (*8*). TCDD has been also implicated as a cause of soft-tissue sarcoma (*48*) and lung cancer (*49*) in humans.

A 10-year mortality study of a population exposed to large quantities of dioxins resulted from an explosion at the Givaudan plant, near Seveso, Italy, on July 10, 1976. This study showed elevated mortality from several types of cancer

---

[4]Porphyria is an abnormality of porphyrin metabolism characterized by urinary excretion of large quantities of porphyrins and by extreme sensitivity of the afflicted subjects to light.

[5]Noninflammatory degenerative disease of nerves.

among the exposed people (*50*). In contrast, data published (*51*) by the Center for Disease Control in Atlanta, Georgia, indicated that no adverse health effects, other than chloracne and other skin diseases, were noted in the Seveso population. However, the time elapsed since this accident is too short to allow definite conclusions to be drawn concerning the possible carcinogenic effects of dioxins in humans.

TCDD is formed during the Kraft process of paper manufacturing, which includes the bleaching of pulp with chlorine. A survey conducted by Greenpeace at Crofton, Vancouver Island, noted (*52*) that the eggs of blue heron colonies in the vicinity of the Crofton paper mill have failed to hatch since about 1987. The implication was that this failure resulted from water pollution with TCDD by the Crofton mill (*52*).

Indeed, TCDD has been found not only in paper mill effluent and sludge, but also, albeit in trace amounts, in chlorine-bleached paper products such as coffee filters, toilet tissue, paper towels, paper plates, and writing paper. In response to public concern, the EPA, the American Paper Institute, and the National Council of the Paper Industry for Air and Stream Improvement initiated a study of 104 U.S. paper mills that use chlorine for bleaching. The conclusion was that the median amounts of dioxin discharged were 6 and 3.5 ppt (parts per trillion) in hardwood and softwood pulp, respectively, and 17 and 0.024 ppt in sludge and wastewater, respectively (*51*).

Considering dioxin toxicity, its tendency to settle with the sediment, and its bioaccumulation potential, concern about the possible environmental impact of even such small amounts in water was justified. Consequently, the EPA proposed 0.014 part per quadrillion as an ambient water quality standard for TCDD.

The tough standards for dioxin, and the clamor of the environmentalists to replace chlorine bleaching with an oxygen bleaching process, were disturbing not only to the paper industry but also to the chloralkali industry that produces chlorine. To alleviate the pressures on the industries, the Chlorine Institute sponsored a dioxin symposium in Banbury Center, Long Island, in October 1990. It is difficult to say whether a consensus was reached among the attending scientists. Some participants declared that dioxin is not as toxic for humans as originally thought and pressured the EPA to revise its standards. In view of this controversy and in consideration of new animal toxicity data, the EPA initiated an extensive study to reevaluate TCDD exposure standards (*53*). In September 1994 the EPA released its long-awaited report. According to this report a new picture of TCDD toxicity emerged. It appears that dioxins and related chemicals are human carcinogens. At doses much lower than those causing cancer, they may cause a wide range of toxic effects in humans: they disrupt normal functioning of the endocrine system, and in consequence affect reproductive function, damage the immune system, and lead to abnormal fetal development (*54*). Their mode of action seems to be related to their ability to interact with cellular receptors, especially estrogen receptors (*55*) (*see* the section **Xenoestrogens** in Chapter 5).

Occurrence and Exposure

Congeners of dioxins, furans, and PCBs differ in their toxicity, depending on the number and position of chlorine atoms. To account for differences in their biological activities, the EPA expresses a compound's mass in terms of its toxicity. The most toxic dioxin, namely TCDD, is used as a reference standard. Thus, the mass of dioxin congeners and related compounds is expressed in terms of TCDD equivalents (TEQs). For instance, if the mass of a particular dioxin congener is equal to that of TCDD, but its toxicity is one tenth of it, then its mass, expressed in TEQ units, will be only one-tenth of its actual weight. According to this system, air emissions of dioxins in the United States amount to 30 lb/yr, 95% being due to waste incineration (mostly municipal and medical waste) (*54*).

Airborne dioxins and furans settle on crops and on water. Those in water settle down with the sediment and hence enter fish via phytoplankton. Those that settle on crops accumulate in the fat of livestock via fodder. Because of their chemical stability and refractivity to biotransformation, they tend to be biomagnified in the food chain. Thus, the general public is exposed to dioxins primarily through consumption of fish, meat, and dairy products. It is estimated that Americans ingest daily, with food, on the average 111 picograms of TEQs. The average dioxin body burden of the U.S. population is 40 ppt. These levels may not be carcinogenic, but they may have adverse effects on the reproductive system. Of main concern is that being readily fat soluble, the milk content of dioxins may be much higher than the average body burden, and nursing infants may be exposed to highly toxic doses (*56*).

Presently there are three estimates of the allowable daily intake (ADI) for TCDD: the EPA's estimate stands at 0.0000064 ng/kg/day, the Food and Drug Administration's at 0.0000572 ng/kg/day, and the Centers for Disease Control's at 0.0000276 ng/kg/day. In other nations the standards are more relaxed; in Germany it is 0.001, in the Netherlands 0.004, and in Canada 0.010 ng/kg/day. The WHO also uses 0.010 ng/kg/day as a safe dose (*53*). The TCDD potential for induction of xenobiotic-metabolizing enzymes (discussed in Chapter 3) may not be related to its carcinogenic and immunosuppressive activities. Nevertheless, the induction of P-450 may be of significance in case of simultaneous exposure to precarcinogens that require metabolic activation.

The Great Lakes

The Great Lakes, which contain 95% of the surface freshwater of the United States and 20% worldwide, constitute a vast economic resource. Because of their enormity, it was thought for decades that they were immune to pollution. In fact, owing to their slow water-replacement rate, the lakes, especially the upper ones, are very sensitive to pollution. The overall annual water outflow from the lakes is less than 1% of their total volume. The flushing times of the individual lakes in years are (*15*):

- Superior, 182;
- Michigan, 10;
- Huron, 21;
- Erie, 2.7; and
- Ontario, 6.

During the 1960s the quality of water in the Great Lakes, especially in the lower lakes, was visibly deteriorating. The most apparent causes were nutrients and phosphate loads from untreated or insufficiently treated sewage and farm runoffs seeping directly, or carried by the tributaries, into the lakes. A natural phenomenon also contributed to this demise; alewives, originally a saltwater fish, which adapted to the freshwater of the lakes, had increased in numbers beyond the carrying capacity of the lakes. As result the fish were dying in large numbers and the dead fish that washed ashore contaminated the beaches. By 1970 Lake Erie had lost or experienced a great reduction of several commercially valuable species of fish; the areas near the shores were covered by algae.

The waters were revived through enforcement of a 1972 U.S.–Canadian agreement that restricted discharge of nutrients and BOD effluents into the Great Lakes.

Toxic Pollution

However, toxic pollution proceeded unnoticed until it was discovered that the fish in many areas had been contaminated with toxic chemicals (such as PCBs and heavy metals) and pesticides (such as mirex).

In response to these findings, a new agreement was signed between the United States and Canada in 1978 and amended in 1987. This document established a new goal: to stop the discharge of toxic chemicals into the lakes. According to the agreement, about 350 hazardous substances should be banned from the lakes. The most critical pollutants were identified as PCBs, polyaromatic hydrocarbons (PAHs), TCDD, tetrachlorodibenzofuran (TCDF), and four pesticides: mirex, DDT, dieldrin, and toxaphene (*15*).

The data on PCBs and dieldrin levels in herring gull eggs from the Great Lakes colonies, shown in Figure 10.2, reflect the changes in the lakes' chemical burden between 1978 and 1986. Although the initial progress in lessening the burden of toxins was considerable, the further advance eventually came to a standstill. This prompted the EPA and Environment Canada to sign amendments to the Great Lakes agreement in 1987, this time focusing attention on the new technologies to be applied in pollution prevention and on the stricter accountability of all parties involved.

Accumulation in Fish

According to a special report in *Chemical and Engineering News* of February 8, 1988 (*15*), "The current overall condition of the lakes is fair to excellent in regard to

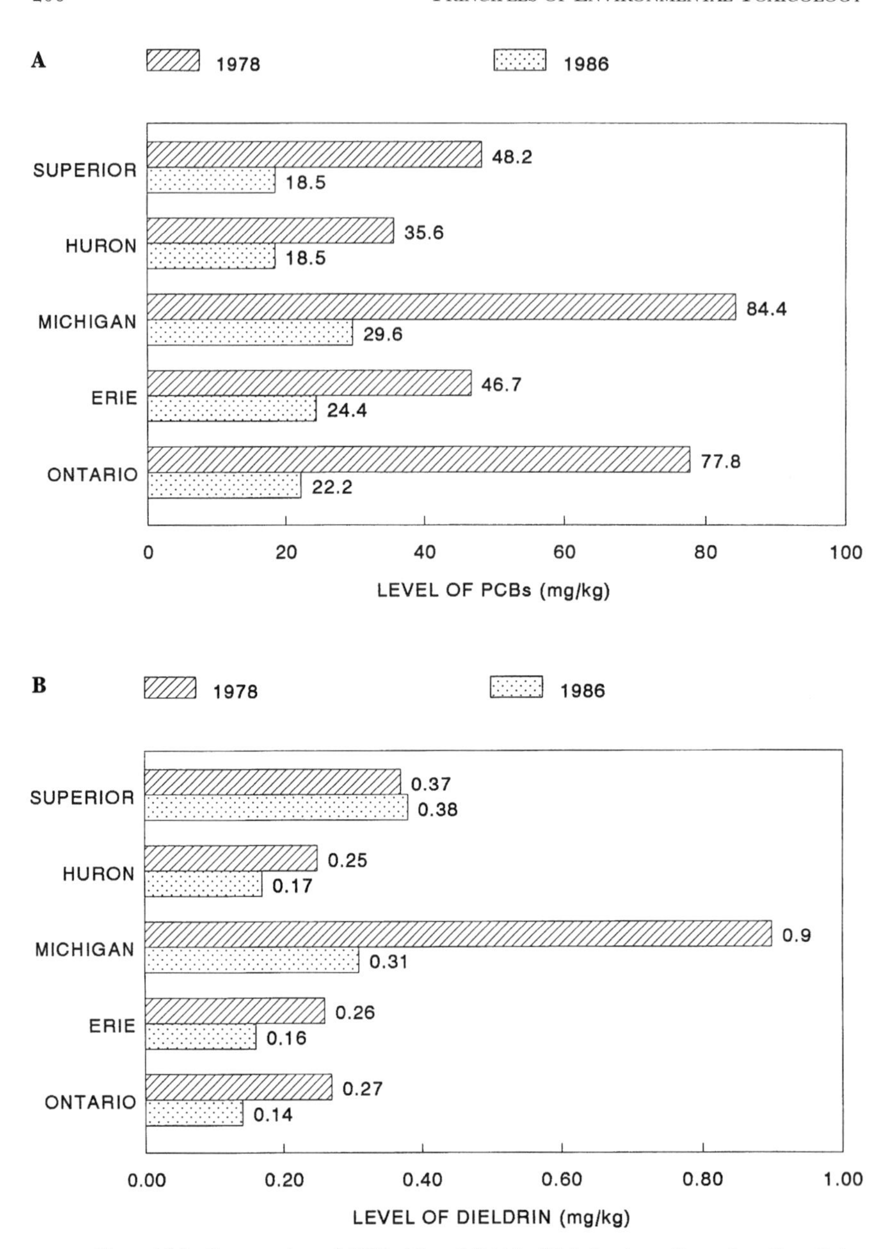

Figure 10.2. Concentrations of PCBs (A) and dieldrin (B) in herring gull eggs from Great Lakes colonies in 1978 and 1986. (Adapted from data in reference 15.)

phosphates, deteriorating in regard to nitrogen, and mixed in regard to toxins." Sediment and sedentary fish from different areas of Lake Ontario were analyzed for certain fluorinated aromatic compounds originating from an abandoned chemical waste dump in Hyde Park at Niagara Falls, New York. The study revealed rapid and uniform distribution of the tracer compounds throughout the lake in sediment and their accumulation in fish in sites remote from the point of origin (*57*).

Cases of skin and liver neoplasia affecting the fish population in several heavily contaminated areas of the Great Lakes have been linked to the presence of a heavy burden of aromatic hydrocarbons in contaminated bottom sediments. Although the causes of the neoplasms have not been fully determined, laboratory experiments with both fish and mice have shown that organic extracts of sediments from the affected waterways have definite carcinogenic potential. In addition, the carcinogenic potential of a number of PAHs present in the sediments has been fully demonstrated in several fish species (*58*).

Fish Consumption

At present a fish consumption advisory recommends that nursing mothers, pregnant women, women who anticipate bearing children, and children under 15 years of age should not eat lake trout over a certain size from Lakes Michigan, Superior, Huron, and Ontario. In addition, people of both sexes and of all ages are advised against eating very large fish from all five lakes.

The toxic pollutants in the Great Lakes also affect fish-eating birds and mammals. In herring gulls the toxicity is manifested principally in reproductive failures due to thinning of the eggshells, whereas in cormorants it is manifested in mutational changes such as crossed bill. Obviously, different chemicals are responsible for cancer in fish than for the toxicity in birds. PAHs, the principal fish carcinogens, are relatively easily metabolized and thus are not biomagnified in the food chain beyond the predatory fish. On the other hand, compounds such as PCBs, dioxins, and chlorinated pesticides, being refractory to biodegradation, persist further in the hierarchy of predatory species. Accumulation of chlorinated organic compounds has also been noted in mink and river otter from the Lake Ontario and Hudson River valley areas (*15*).

Cleaning up toxic chemicals from the Great Lakes is difficult because of the variety of pollution sources. Even if the discharges by the industrial and municipal point sources and the nonpoint urban and rural runoffs were reduced to zero, there remain the problems of atmospheric depositions and leachates from the hazardous dumps. The EPA and Environment Canada undertook a joint project to determine quantities of airborne toxins over the United States and Canada. They estimated that the deposition of airborne toxins is responsible for 90% of pollutants in Lake Superior, 63% in Lake Huron, 57% in Lake Michigan, 7% in Lake Erie, and 6% in Lake Ontario (*15*). The sources of the pollutants are numerous and frequently very distant from the points of deposition. Evaporation from contaminated sewage sludge deposited on land and from open lagoons of toxic waste, exhausts from municipal and toxic waste incinerators, and exhausts from coal-fired plants contribute to the pollution of the lakes. In addition, airborne pesticides, banned in the United States but still used in Latin America, are transported by the wind and deposited into the lakes.

The other problem is leaching of toxic compounds from the abandoned hazardous waste sites. In the past the chemical waste was deposited haphazardly, without regard to the geological structure of the land and the proximity of a lake or a river. There were no liners and no leachate collecting systems to prevent the leachate from flowing along the way of least resistance. The removal of hazardous waste from old dumps or at least confinement of the leachates is difficult, expensive, and time-consuming. The fact that along a 3-mile stretch of the Niagara River alone there are 164 abandoned chemical waste sites (*15*) exemplifies the scope of the problem. It appears that if the present slow rate of progress of the Superfund cleanup continues, the problem will persist for a considerable time to come.

The Great Lakes is just one example of despoiled waters in the United States. There are many other areas of concern, including Long Island Sound and those already discussed: Galveston Bay, the Chesapeake Bay, the lower Mississippi River, the upper Hudson River, and the Hudson–Raritan estuary.

Europe

Water pollution is also an acute problem in Europe. An example of extensive river pollution is the Rhine River in Germany. Although dissolved oxygen levels have increased considerably since the 1970s, salts, chemicals, metals, oils, pesticides, and thermal discharges from industry and power plants remain high. Despite some improvement, the chemical burden of the river is so high that dredged sediment from Dutch harbors is considered to be a hazardous waste (*59*).

However, with the awakening of cognizance among the European governments and the population at large, of the economic consequences of environmental despoilment, with the investment in new antipollution technologies, and with stationary or declining populations (as is now the case in some European countries), there are good chances for environmental improvement. The birth and relative popularity of the Green Party in the European Community shows societal concern with the environment.

The situation in Eastern Europe is much worse than that in Western Europe. For example, half of the Polish communities that line the Vistula River, including Warsaw, discharge inadequately treated sewage into the river. In addition, industrial discharges have made the water in many sections of the Vistula unsuitable even for industrial use; it may corrode the plants' machinery (*60*).

The soil in Silesia, which is the center of Polish heavy industry, is polluted with heavy metals (zinc, mercury, cadmium, and lead) to such an extent that five villages have to be relocated and the government is considering a ban on agriculture in certain areas (*61*). This environmental deterioration is not limited to Poland. It occurs in most of the Eastern European countries. The worst contaminated area is the coal-rich industrial zone comprising the southwestern part of East Germany, the western Czech Republic (Bohemia), and the southern part of Poland.

The environmental degradation took its toll on people's health. In the highly polluted regions of the Czech and Slovak Republics, the frequency of respiratory diseases among preschoolers and school-age children was five times and three times higher, respectively, than it was among the children from the less-polluted western region (*62*). Similarly, it has been noticed in Poland that the rates of chronic bronchitis were three times higher, and asthma four times higher, among army recruits from the areas heavily polluted by sulfur dioxide than among recruits from the unpolluted areas (*62*). Life expectancy at birth during the period 1985–1990 was 5% lower in Eastern than in Western Europe. On the other hand, infant mortality was nearly twice as high in the eastern countries (Poland, Czechoslovakia, and Hungary) than in West Germany (*58*). The United States is currently contributing funds and technical assistance to clean up water and to retrofit industrial plants with modern clean-coal technologies.

Heat Pollution

Power plants, conventional as well as nuclear, and the steel industry use large amounts of water for cooling purposes. The released water carries heat from the plants into rivers or lakes, and this heat increases the ambient water temperature in the vicinity of the release point.

The elevated temperature stimulates the metabolism of aquatic organisms, which in turn increases the demand for oxygen. At the same time, the amount of dissolved oxygen decreases with increasing temperature. Thus, the effect of heat pollution is similar to that of BOD contaminants or nutrients.

Some aquatic species have difficulty adapting to the warmer environment. Other species adapt to the warmer water and congregate around discharge points in winter. If the plants are shut down temporarily, massive fish kills from temperature shock result.

Pollution of Groundwater

Groundwater is an important natural resource. In the United States the use of groundwater increased from 34 billion in 1950 to 88 billion gallons per day in 1980. Of the latter amount, 54 billion gallons was used for irrigation. The rest was used for industrial purposes and as drinking water. About half of the U.S. population depends on groundwater for drinking. Thus, preservation of clean groundwater is of utmost importance.

Although there are numerous sources of contaminants, they are all related to three potential roots:

1. water-soluble products that are stored or spread on the land surface,
2. substances that are deposited or stored in the ground above the water table, and
3. material that is stored, disposed of, or extracted from below the water table.

Agricultural pollutants and waste disposed on land belong to the first category; waste disposed in landfills, leaking septic tanks, and leaking underground storage tanks, to the second one; and waste disposed in deep wells and waste originating from mining activities to the third.

Essentially, all chemicals that contact the ground, such as fertilizers and pesticides spread on the fields, especially if they are water soluble, present a potential hazard of groundwater contamination. Extensive contamination of groundwater is also caused by animal feedlots. Although the feedlots occupy relatively small areas, they provide an enormous amount of waste that leaches nitrates, phosphates, ammonia, chlorides, and bacteria into groundwater. Certain irrigation practices, such as use of automatic fertilizer feeders that are attached to irrigation sprinkler systems, may also contribute to groundwater contamination. This is because when the irrigation pump is shut off, water flows back into the well, siphoning the fertilizer from the feeder into the well (*63*).

Waste Disposal Sites

In addition to agriculture, waste dumps are also major pollutants of groundwater. The Resource Conservation and Recovery Act (RCRA) of 1976 prescribes structural features to prevent leaching of chemicals from toxic waste disposal sites. However, according to a 1985 EPA accounting, there are 19,000 old abandoned hazardous waste dumps (*64*).

These dumps, established before regulation of disposal sites was enacted, are frequently located on sites with little commercial value, such as marshes and old gravel or strip mining pits. Such sites are most unsuitable for disposal, as they provide an easy conduit for leachate.

The long-term effectiveness of the plastic and clay liners used to confine the leachate in modern sanitary and toxic disposal sites is questionable. Evidence is accumulating that sooner or later tears will develop in plastic liners and cause oozing of leachate. New research (*65*) indicates that clay liners, although impervious to leachate, may be penetrated by chemicals through diffusion.

According to a new epidemiological study, toxic waste sites may also represent public health hazards other than those caused by the contamination of groundwater. Review of New York State Department of Health data revealed a link between an elevated risk of congenital malformations among the newborn and their mothers' residential proximity to a toxic waste landfill; children borne to women residing within a mile from hazardous waste sites had, on the average, a 20% higher frequency of congenital malformations than the controls whose mothers resided elsewhere (*66*). In addition, the magnitude of the risk could be correlated with the presence of a chemical leakage from the waste site (*66*).

Another source of groundwater contamination is underground storage tanks. Of the 1.4 million underground gasoline storage tanks, 70,000–100,000 are estimated to be leaking (*64*).

Deep wells, used by some industries as a relatively inexpensive and supposedly environmentally safe method of chemical waste disposal, also cause some concern. Although in this procedure liquid waste is injected through wells below the groundwater aquifer, chemicals have been observed leaking into groundwater through cracks in the rock.

Contamination by Leaching

Leaching of pesticides into groundwater cannot be ignored. In Long Island, New York, aldicarb, which was used to control potato pests, leached into a groundwater aquifer and contaminated a local source of drinking water (*8*).

The Office of Technology Assessment reported (*64*) in October 1984 that incidences of groundwater contamination have been found in every state and that a number of organic and inorganic chemicals were detected in various groundwater supplies. Many of these contaminants are known toxins and carcinogens. In addition, some microorganisms and radioactive contaminants were found.

The problem of groundwater contamination is magnified by the fact that groundwater flows extremely slowly (about 1–10 ft per day). Thus, in comparison to surface water, there is little mixing and dispersal of contaminants.

The link between the use of contaminated groundwater and any specific disease cannot easily be established. However, one case of such a correlation has been reported. In Woburn, Massachusetts, the groundwater aquifer supplying drinking water became contaminated with trichloro- and tetrachloroethylene. A statistically significant increase was reported (*67*) in childhood leukemia, birth defects, and pulmonary and urinary infections related to immunosuppression.

Airborne Water and Land Pollution

Airborne pollutants may be divided into three categories: pollutants that cause changes in acidity, nutrients, and toxins.

Acid Deposition

Sulfur dioxide from coal combustion is converted in the atmosphere to sulfuric acid (Chapter 8). The sulfuric acid is then driven by the wind and eventually comes down to the earth, either directly (dry precipitation) or with rain or snow (wet precipitation, also referred to as acid rain), many miles from its origin. It is estimated that about one-third of sulfur deposition in the Eastern states originates from sources 300 miles away, one third from sources 120–300 miles away, and the rest from sources within 120 miles (*68*). Table 10.6 shows EPA estimates of annual industrial emissions of SO_2 in the United States in 1983.

Nitrogen oxides also contribute to acid deposition in the form of nitric and nitrous acids. Although automobiles produce nearly half of the total NO_x emitted, their contribution to acid rain is less significant than that of stationary sources because their emissions occur at ground level and are not likely to be carried for long distances.

Effect on Freshwater

Acid deposition lowers the pH of lakes, rivers, and soil. Thirteen rivers in Nova Scotia, at least 1600 lakes in Ontario, and an unspecified number of watersheds in New England and in upstate New York are practically devoid of fish as a result of their high acidity (*68*).

For years the problem of acid rain had political overtones. The Reagan administration maintained that there was not enough evidence connecting SO_2 emission with environmental damage and insisted that more study was needed before any restrictions should be imposed. On the other hand, most of the scientific community on both sides of the border supported the Canadian position that the cause–effect relationship between SO_2 emission and the deterioration of the environment is a well-established fact.

The Freshwater Institute in Manitoba initiated a research project (*69*) whereby a small lake in Ontario was purposefully acidified over an 8-year period, from the lake's original pH of 6.8 down to 5.0. At pH 5.9 the population of a shrimp species decreased considerably, another species of crustacean disappeared, and fathead minnow stopped reproducing. At pH 5.4 all fish stopped reproducing (*69*).

Table 10.6. Estimates of 1983 Industrial SO_2 Emission in the United States

Source	*Amount Released (million tons)*	*Percent Contribution*
Electric utilities	13.9	67
Industrial boilers	4.1	20
Smelters	1.1	5
Other	1.7	8
Total	20.8	100

SOURCE: Adapted from data in reference 67.

Effect on Forests and Soil

Acid rain is implicated in the destruction of forests. Unpolluted rain generally has a pH of 5.6, but soil has the capacity to neutralize it. However, the buffering capacity of soil may be exceeded when too much acid precipitates. Although no definitive cause–effect relationship between acid precipitation and damage to forests has been established, there is enough evidence linking high soil acidity with damage to trees. Most forests' damage occurs in the areas downwind of concentrated sources of emission of sulfur dioxide and nitrogen oxides; the damage is greatest at high elevations where very acidic fog lingers around mountain tops.

Damage to trees has been noted in several areas in California (the San Bernardino National Forest, the Laguna Mountains, the Sierra Nevada, and the San Gabriel Mountains), in the eastern United States, and also in Germany, Poland, Czechoslovakia, and Scandinavia. Since ozone is a known plant toxin, it is difficult to distinguish between damage inflicted by ozone drifting from the cities and that caused by acid deposition; most likely both factors play a substantial role, either by directly damaging the trees or by predisposing them to natural blight, such as infections, root rot, insects, and fungi.

It is believed that acid rain leaches Ca, Mg, and K out of the soil. Although this leaching may cause the availability of these cations to increase temporarily, eventually they are washed out, and in the long run a nutrient deficiency may occur. In addition, sufficiently high concentrations of aluminum may be released from the minerals to be toxic to plants (*70*). Indeed, deficiencies of Ca, Mg, K, and possibly Na, as well as increased concentrations of soluble Al, Mn, Fe, and other toxic metals, have been demonstrated in acidic soil (*71*). The soluble aluminum in water is toxic to fish because it precipitates in the gills and inhibits respiration. It may also have human health effects, as aluminum has been implicated as playing a role in Alzheimer's disease.

In addition to damaging trees and aquatic life, acid rain damages galvanized structures, and marble edifices and monuments. The damage to galvanized structures is due to zinc being dissolved out of the surface of the structure. Because zinc is always contaminated with very toxic cadmium, the runoff from such structures adds to the toxic pollution of soil and water.

Airborne Nutrients

Airborne transport of nutrients has been shown to be a significant factor in pollution of the Chesapeake Bay (*see* the discussion earlier in this chapter). Similar situations of airborne nitrate and ammonia deposition were observed in other watersheds. Because NO_x emissions are expected to increase in the future, the problem of aerial transport and deposition of nitric acid deserves special attention.

Airborne Transport of Toxins

In addition to SO_2 and NO_x, other chemicals (some of them toxic) and numerous metals are carried aloft with the wind. They come down with rain or snow to pollute soil and water.

The best evidence for this type of airborne pollution was brought about by the 1982 discovery of an insecticide, toxaphene, in fish in a lake on Isle Royale in Lake Superior (*15*). Isle Royale is a national park kept in a wild and unspoiled state. There is no industry, no agriculture, and no human settlements. Thus, the only explanation for the chemical pollution of the park is airborne transport (*see* the section on the Great Lakes earlier in this chapter).

At present there are strong indications that most of the toxic materials found in Lake Superior come from the atmosphere. The same applies to Lakes Michigan and Huron. The sources of these contaminants may be hundreds of miles away. Some chemicals such as DDT, which have not been used in the United States for many years, are found in the Great Lakes. Because they are still used in Latin America, airborne transport is suspected.

Vaporization of organic chemicals (such as pesticides) from soil and water is affected by external factors as well as by the physicochemical properties of chemicals. The external factors are temperature, type of soil, water content of soil, and wind velocity over the evaporating surface. Low water solubility and high vapor pressure of a chemical favor vaporization. The degree of adsorption of a chemical to the soil and the ease of its desorption by water molecules also play important roles.

As described earlier in this chapter, the sources of other airborne pollutants may be numerous: municipal waste incinerators (the main source of lead, cadmium, and mercury), the open lagoon treatment of toxic waste by aeration, sewage sludge incineration or disposal on land, wood fireplaces, etc. (*15*).

Another global problem related to airborne transport of pollutants is contamination of oceans caused by incineration of toxins at sea. In 1969, the West German giant chemical corporations introduced the practice of burning their toxic waste in specially built incinerator ships. This practice was meant as a better alternative to direct dumping of toxins into the sea. The attraction of this technology was that there was no need for political maneuvering to overcome the objections of communities against toxic waste incinerators in their vicinities; after all, "fish do not vote" (*72*).

The airborne toxins resulting from incomplete combustion or formed during the process of combustion cause considerable damage to the marine life of the North Sea and the eastern Atlantic. Some sources suggest that the high seal mortality from a viral infection that occurred in 1988 in the North Sea may have been a result of water pollution with chemicals that affected the seals' immune systems.

In the United States the practice of burning hazardous waste in the Gulf of Mexico began with a permit from the EPA in 1974 and was continued occasionally until 1983. However, the practice was then discontinued because of public pressure (*73*).

With the lifting of the Iron Curtain, the West was allowed to take a good look at the environmental devastation of Eastern Europe caused by 45 years of uncontrolled pollution. The results of this total environmental neglect are horrifying: polluted air, impaired human health, dying lakes and rivers, destroyed forests, and despoiled soil. Perhaps this is a warning of what may happen if short-term economic gains are allowed to take precedence over protection of the environment. It may also be a practical lesson for those who claim that more research is needed to prove that acid precipitation damages lakes, rivers, and forests.

References

1. World Resources Institute, International Institute for Environment and Development in collaboration with U.N. Environment Programme. *World Resources 1988–89, Freshwater;* Basic Books: New York, 1988; Chapter 8, p 127.
2. World Resources Institute, International Institute for Environment and Development in collaboration with U.N. Environment Programme. *World Resources 1992–93, Freshwater;* Oxford University: New York, 1992; Chapter 11, p 159.
3. Florini, K. L.; Krumbhar, G. D., Jr.; Silbergeld, E. K. *Legacy of Lead: America's Continuing Epidemic of Childhood Lead Poisoning;* Environmental Defense Fund: Washington, DC, 1990.
4. Mielke, H. W. *Water Air Soil Pollut.* **1991,** *57–58,* 111.
5. Mielke, H. W. *Appl. Geochem.* **1993,** Suppl. Issue No. 2, 257.
6. Mielke, H. W. *Environ. Geochem. Health* **1994,** *16*(3/4), 123.
7. Postel, S.; Heise, L. In *State of the World 1988;* Brown, L. R., Ed.; W. W. Norton: New York, 1988; Chapter 5, p 83.
8. Menzer, R. E.; Nelson, J. O. In *Cassarett and Doull's Toxicology;* Klaassen, C. D.; Amdur, M. O.; Doull, J., Eds.; MacMillan: New York, 1986; Chapter 26, p 825.
9. World Resources Institute, International Institute for Environment and Development in collaboration with U.N. Environment Programme. *World Resources 1988–89, Food and Agriculture;* Basic Books: New York, 1988; Chapter 17, p 271.
10. World Resources Institute, International Institute for Environment and Development in collaboration with U.N. Environment Programme. *World Resources 1992–93, Forests and Rangelands;* Oxford University: New York, 1992; Chapter 8, p 111.
11. World Resources Institute, International Institute for Environment and Development in collaboration with U.N. Environment Programme. *World Resources 1992–93, Oceans and Coasts;* Oxford University: New York, 1992; Chapter 12, p 175.
12. Durning, A. T.; Brough, H. B. In *State of the World 1992;* Brown, L. R., Ed.; Worldwatch Institute: Washington, DC, 1992; Chapter 5, p 66.
13. World Resources Institute, International Institute for Environment and Development in collaboration with U.N. Environment Programme. *World Resources 1992–93, Food and Agriculture;* Oxford University: New York, 1992; Chapter 7, p 93.
14. Long, J. *Chem. Eng. News* March 7, 1988, p 7.
15. Hileman, B. *Chem. Eng. News* February 8, 1988, p 22.
16. Carson, R. *Silent Spring*; Houghton Mifflin: Boston, MA, 1962.
17. Murphy, S. D. In *Cassarett and Doull's Toxicology;* Klaassen, C. D.; Amdur, M. O.; Doull, J., Eds.; MacMillan: New York, 1986; Chapter 18, p 519.

18. Hanson, D. *Chem. Eng. News* June 24, 1991, p 19.
19. Hanson, D. *Chem. Eng. News* August 17, 1992, p 19.
20. Hanson, D. *Chem. Eng. News* May 2, 1994, p 6.
21. Hileman, B. *Chem. Eng. News* November 16, 1992, p 6.
22. Hileman, B. *Chem. Eng. News* January 31, 1994, p 19.
23. Hardell, L.; Erikson, M.; Lenner, P.; Lundgren, E. *Br. J. Cancer* **1981,** *43,* 169.
24. Blair, A.; Grauman, D. J.; Lubin, J. H.; Fraumeni, J. F., Jr. *J. Natl. Cancer Inst.* **1983,** *71*(1), 31.
25. Wolff, M. S.; Toniolo, P. G.; Lee, E. W.; Rivera, M.; Dubin, N. *J. Natl. Cancer Inst.* **1993,** *85,* 618.
26. Ames, B. N.; Magaw, R.; Swirsky Gold, L. *Science (Washington, D.C.)* **1987,** *236,* 271.
27. Hileman, B. *Chem. Eng. News* July 5, 1993, p 3.
28. Mayers, J. P.; Colborn, T. *Chem. Eng. News* January 7, 1991, p 40.
29. Hayes, H. M.; Tarone, R. E.; Cantor, K. P.; Jessen, C. R.; McCurnin, D. M.; Richardson, R. C. *J. Natl. Cancer Inst.* **1991,** *83,* 1226.
30. Davis, J. R.; Brownson, R. C.; Garcia, R.; Bentz, B. J.; Turner, A. *Arch. Environ. Contam. Toxicol.* **1993,** *24,* 87.
31. Gold, E.; Gordis, L.; Tonascia, J.; Szklo, M. *Am. J. Epidemiol.* **1979,** *109,* 309.
32. Hanson, D. *Chem. Eng. News* August 17, 1992, p 19.
33. Hileman, B. *Chem. Eng. News* March 5, 1990, p 26.
34. Schroeder, R. C. In *The Earth's Threatened Resources;* Editorial Research Reports; Gimlin, H., Ed.; Congressional Quarterly: Washington, DC, 1986; p 43.
35. *The Global Ecology Handbook;* Corson, W. H., Ed.; Beacon: Boston, MA, 1990; Chapter 8, p 135.
36. Findley, R. W.; Farber, D. A. *Environmental Law;* West Publishing: St. Paul, MN, 1988; Chapter 6/2, p 323.
37. Fisher, D.; Cesaro, J.; Mathew, T.; Oppenheimer, M. *Polluted Coastal Waters: The Role of Acid Rain;* Environmental Defense Fund: New York, 1988.
38. Hanson, D. *Chem. Eng. News* May 31, 1993, p 6.
39. Postel, S. In *State of the World 1987;* Brown, L. R., Ed.; W. W. Norton: New York, 1987; Chapter 9, p 157.
40. National Research Council. *Toxicity Testing;* National Academy Press: Washington, DC, 1984.
41. Potts, A. In *Cassarett and Doull's Toxicology;* Klaassen, C. D.; Amdur, M. O.; Doull, J., Eds.; MacMillan: New York, 1986; Chapter 17, p 478.
42. Waldbott, G. L. *Health Effects of Environmental Pollutants,* 2nd ed.; C. V. Mosby: St. Louis, MO, 1978; Chapter 12, p 141.
43. Slemr, F.; Langer, E. *Nature (London)* **1992,** *355,* 434.
44. Clark, S. L. *Lurking on the Bottom: Heavy Metals in the Hudson–Raritan Estuary;* Environmental Defense Fund: New York, 1990.
45. Armstrong, N. E. In *Introduction to Environmental Toxicology;* Guthrie, F. E.; Perry, J. J., Eds.; Elsevier Science: New York, 1980; Chapter 19, p 249.
46. Clesceri, L. S. In *Introduction to Environmental Toxicology;* Guthrie, F. E.; Perry, J. J., Eds.; Elsevier Science: New York, 1980; Chapter 17, p 227.
47. Zurer, P. *Chem. Eng. News* November 21, 1988, p 5.
48. Hardell, L.; Erikson, M. *Cancer (Philadelphia)* **1988,** *62,* 652.
49. Blair, A.; Grauman, D. J.; Lubin, J. H.; Fraumeni, J. F. *J. Natl. Cancer Inst.* **1983,** *71*(1), 31.
50. Bertazzi, P. A.; Zocchetti, C.; Pesatori, A. C.; Guercilena, S.; Sanarico, M.; Radice, L. *Am. J. Epidemiol.* **1989,** *129*(6), 1187.
51. Stinson, S. *Chem. Eng. News* July 24, 1989, p 28.
52. von Stackelberg, P. *Greenpeace* **1989,** *14*(2), 7.

53. Hanson, D. *Chem. Eng. News* August 12, 1991, p 7.
54. Hileman, B. *Chem. Eng. News* September 19, 1994, p 6.
55. Colborn, T.; von Saal, F. S.; Soto, A. M. *Environ. Health Perspect.* **1993,** *101*(5), 378.
56. Schmidt, K. F. *Science News* **1992,** *141,* 24.
57. Jaffe, R.; Hites, R. A. *Environ. Sci. Technol.* **1986,** *20,* 267.
58. Black, J. J. In *Toxic Contaminants in Large Lakes;* World Conference on Large Lakes, Mackinac '86; Schmidtke, N. W., Ed.; Lewis Publishers: Chelsea, MI, 1988; Vol. I, p 55.
59. Meadows, D. H.; Meadows, D. L.; Randers, J. *Beyond the Limits;* Chelsea Green: Post Mills, VT, 1992; pp 88 and 99.
60. Lasota, J. P. *Sciences (N.Y.)* **1987,** *July–August,* 23.
61. Ember, L. R. *Chem. Eng. News* April 16, 1990, p 7.
62. World Resources Institute, International Institute for Environment and Development in collaboration with U.N. Environment Programme. *World Resources 1992–93, Central Europe;* Oxford University: New York, 1992; Chapter 5, p 57.
63. *EPA Handbook, Ground Water;* U.S. Environmental Protection Agency, Center of Environmental Research Information: Cincinnati, OH, 1990; Vol. 1, Chapter 5, p 94.
64. Thompson, R. In *The Earth's Threatened Resources;* Editorial Research Reports; Gimlin, H., Ed.; Congressional Quarterly: Washington, DC, 1986; p 121.
65. Johnson, R. L.; Cherry, J. A.; Pankow, J. F. *Environ. Sci. Technol.* **1989,** *23,* 340.
66. Geschwind, S. A.; Stolwijk, J. A.; Bracken, M.; Fitzgerald, E.; Stark, A.; Olsen, C.; Melius, J. *Am. J. Epidemiol.* **1992,** *135*(11), 1197.
67. Bayers, V. S. *Cancer Immunol. Immunother.* **1988,** *27,* 77.
68. Thompson, R. In *The Earth's Threatened Resources;* Editorial Research Reports; Gimlin, H., Ed.; Congressional Quarterly: Washington, DC, 1986; p 1.
69. Schindler, D. W.; Mills, K. H.; Malley, D. F.; Findlay, D. L.; Shearer, J. A.; Davies, I.; Turner, M. A.; Linsey, G. A.; Cruikshank, D. R. *Science (Washington, D.C.)* **1985,** *288*(4706), 1395.
70. Hutchinson, T. C. In *Effects of Acid Precipitation on Terrestrial Ecosystems, NATO Conference on Acid Precipitation on Vegetation and Soils, Toronto, 1987;* Hutchinson, T. C.; Havas, M., Eds.; Plenum: New York, 1978; p 481.
71. Rorison, I. H. In *Effects of Acid Precipitation on Terrestrial Ecosystems, NATO Conference on Acid Precipitation on Vegetation and Soils, Toronto, 1987;* Hutchinson, T. C.; Havas, M., Eds.; Plenum: New York, 1978; p 283.
72. Asmus, P.; Johnston, R. *Greenpeace* **1988,** *13*(2), 6.
73. Thompson, R. In *The Earth's Threatened Resources;* Editorial Research Reports; Gimlin, H., Ed.; Congressional Quarterly: Washington, DC, 1986; p 101.

11 Pollution Control

Clean-Coal Technology

Coal is now used mainly as fuel for the production of electricity. About 30% of the total electricity in the United States and (on the average) worldwide is produced by coal-firing plants; in China this figure is 80% (*1*). Because of the ample supply of available coal, dependence on coal as an energy source will probably remain high for some time to come.

However, coal is the most polluting of all fuels; its main pollutants are sulfur dioxide and suspended particulate matter (SPM). Depending on its origin, coal contains between 1 and 2.5% or more sulfur. This sulfur comes in three forms: pyrite (FeS_2), organic bound sulfur, and a very small amount of sulfates (*2*). Upon combustion, about 15% of the total sulfur is retained in the ashes. The rest is emitted with flue gases, mostly as SO_2 but also, to a lesser extent, as SO_3. This mixture is frequently referred to as SO_x (*2*).

The three basic approaches to the control of SO_x emission are prepurification of coal before combustion, removal of sulfur during combustion, and purification of flue gases.

Prepurification

The first approach, referred to as a *benefication* process, is based on a difference in specific gravity between coal (sp gr = 1.2–1.5) and pyrite (sp gr = 5). Although the technical arrangements may vary, in essence the procedure involves floating the crushed coal in a liquid of specific gravity between that of pure coal and that of pyrite. Coal is removed from the surface while pyrite and other minerals settle to the bottom. Coal benefication can reduce sulfur content by about 40% (*2*).

Although gravity separation is presently the only procedure in use, research is in progress on microbial purification of coal. A research project conducted by the Institute of Gas Technology, with funding from the U.S. Department of Energy, is aimed at the development of genetically engineered bacteria capable of removing organic sulfur from coal. Inorganic sulfur can be removed by the naturally occurring bacteria *Thiobacillus ferrooxidans, Thiobacillus thiooxidans*, and *Sulfolobus acidocaldarius* (*3*).

Clean Combustion

Gasification Combined Cycle

The two procedures for the clean combustion of coal are the coal *gasification combined cycle* (GCC) and fluidized-bed combustion. GCC involves conversion of coal to methane by the following procedure (*4*).

Preheating coal to 500–800 °C removes the volatile components of coal in the form of methane. The remaining char is treated with steam at temperatures above 900 °C to produce water gas (a mixture of CO and H_2). The water gas is then converted to methane and carbon dioxide according to the following equations.

$$C + H_2O \rightarrow CO + H_2 \tag{11.1}$$

$$C + 2H_2 \rightarrow CH_4 \tag{11.2}$$

$$CO + H_2O \rightarrow H_2 + CO_2 \tag{11.3}$$

$$\text{Net: } 2C + 2H_2O \rightarrow CH_4 + CO_2 \tag{11.4}$$

Methane is burned directly to drive a turbine. The excess heat is recovered to produce steam, which drives a steam turbine. This dual action led to the name "combined cycle" (*5*).

The formation of methane from coal is a reductive process in which sulfur is also reduced to H_2S. Elemental sulfur may be recovered by reacting H_2S with SO_2.

$$SO_2 + 2H_2S \rightarrow 2H_2O + 3S \tag{11.5}$$

By 1991 only one 100-MW demonstration plant using the GCC process existed in the United States; it is operated by California Edison Company in the Mojave Desert. The plant seems to be very successful in removing sulfur and in meeting the toughest environmental standards (*6*).

Fluidized-Bed Combustion

In the *fluidized-bed combustion* procedure, pulverized coal is mixed with limestone ($CaCO_3$). This mixture is ignited and held in suspension by a stream of hot air from below. The heat produced and the velocity of air give the appearance of a boiling fluid to the mixture. Sulfur reacts with limestone to form $CaSO_4$.

An added advantage of this process is the extremely efficient heat transfer as the boiler tubes are immersed directly in the fluidized bed. This, in turn, allows the combustion temperature to remain relatively low (730–1010 °C, as compared to 1510–1815 °C for conventional units burning pulverized coal). Low combustion temperature reduces formation of NO_x.

As of 1986, there were more than 40 fluidized-bed combustion boilers in operation or under construction in the United States (*6*).

Purification of Flue Gases

Desulfurization

Desulfurization can be achieved by the use of scrubbers. The two types of scrubbers are nonregenerative and regenerative. The use of scrubbers increases the cost of electricity by about 20–30% and uses 5–15% of the plant energy output (*2*).

Nonregenerative Scrubbers

In nonregenerative scrubbers, flue gases are guided through a slurry of limestone; SO_x reacts with $CaCO_3$ to form $CaSO_3$ and $CaSO_4$. The drawback of the limestone scrubbers is that large amounts of sludge accumulate. This sludge has to be disposed of on land, usually in lagoons. Leaching from such lagoons creates the danger of groundwater contamination. In addition, occasional operating problems can put the scrubbers temporarily out of commission.

Regenerative Scrubbers

Regenerative scrubbers recycle the SO_2-trapping reagent and produce sulfur products of commercial value. The Wellman–Lord process uses sodium sulfite, which reacts with SO_2 to produce sodium bisulfite.

$$Na_2SO_3 + SO_2 + H_2O \rightarrow 2NaHSO_3 \qquad (11.6)$$

The reaction is reversed by treating the sodium bisulfite with steam in the presence of alkali to produce sodium sulfite and SO_2. This sulfur dioxide may be converted to elemental sulfur, liquid SO_2, or sulfuric acid (*2*).

Suspended Particulate Matter (SPM)

Another concern is removal of particulate matter from flue gases. The particles emitted in the process of coal combustion are fly ash, soot, and smoke. *Fly ash* consists mostly of mineral matter contained in the coal and altered by high temperatures, whereas *soot* consists of fine, unburned carbon particles. In practice, depending on the completeness of combustion, fly ash may contain varying amounts of admixed soot particles. *Smoke* is a mixture of soot and condensed tar vapors. Because smoke results from incomplete combustion, the combustion technique is an important factor in eliminating smoke. The use of pulverized coal and a thorough mixing of fuel with an excess of air, as in modern boilers, eliminates smoke and soot (*2*).

The behavior of fly ash depends on the size of the particles. Large particles precipitate on impact with each other and any obstruction encountered, whereas small ones are propelled by the gases. The very small particles, on the order of magnitude of molecules, behave like gas particles (i.e., they move like molecules and frequently collide). The removal of small and very small (less than 1 mm in diameter) particles is important, as they are the most damaging to human health.

Particle-Removal Techniques

The four techniques for removal of particles from flue gases are filtration, centrifugal separation, use of wet collectors, and electrostatic precipitation.

Filtration involves either bags, mats, or columns. The efficiency of these devices for all sizes of particles is about 99%. However, they become partially plugged with time and require progressively increased gas pressures, which consume energy. They are also sensitive to corrosion and high temperatures.

Centrifugal separators are inexpensive and highly efficient. The gas enters a conical vessel at the top and is forced into a rotating motion. Particles are thrown by centrifugal force against the walls and slide down into a collecting compartment.

Wet collectors involve a variety of arrangements whereby the gas passes through a water spray and the particles are washed out. Although wet collectors are very efficient, especially for the removal of small particles, they produce a large amount of sludge and they lower the flue gas temperature.

Electrostatic precipitators are very efficient for the removal of 0.05- to 200-μm particles. They have a low operating cost but are expensive to install. Their operation is based on the passage of the gas through an electric field, whereby the particles become charged and migrate to collecting electrodes.

Reference 4 provides a more complete description of technical arrangements to control emission of particulates.

The efficacy of removal of air pollutants by different clean-coal technology systems is compared in Table 11.1. The advanced technologies, GCC and fluidized-bed combustion, compare favorably with flue purification systems. In ad-

Table 11.1. Efficacy of Clean-Coal Technology Systems for Removal of Air Pollutants

System	*SO_2 Removal (%)*	*Emission (lb/10^6 BTU)*	
		NO_x	*Particulates*
Pulverized coal with flue gas purification	90–98	0.5–0.6	0.03
Fluidized bed	90–95	0.2	0.01
GCC	90–99	0.1–0.3	None

SOURCE: Adapted from data in reference 5.

dition, GCC and fluidized-bed combustion operate at a lower cost than flue gas purification systems. They eliminate problems such as sludge and solid waste buildup and malfunctions caused by filter clogging. However, neither of these clean-coal technologies helps to abate CO_2 emission.[1]

Control of Mobile-Source Emission

Control of pollution from mobile sources (i.e., cars, trucks, and buses) involves several aspects: exhaust emission, volatile organic compound (VOC) emission, and rubber and asbestos emission from tires and brakes, respectively.

A point of concern is also emissions of carbon dioxide from motor vehicles. Although the amount of carbon dioxide from combustion of gasoline per unit of heat produced is less than that from coal combustion, nevertheless 19 lb of carbon dioxide (corresponding to 5.3 lb of carbon) are released per gallon of gasoline consumed. Globally, motor vehicles contribute 14% to the total carbon dioxide released into the atmosphere; in the United States, motor vehicles are responsible for 25% of national carbon dioxide emissions (*7*).

Exhaust Emission

The main exhaust pollutants are carbon monoxide (CO), hydrocarbons (also referred to as VOCs), lead, and nitrogen oxides (NO_x). Both CO and hydrocarbons result from incomplete combustion of fuel. This problem can be remedied by the use of catalytic converters and by strict adjustment of combustion conditions. The pollutant most difficult to control is NO_x, because it originates mostly from

[1]Although all fossil fuels produce CO_2 on burning, the amount of this gas produced per unit of heat generated varies. Thus, compared to natural gas, oil produces 1.35 times and coal 1.8 times the amount of CO_2 per British thermal unit (Btu). Of all coal combustion processes, the fluidized-bed process probably produces the least amount of CO_2 per Btu because of its efficient heat transfer.

the combustion of nitrogen from the air and not from the fuel. NO_x control technology will be discussed later.

Control Systems

Catalytic converters, which consist of a platinum or platinum–palladium catalyst spread on an alumina substrate, promote oxidation of unburned hydrocarbons and of CO. Catalysts are sensitive to inactivation by lead. Thus, use of unleaded gasoline (0.05 g of Pb per gallon, as compared to 2 g of Pb per gallon in leaded gasoline) is essential for proper functioning of catalytic converters. As an added advantage, lead pollution is considerably curtailed.

Additional pollution control is achieved by a computer-controlled electronic system that monitors exhaust gas composition. The same system adjusts the fuel–air ratio and spark advance as needed to minimize pollution.

Alternate Fuels

Use of alternate, less-polluting fuels is presently under consideration in the United States. This change would both combat pollution and reduce dependence on imported oil. The following possibilities are considered:

1. replacement of gasoline with compressed or liquefied natural gas;
2. replacement of gasoline with alcohols, methanol, or ethanol;
3. use of oxygen-containing additives in gasoline (the additives, called oxygenates, effect more efficient combustion);
4. reformulation of gasoline to decrease evaporation of VOCs during refueling; and
5. a combination of some of the above (*8, 9*).

Natural gas is a clean-burning fuel and produces the least amount of carbon dioxide per energy unit of all fossil fuels. The drawbacks of its use for motor vehicle propulsion are that it requires a change of the motor vehicle fuel system and generation of a new fuel supply network. Presently, at the worldwide production rate of 70,770 petajoules/year, the known global reserves of natural gas are estimated to last for about 60 years. Should all automotive gasoline in North America (the United States and Canada) be replaced by natural gas, the production rate would have to increase by 21,482 petajoules/year (the present level of use of oil for motor vehicle fuel in North America) to a total of 92,252 petajoules/year. Under such circumstances, the world natural gas reserves would still last for about 45 years; however, the United States would have to import more than 90% of it. Moreover, these calculations do not take into consideration the expected growth of the motor vehicle fleet and the fact that other nations may have similar ideas.

Another potential fuel is methanol. Methanol is an efficient fuel, being used extensively in racing cars. Compared to gasoline, methanol combustion produces fewer VOCs and less carbon dioxide. However, these environmental benefits are offset by emissions of carcinogenic formaldehyde and increased emissions of nitrogen oxides (*10*). An important factor to consider is the feedstock for manufacturing methanol. Coal is not practicable because the process of preparation of methanol from coal is accompanied by emissions of carbon dioxide. Adding carbon dioxide emissions from methanol synthesis and methanol combustion, there would be an 80% increase over emissions from combustion of gasoline. The most economical feedstock for methanol production is natural gas; however, the economics and practicality of converting natural gas to methanol instead of using it directly as a fuel may be questioned. Other feedstocks to be considered are wood and agricultural waste.

Ethanol may be used as a fuel by itself or as a 10% blend with gasoline as *gasohol*. Pure ethanol and gasohol are cleaner-burning fuels than gasoline. However, there are drawbacks in using pure ethanol made of corn:

1. There is not enough corn in the world to satisfy the appetite of the United States for automotive fuel, let alone to provide food for the hungry world and fuel for the United States.
2. The energy balance is unfavorable, that is, it takes almost as much energy to cultivate the soil, harvest the grain, and distill the ethanol as is gained by combustion of ethanol (*11*).

Although the amount of carbon dioxide emitted by combustion of ethanol is balanced by the carbon dioxide assimilated by the growing corn, the overall carbon dioxide balance is unfavorable. Emissions of carbon dioxide associated with feedstock production, and product distillation (*12*), far outweigh the assimilation capacity of the growing corn. Other feedstock, such as plant material or municipal waste, may provide better options for ethanol production (*13*).

Oxygenates are gasoline fuels blended with oxygen-containing additives to provide for cleaner burning gasoline. The additive most frequently used is methyl *tert*-butyl ether (MTBE); another one is ethanol. The disadvantage of ethanol is that it increases volatility of the gasoline, thus augmenting emissions of VOCs during refueling; this defeats the purpose of reformulated gasoline. The Clean Air Amendments of 1990 require that large cities that are unable to meet national ambient air quality standards (NAAQS) ozone limits must use gasoline reformulated to low-volatility standards. This requirement makes the value of ethanol as an additive to reformulated gasoline highly problematic (*14*).[2]

An Alternative Fuel Council has been established as an advisory body to the Department of Energy. This organization is scrutinizing arguments for and

---

[2]The purpose of reformulating gasoline is to decrease emissions of VOCs, NO_x, and toxic compounds, such as carcinogenic benzene and other aromatics. The addition of oxygenates is meant to reduce emission of CO.

against alternative motor fuels, taking into consideration not only fossil fuels but also fuels of the future such as hydrogen and electricity.

Volatile Organic Compounds

The three sources of VOC emission other than exhaust emission are fuel evaporation through the fuel tank and carburetor vents, escape of crankcase gases, and fuel evaporation while refueling.

Controls in the Vehicle

Fuel evaporation through tank and carburetor vents is controlled by connecting the carburetor and the fuel tank to an activated charcoal container. The charcoal traps the fuel vapors while the car is at rest and releases them into the induction system while the engine is running.

During the compression stroke of the engine, some gasoline vapors escape through the piston ring gaps into the crankcase, and hence through the breather tube into the atmosphere. To prevent this, the breather is connected to the intake manifold. Part of the air sucked into the air cleaner is used to purge the crankcase, thus sweeping the gases into the intake manifold. This system is known as positive crankcase ventilation (PCV).

Controls at the Gas Tank

When gasoline is pumped into a partially empty gas tank, the vapors contained in the tank are displaced and forced into the air. Approximately 1.27 billion pounds of VOCs escape annually into the atmosphere during refueling in the United States alone (*15*).

This level of emission can be reduced by the installation of a stage II vapor recovery system at the gas pump. The simplest installation, called a vapor balance system, consists of a rubber boot on the filler nozzle connected by a hose to the underground tank. When the boot tightly covers the car's filler neck, the displaced vapors are forced into the underground tank.

A modification of the vapor balance system is a vacuum-assisted system, in which there is no need for airtight contact between the boot and the filler neck because a pump-generated vacuum pulls vapors from the vehicle tank into the underground tank. With this arrangement a larger volume of air and vapors is drawn into the underground tank than the volume of fuel delivered to the car. Thus the underground tank must be vented, and venting requires installation of an additional vapor-trapping device at the vent.

A hybrid system uses only a slight vacuum, created by the Venturi effect of a gasoline sidestream. The vacuum is not as strong as that created by a pump in the vacuum-assisted system. Because little excess air is drawn into the underground tank, a balance is maintained between volume of fuel delivered and vapor displaced. The use of stage II vapor recovery has not yet been implemented in most states.

Control of Nitrogen Oxides

Abatement of NO_x emission is difficult to achieve because NO_x originates from the air. Thus it is a byproduct of any combustion process, regardless of the fuel used. Both stationary and mobile sources contribute to NO_x pollution, but their environmental impact differs somewhat. Mobile sources are responsible primarily for urban smog, whereas stationary sources contribute primarily to acid and nutrient precipitation. Successful abatement of photochemical smog formation depends more on the control of NO_x emission than on that of hydrocarbons, because hydrocarbons of natural origin are abundant in the ambient air, and thus the anthropogenic contribution is of less importance (*16*).

Combustion Conditions

Control of NO_x emissions from stationary sources depends on proper adjustment of combustion conditions. NO_x is formed in appreciable amounts only at temperatures above 1400–1500 °C, and it decomposes slowly with cooling. Therefore, control measures involve a low flame temperature and slow cooling of the flue gases. Decreasing excess air in combustion gases is also useful in reducing NO_x formation.

Practical methods of lowering the combustion temperature involve partial recirculation of flue gases back into the combustion chamber, addition of moisture to the combustion air in the form of steam or water spray, and a two-stage combustion process in which the fuel is burned initially with insufficient air. The resulting gases, CO and hydrocarbons, are then mixed with additional air; complete combustion is achieved in the second stage.

Control Systems

The conventional methods used for purification of flue gases will not remove NO_x. Wet scrubbers do not work because of the low water solubility of NO_x. However, a new method, referred to as selective catalytic reduction (SCR), which can reduce NO_x by 85 to 90%, has been developed and is now widely used in Europe and Japan. The process involves addition of ammonia to a nitrogen-oxide-containing exhaust stream and passage of this mixture over a catalyst. Three types of catalysts are in use: platinum, vanadium pentoxide on a titanium dioxide support, and zeolite catalyst (*17*).

Control of NO_x emissions from automobiles and trucks involves the use of three-way catalytic converters. In addition to a platinum–palladium catalyst that oxidizes CO and hydrocarbons, these converters contain a rhodium catalyst that reduces NO_x to N_2. As mentioned before, successful operation of catalytic converters requires meticulous control of combustion conditions; to ensure this control, proper maintenance is essential.

Energy Conservation

One potentially significant method of air pollution control is energy conservation. Improving automobile fuel efficiency by just 1 gallon per mile will save, in the United States, approximately 420,000 barrels of oil per day. This corresponds to about 3% of the daily consumption in 1988 (*18a*).

As indicated by data presented in Figure 11.1, a private automobile is the least energy efficient mode of urban transportation. More reliance on public transportation rather than on private automobiles (especially within the city), and strict enforcement of speed limits, may save a considerable amount of energy. The relationship between driving speed and fuel consumption is shown in Figure 11.2.

Energy conservation, in addition to its positive impact on the environment and human health by reducing air pollution, is also economically sound. Energy-efficient electrical appliances and lighting, and thermally insulated houses, help to conserve energy and thus to reduce pollution from stationary sources.

Both the use of energy and its production have an environmental impact.

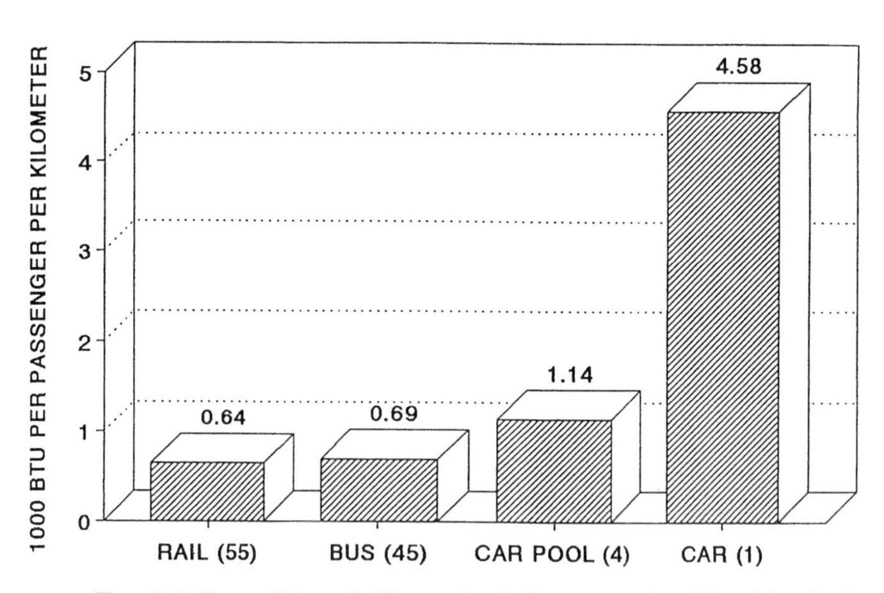

Figure 11.1. Energy efficiency of different modes of urban transportation. (Adapted from data in reference 18b.)

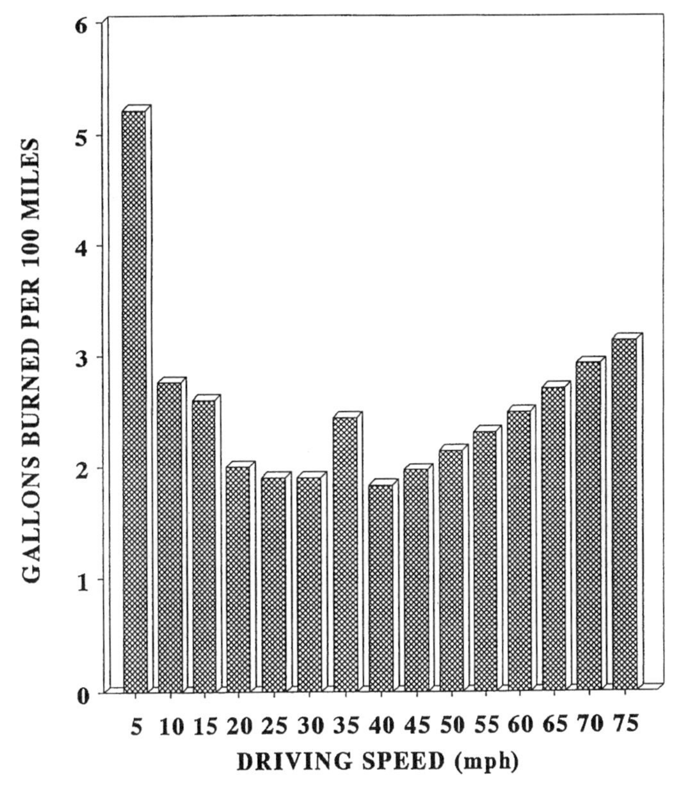

Figure 11.2. Driving speed and fuel consumption. (Data courtesy of Ford Motor Company, Dearborn, MI.)

Practices such as off-shore drilling for oil and transport of oil by ships create environmental hazards that can be reduced by energy conservation.

Energy conservation and reforestation of the earth are significant means of combating the greenhouse effect and the climatic changes associated with it.

Wastewater Treatment

Wastewater treatment is divided into four stages: primary, secondary, tertiary, and advanced. Because of the cost, not every plant includes all four stages. However, primary and secondary treatments are required by law for all communities in the United States.

Before entering the primary stage, sewage usually passes through grit chambers, where large nonputrescible solids (such as grit, stones, and pieces of lumber) are removed by sedimentation and screening through grills.

Primary treatment involves retention of sewage for 1–3 hours in settling tanks equipped with surface skimmers. The heavy solid particles settle to the bottom and those that float to the surface are skimmed off. This treatment removes 25–40% of the BOD contaminants. Primary treatment sludge is digested by anaerobic bacteria. The residue, which contains no (or very little) putrescible matter, is disposed of in a landfill. The water leaving primary treatment contains putrescible matter in solution or in colloidal suspension.

During secondary treatment, the putrescible materials are digested by aerobic microorganisms. To facilitate aerobic digestion the wastewater is aerated or oxygenated vigorously. This stage removes 85–99% of the BOD contaminants. The bacterial mass settles to the bottom and is collected as sludge. Secondary treatment sludge is rich in nutrients and formerly was dried and sold as a fertilizer. This practice has been discontinued in most areas because heavy metals, present at low concentrations in the sewage, are concentrated in the sludge and make its use as fertilizer hazardous.[3] Secondary treatment sludge was then disposed of in landfills. However, because of the shortage of landfill sites and increasing cost of solid waste disposal, the trend was again reversed. The sludge may now be sold for fertilizer if it conforms to the standards of the Environmental Protection Agency (EPA) with respect to the content of metals and polychlorinated biphenyls.

Tertiary treatment is designed to remove bacteria that remain suspended in the now-purified water leaving secondary treatment tanks. This removal may be accomplished by a combination of long-time retention in shallow oxidation ponds (where aeration is achieved by growing algae or by mechanical means), by filtration through sand, or by a combination of both methods.

The purpose of advanced treatment is to remove nutrients (such as phosphates, nitrates, and ammonia) and to remove salts and specific compounds that may be present in the wastewater of certain localities. Phosphates are best removed by precipitation with lime, whereas nitrates and ammonia may be converted to elemental nitrogen by anaerobic or aerobic microorganisms, respectively. Further processes, such as filtration (through activated charcoal or ion exchangers) and chlorination, are sometimes used. A schematic representation of a modern wastewater purification plant is depicted in Figure 11.3. **1,** At the pump station, large debris is removed from the sewage by bar screens, and hammer mills grind it to a size that can be handled by the 600-hp pumps. These pumps lift the sewage to a higher level and thus make possible a gravity flow through most of the remaining processes. **3,** At the grit chamber, heavier (mainly inert) solids such as sand and gravel settle to the bottom and are drawn off for disposal. **3,** The equalization tanks act as a "balancing reservoir". They accept excess flows and guarantee that only flows of a reasonably uniform volume and composition enter the downstream secondary process. **4,** In the oxygen-transfer

---

[3]Cadmium, a toxic metal and a carcinogen, is particularly dangerous because it is very easily assimilated from the soil by plants.

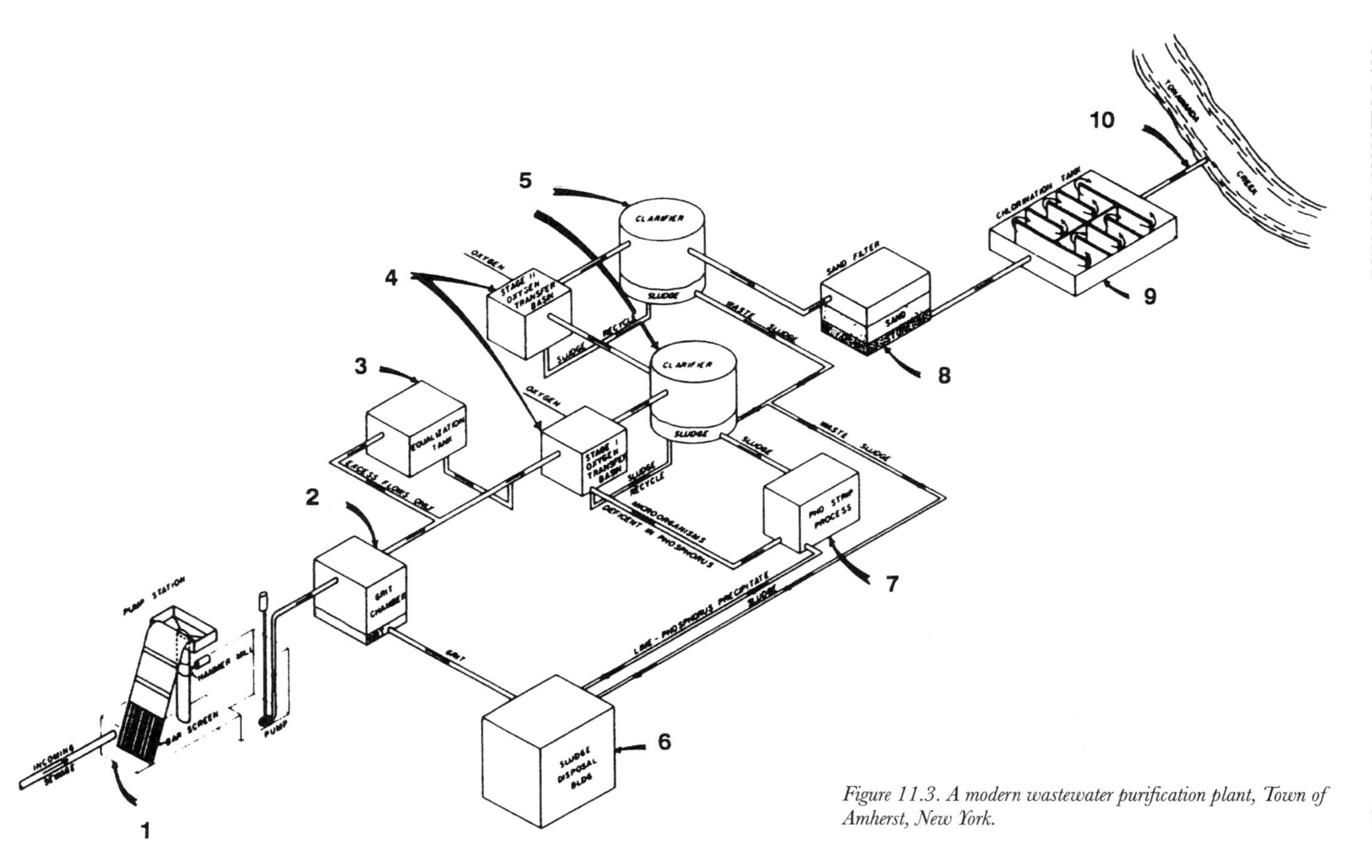

Figure 11.3. A modern wastewater purification plant, Town of Amherst, New York.

basins, microorganisms consume the organic matter and stabilize the nitrogen in the sewage. Oxygen, required for the respiration of the microorganisms, is supplied in the form of pure oxygen from the on-site cryogenic oxygen-supply system. **5,** Sludge, which consists mainly of microorganisms with some inert material, is slightly denser than water and therefore settles to the bottom of the clarifiers and is drawn off. Most of it is recycled to the oxygen-transfer basins to maintain the "biological" secondary process. Some is conveyed to the Pho-strip process, where phosphorus is removed. The excess is wasted to the sludge building for disposal. **6,** The grit, lime–phosphorus precipitate, and organic sludge, which are the end products of the treatment processes, are concentrated and incinerated in the sludge-disposal building. The inert ash will then be trucked to a landfill site for disposal. **7,** When oxygen is absent, microorganisms release phosphorus to the surrounding solution, and it can then be removed by precipitation with lime. The microorganisms, stripped of phosphorus, are returned to the secondary reactor. The lime–phosphorus precipitate is conveyed to the sludge building for disposal. **8,** The sand filter removes any remaining suspended matter. **9,** The chlorination tank, a mazelike structure, provides sufficient contact time to allow chlorine to kill any disease-causing organisms that may be present. **10,** The treated effluent, which is environmentally safe, is discharged into Tonawanda Creek. (Courtesy of Town of Amherst, New York, Water Pollution Control Facility.)

Waste Disposal and Recycling

Currently, many industrialized countries are facing garbage crises. In the United States, the amount of garbage rose from 87.5 million tons in 1960 to 157.7 million tons in 1986 (*19*), a 1.8-fold increase, whereas the population increased in the same time by a factor of only about 1.3. These figures indicate that the increase in the amount of waste results not only from the growth of the population: waste production per capita has also increased. By the year 2000, the amount of garbage produced in the United States is predicted to be 192.4 million tons per year (*19*). The composition of American trash is as follows (*20*):

- paper and paperboard, 36%;
- yard waste, 20%;
- food, 9%;
- metals, 9%;
- glass, 8%;
- plastics, 7%;
- wood and fabric, 6%;
- rubber and leather, 3%; and
- other inorganic substances, 2%.

In industrialized countries, 30% by weight and 50% by volume of the total trash is packaging material (*21*). This amount translates to 47.3 million tons of discarded packaging materials in the United States in 1986. Not only is the amount of packaging increasing; the material used for packaging has changed. Paper, glass, and metal are being replaced by plastics. The tendency for over-packaging is more pronounced in the United States than anywhere else. Nearly 10% of the money spent on food and beverages goes for packaging. The U.S. Department of Agriculture estimates that the amount of money spent on packaging food is more than farmers earn for producing this food (*21*).

Methods of Trash Disposal

Historically, landfill disposal was the most common method of disposing of trash. With the advent of the industrial revolution and with the associated growth of cities, municipal authorities adopted responsibility for collecting and disposing of trash. Originally it was thrown on heaps or deposited in pits. Presently, 90% of the refuse in the United States is disposed of in landfills.

Hazardous-Waste Landfills

Recognition of the danger of groundwater contamination by leachates from hazardous-waste landfills led to a federal law, designated as the Resource Conservation and Recovery Act (RCRA). This law required the operators of hazardous-waste dumps to provide double clay or plastic liners as well as a leachate-collecting system. Moreover, groundwater in the vicinity of hazardous-waste dumps must be monitored. However, the law did not impose any restrictions on municipal-waste landfills.

A recent survey of available data on 58 municipal and hazardous-waste disposal sites indicated that toxic chemicals were present in leachates from all sites considered in the study. Although the composition of chemicals varied depending on the type of site, their carcinogenic potential was similar in all cases (*22*).

Since 1991 the EPA guidelines for municipal solid-waste dumps have prescribed standards for location, design, operation, and closure (*23*). The design requires double clay or plastic liners, a leachate-collecting system, and monitoring of groundwater for 45 organic chemicals and 10 metals. Thus the newly designed municipal waste dumps do not differ substantially from the hazardous-waste dumps.

Figure 11.4 shows a schematic representation of a cross section of the bottom of a modern hazardous-waste dump. The collected leachate is disposed of through a municipal wastewater purification plant either directly or (as in the case of a hazardous-dump leachate) after preliminary biological and physical treatment. When it is retired from service, the landfill is capped with clay or plastic to prevent spilling of the leachate over the top.

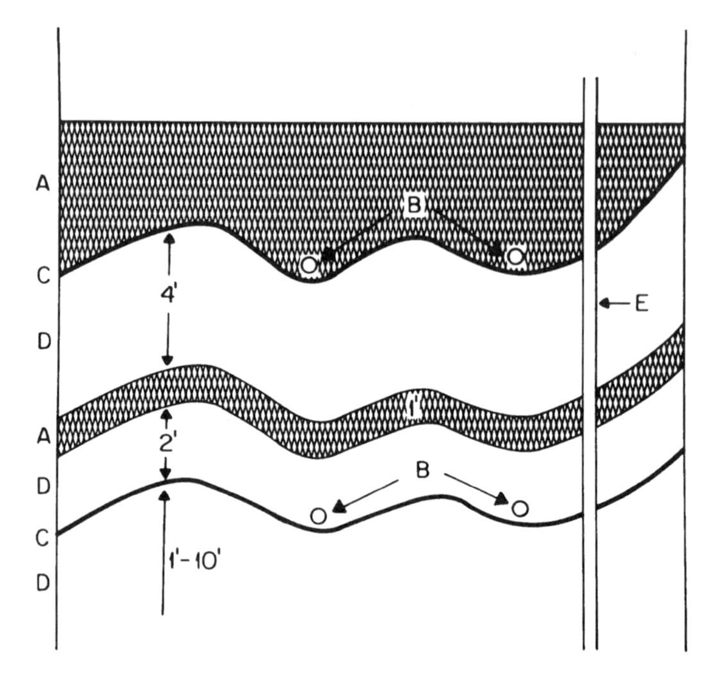

Figure 11.4. Cross section of the bottom of a modern sanitary landfill. A: Protective stone drainage layer. B: Leachate-collecting system, perforated pipes. C: High-density polyethylene liner (0.8 inch). D: Clay compacted to a permeability of 10^{-7} cm/s. E: Groundwater-monitoring well. (Courtesy of BFI waste management and Cecos waste disposal, Niagara Falls, New York.)

A 1987 study of clay liners was conducted for the EPA by a private company (*24*). The study revealed that the clay liners, even those conforming to EPA specifications (permeability no more than 10^{-7} cm s^{-1}) will, after 15 years, produce a steady leachate of 90 gallons per acre per day. The most recent study by an American–Canadian team (*25*) disclosed that organic chemicals can penetrate the clay liner by diffusion. This mechanism will allow passage of these chemicals through a 3-ft clay liner in 5 years.

Plastic liners, so-called flexible-membrane liners or FMLs, will develop leaks sooner or later because of the pressure of tons of garbage. They may contain pinholes formed during manufacturing or during gluing or welding together of the plastic sheets. Both systems need a way to prevent clogging of leachate-collecting pipes by silt, mud, slime buildup, or chemical precipitation.

According to some authorities, half of the cities in the United States will exhaust their landfills by 1990. Moreover, the cost of landfill disposal is increasing rapidly. For instance, in Minneapolis the cost of landfill disposal increased in 6 years from \$5 to \$30 per ton. Philadelphia, which ships its garbage to Ohio or Virginia for disposal, paid \$90 per ton in 1988 (*22*).

Incineration

Another method of trash disposal is incineration. Incineration does not dispense entirely with the need for landfills, but it reduces the volume of trash by 90% and its weight by 70%. Trash incineration has been in use since 1874. However, the old incinerators have been largely retired because of their inability to meet present air quality standards.

A new trend is controlled waste-to-energy incineration. The heating value of trash is about 1/3 that of coal. In addition, the flue gases generated are very low in SO_2. Some modern plants segregate the garbage by removing undesirable materials and separating iron for recycling. A 1978 federal law requires electric utilities to purchase, at a fair price, electricity generated by small producers. This potential profit is an additional incentive for municipalities to invest in waste-to-energy incinerators.

In 1986 there were 62 waste-to-energy plants in operation in the United States and 65 others under construction or in planning. The burning of trash is expensive, but at least part of the cost may be offset by the energy sold. Thus, it may be less expensive than landfill disposal. Frequently, waste-to-energy plants use the excess heat (in the form of hot steam) that remains after generating electricity to heat plants or residential dwellings. This is referred to as cogeneration. The term "cogeneration" applies whenever fuel is burned to produce electricity and the excess heat is used either for space heating or to provide mechanical power.

In principle, waste-to-energy incineration appears to be a good idea. However, the exhaust gases and residual ash present serious problems. Chlorine-containing compounds, such as polyvinyl plastics and bleached paper, form dioxins and furans on combustion. These toxins may be emitted into the air or retained in the ashes, depending on conditions such as temperature of combustion, cooling process, and adsorption to fly ash particles. Toxic metals such as lead, cadmium, arsenic, and mercury likewise may become air pollutants or a landfill hazard (*26*). Little is known about the chemistry of combustion, and high levels of dioxins detected in the milk of nursing mothers are attributed to the pollution caused by trash incinerators (*21*).

According to a newly released report by the EPA, incineration of medical wastes is responsible for more than half of estimated U.S. dioxin emissions, and incineration of municipal wastes for another 30%. In contrast, hazardous-waste incineration produces only 0.4% of dioxins (in terms of TEQ [tetrachlorodibenzo-*p*-dioxin equivalents]) (*27*).

Proponents of waste-to-energy incineration maintain that the emission of dioxins can be controlled by filtering the exhaust gases. Opponents argue that the toxins trapped on the filters have to be discarded somewhere and will end up in landfills. Another concern is that the filters do not retain very small particles (smaller than 2 μm). These particles are able to penetrate deeply into the lungs and thus present a health hazard. In addition, incinerator ashes (which are deposited in landfills) are spiked with toxic metals. Metals are more concentrated in ashes than in the trash being incinerated. Moreover, ashes are a conglomeration of small particles and as such have a large surface area. This large surface area facilitates the leaching of metals and other toxins.

New standards for municipal waste incinerators were proposed by the EPA. Accordingly, more-effective scrubbers are now required to replace the previously used spray dryers and electrostatic precipitators. The compliance threshold with the new regulations was lowered from 250 to 40 tons of wastes incinerated per day (*28*). The new rules took effect in September 1995.

Problems with Plastics

The 42 different polymers designated as plastics can be divided into two general classes: thermoplastics and thermosetting plastics. Thermoplastics, which constitute 87% of all plastics sold, are recyclable in principle (i.e., they can be melted down and remolded). Thermosetting plastics, on the other hand, once molded, cannot be remelted into the virgin resin.

Thermoplastics, which include polyethylene, polypropylene, polystyrene, poly(vinyl chloride) and poly(ethylene terephthalate) (PET), are used mostly in packaging. Plastic packaging, which accounts for 25% of the total use of plastics, represents the largest share of the market for plastics. The second largest use is building materials, with 20% of the market share (*29*).

Environmental Persistence

The major objection to plastic packaging is that it is neither bio- nor photodegradable, and thus will persist in landfills for centuries. Although plastics can be incinerated and have the highest heating value of all materials in the waste stream, some of them, such as poly(vinyl chloride), form toxic dioxanes and furans when burned. Separation of plastics for incineration into safe-burning and toxic is not economically feasible.

The argument against plastics based on their persistence in landfills for 400 years or so was weakened by research into the composition and biodegradability of waste deposited in landfills. It has been found (*30*) that the total landfill refuse retained its original weight, volume, and form even after being buried for 25 years.

However, this study was done on landfills in Arizona, where extremely dry conditions prevail. This environment made the survival of anaerobic bacteria, which are needed for the digestion of the waste, problematic. Under different climatic conditions, enough moisture is present to make bacterial fermentation possible. Indeed, subsequent investigation of Fresh Kills, the world largest landfill in Staten Island, NY, confirmed this point. Fresh Kills landfill covers an area more than 1200 hectares and consists of dry and wet areas. The decomposition of paper in the wet areas was found to be considerably faster than in the dry areas (*31*). Another study (*32*) indicates that the average volume of waste in landfills decreases by 7% per decade.

To overcome the antiplastic sentiment of citizen groups, some manufacturers have developed biodegradable plastics in which chains of polyethylene are linked by short segments of starch. The alleged advantage of this type of plastic is that the starch links are digested by bacteria, and this digestion breaks the integrity of plastic sheets and reduces their volume. Although this bacterial digestion may offer an advantage by saving marine species and waterfowl from suffocation by plastics discarded into water, it will leave powderized polyethylene in the environment. The consequences of this residue are still unknown.

The term "biodegradable" is rather misleading because no standards have been set with regard to the time span within which the degradation must occur. Considering that biodegradation in landfills is an extremely slow process, the environmental benefits of biodegradable plastics are questionable.

Plastics discarded into waterways represent a real hazard to aquatic species and waterfowl. Although the United States ratified the international convention that prohibits discharge of refuse from ships, this law is difficult to enforce.

Recycling of Plastics

Under pressure from environmental groups and from some local and state governments, the producers of plastics began to investigate recycling possibilities. Industries dedicated to sorting, cleaning, and shredding discarded plastic products were developed. Sorting of plastics by their chemical nature is now facilitated by the following numeric coding system:

1. poly(ethylene terephthalate) (PET),
2. high-density polyethylene (HDPE),
3. poly(vinyl chloride) (PVC),
4. low-density polyethylene (LDPE),
5. polyethylene (PE),
6. polystyrene (PS), and
7. composite plastics.

Presently, many communities include plastic packaging in their recycling program for glass, aluminum, and paper. Also, supermarkets place bins for collection of plastic bags for recycling. Despite this effort, the rate of plastic recycling is much below that of other materials. The plastics industry tries to make us believe that most of the plastic packaging is recycled. In fact, the rate of plastics recycling is now only 4.8 to 6.5% (up from 1% in the late 1980s), and the production of plastics from virgin materials is still outpacing recycling by almost 10 to 1 (*33*).

Recycling efforts are further complicated by the fact that many products are actually a composite of several resins. Some plastics are combined with other materials, such as paper or aluminum foil, which make them unsuitable for recycling. Some plastic products, such as HDPE milk jugs or PET soda bottles, are recycled by being shredded and used for fill in pillows and jackets, or as packing material. Although this approach is better than direct disposal after a single use, it can be considered only a postponement of the problem. Eventually these products, too, will end up in dumps. Manufacture of durable goods, such as plastic lumber or outdoor furniture, from discarded plastics may be a better idea.

Difficulties in recycling composite products may be solved by technological advances. For instance, a technology has been developed for recycling composite soda bottles. These bottles consist of four components: a PET body, an HDPE base, an aluminum cap, and a paper label. After the components are separated, PET and HDPE chips are sent for remelting and recycling into new plastic products (*34*).

A new recycling trend, called *feedstock* or *chemical recycling*, is now emerging in Europe, especially in Germany. The process involves depolymerization of plastics to original components from which it was synthesized. The technology of chemical depolymerization of individual types of plastics is well-developed and presents no technical problems. More difficult, although not impossible, is breaking down mixed plastics to basic oil feedstocks (*35*). Presently, a consortium of German chemical companies is starting chemical recycling of mixed plastics on a commercial scale. Mixed-plastics recycling dispenses with the necessity of sorting and cleaning the individual types of plastics, thus reducing the cost of recycling from $1,765 to $190 per ton of waste (*36*).

Recycling

Recycling as much as possible may be the best way to handle the garbage crisis. The advantages of recycling lie not only in diminishing the solid waste stream, but also in conservation of virgin resources such as trees and ores, conservation of energy, and reduction of air and water pollution. In the United States, consumption of raw materials has doubled in the last 35 years (*29*). With finite availability of resources, this rate of consumption is not sustainable for a prolonged period. Moreover, worldwide use of resources may be expected to rise as the developing nations strive to achieve living standards comparable to those of the industrialized nations.

Table 11.2 shows the environmental benefits of recycling, in terms of energy savings and pollution reduction, as compared to production of the same materials from virgin resources.

In the past the record of recycling in the United States was not impressive as compared to the record in some European countries. In 1987, 28% of aluminum, 27% of paper,[4] and 10% of glass were recycled in the United States.

---

[4]Although more paper products are now made from recycled paper, the designation "recycled" does not necessarily mean what the public expects. The paper made from the mill's waste, which would be otherwise discarded, is referred to as preconsumer recycled as opposed to postconsumer recycled, that is, made of paper that has been used.

Table 11.2. Percent Reduction of Energy Use and Pollution with Recycled Products

Product	*Energy Use*	*Air Pollution*	*Water Pollution*
Aluminum	90–97	95	97
Steel	47–74	85	76
Paper	23–74	74	35
Glass	4–32	20	—[a]

[a]The dash means "not reported".
SOURCE: Adapted from data in reference 22.

The corresponding figures in the Netherlands were 40, 46, and 53%; and in West Germany they were 34, 40, and 39% (*21*). This hesitation may have been due to a lack of serious environmental concerns by the former federal administrations and to their obsession with the idea that the government should not interfere with market forces. Thus, the extent of recycling depended solely on market demand for the recyclable materials. Lately, with some prodding by the federal and state governments, the recycling effort in the United States is gaining momentum. For municipalities, the extra benefit of recycling is that it is less expensive than dumping or incineration and in some cases may even be profitable.

Conflict of interest between the recycling and incineration industries sometimes interferes with progress. Private companies that contract to build and operate waste-to-energy incinerators require that communities obligate themselves to supply a steady stream of burnable waste. This obligation obviously reduces motivation for recycling.

Although recycling offers many advantages over dumping and incineration, it also takes its toll on the environment. For instance, removal of ink from newsprint releases a wide variety of hydrocarbons into the wastewater (*29*).

Last, but not least, is the problem of reducing the production of waste. In the past, not much effort has been expended toward reducing the waste stream. However, since the late 1980s the EPA has been moving forward, albeit slowly, to develop a policy of waste reduction (*20*). Certainly, much could be done in this area by reducing unnecessary and frequently redundant packaging.

In summary, the strategy to combat the garbage crisis and its associated environmental degradation should include the following steps:

1. reduction of the waste stream;
2. recycling of glass, metals, paper, and plastics;
3. composting of organic matter (yard and food waste);
4. incineration of the remainder; and
5. burying of the ashes.

Hazardous Waste

Superfund Projects

According to EPA estimates, 1163 hazardous-waste sites are on the priority list for urgent cleanup under Superfund legislation (Chapter 14). Another 30,000 remain to be evaluated; however, estimates by the General Accounting Office go as high as 130,000–425,000 sites (*37*).

Many hazardous-waste sites have been covered over, and subsequently housing developments, schools, or recreation areas were built on them. Such unidentified sites may be discovered only after health problems arise in those areas. This was the case with the Love Canal, where a housing development was erected on an abandoned chemical dump (*38*). Eventually the whole neighborhood had to be evacuated after toxic leachates began to seep into basements and an unusually high incidence of health problems was identified.

Since 1980 the EPA has begun cleanup at 257 sites; by 1989 cleanup was completed at 48 sites. In addition to this slow progress, there is criticism concerning the quality of the results. In many cases the cleanup procedure involved containment rather than detoxification or incineration of the toxic waste. In the short term, containment is a less expensive procedure than complete destruction or detoxification. However, in the long run it may turn out to be more expensive. As has been discussed, no clay or plastic liners will contain leachates permanently. Eventually another treatment of the contained sites will be necessary.

International Export of Hazardous Waste

With the increasing generation of toxic waste in the industrialized world, and with the increasing cost of its disposal, many industries found it profitable to ship their toxic waste to financially strapped developing countries. The amount of hazardous waste generated in the United States rose from 25 million tons/year in 1970 to 500 million tons/year in 1989. Another 40 million tons was generated annually by the other countries of the Organization for Economic Cooperation and Development (OECD) (*39*). At the same time the cost of disposal increased, between 1976 and 1991, from $10 to $250 per metric ton for disposal as landfill, and from $50 to $2600 per metric ton for incineration. In contrast, a metric ton of hazardous waste could be disposed of in developing countries for $5 to $50 (*39*).

The export of toxic waste to developing countries has been severely criticized by environmental groups on both ethical and environmental grounds. The feeling was that it is highly immoral to dump our toxic waste on impoverished people who lack the technical knowledge of how to handle the waste safely. In addition, the developing countries have enough of their own, difficult-to-solve, environmental problems to be burdened with the hazardous byproducts of our extravagant lifestyle. Greenpeace estimated the volume of hazardous waste shipped to developing countries between 1986 and 1988 at more than three million tons (*39*).

Cases of illegal dumping, or attempted dumping, aroused the international community against the practice of unregulated and uncontrolled trade in toxic waste that frequently exploited the poor for the profit of the rich. Accordingly, an international conference was convened, under the auspices of the United Nations Environment Program (UNEP), in Basel, Switzerland, in March 1989. Delegates from 116 countries drafted a treaty titled The Basel Convention on Control of Transboundary Movements of Hazardous Waste and Their Disposal. In essence, the postulates of the treaty were:

1. Establishment of notification procedures before the export of hazardous waste may be permitted.
2. A written consent of the importing country, and of the transit countries involved, must be obtained before the shipment can take place.

The treaty was hailed by UNEP executive director Mostafa Tolba as a significant advance toward sharp reduction of transboundary movement of toxic waste. Tolba said: "The ultimate goal is to make the movement of hazardous waste so costly and difficult that industry will find it more profitable to cut down on waste production, and reuse or recycle what waste they produce" (*40*). Greenpeace, on the other hand, disapproved of the treaty on the grounds that it gave, de facto, a seal of approval to the trade in hazardous waste that should be outlawed altogether. Ernst Klatte of Greenpeace put it this way: "This convention risks involving developing countries in solving the waste problem of industrialized countries" (*40*).

The United States participated in the Basel conference but so far has not ratified the treaty. However, a bill designed to curb the transboundary movement of hazardous waste is now under consideration by the U.S. Congress. In March 1994 the Clinton administration recommended that the Congress adopt the postulates of the Basel Convention.

Storage in Concrete Silos

An innovative concept for the cleanup of hazardous waste is excavation of the waste and storage in aboveground concrete silos. The waste can be safely stored in this way until technology for its detoxification or destruction is developed. This type of cleanup has been suggested but has not as yet been implemented (*41*).

References

1. World Resources Institute, International Institute for Environment and Development in collaboration with U.N. Environment Programme. *World Resources 1992–93, Energy;* Basic Books: New York, 1988; Chapter 7, p 109.

2. Schobert, H. H. *Coal: The Energy Source of the Past and Future;* American Chemical Society: Washington, DC, 1987.
3. Haggin, J. *Chem. Eng. News* August 29, 1988, p 36.
4. Matteson, M. J. In *Introduction to Environmental Toxicology;* Guthrie, F. E.; Perry, J. J., Eds.; Elsevier Science: New York, 1980; Chapter 31, p 404.
5. Braustein, L. *Electric Perspectives (New York)* Fall 1985, p 6.
6. Thompson, R. In *Earth's Threatened Resources;* Gimlin, H., Ed.; Congressional Quarterly: Washington, DC, 1986; Editorial Research Reports; p 1.
7. MacKenzie, J. J; Walsh, M. P. *Driving Forces: Motor Vehicle Trends and Their Implications for Global Warming, Energy Strategies and Transportation Planning;* World Resources Institute: New York, 1990.
8. Higgin, J. *Chem. Eng. News* August 14, 1989, p 25.
9. Renner, M. In *State of the World 1989;* Brown, L. R., Ed.; W. W. Norton: New York, 1989; Chapter 8, p 132.
10. Lenssen, N.; Young, J. E. *World Watch (Washington, D.C.)* **1990,** *3*(3), 19.
11. Chang, T. Y.; Hammerle, R. H.; Japar, S. M.; Salmeen, I. T. *Environ. Sci. Technol.* **1991,** *25*(7), 1190.
12. Fisher, D. C. *Reducing Greenhouse Gas Emissions with Alternate Transportation Fuels;* Environmental Defense Fund: Oakland, CA, 1991.
13. Lynd, R. L.; Cushman, J. H.; Nichols, R. J.; Wyman, C. E. *Science (Washington, D.C.)* **1991,** *251,* 1318.
14. Anderson, E. W. *Chem. Eng. News* November 2, 1992, p 7.
15. *Compliance of Selected Areas with Clean Air Act Requirements;* American Lung Association: New York, 1988.
16. Abelson, P. H. *Science (Washington, D.C.)* **1988,** *241*(4873), 1569.
17. Farrauto, R. J.; Heck, R. M.; Speronello, B. K. *Chem. Eng. News* September 7, 1992, p 34.
18. (a) World Resources Institute, International Institute for Environment and Development in collaboration with U.N. Environment Programme. *World Resources 1990–91, Energy;* Oxford University: New York, 1990; Chapter 9, p 141. (b) Lowe, M. D. In *State of the World 1991;* Brown, L. R., Ed.; W. W. Norton: New York, 1991; Chapter 4, p 56.
19. Church, G. J. *Time* September 5, 1988, p 81.
20. Ember, L. *Chem. Eng. News* July 3, 1989, p 23.
21. Pollock, C. In *State of the World 1987;* Brown, L. R., Ed.; W. W. Norton: New York, 1987; Chapter 6, p 101.
22. Brown, K. W.; Donnelly, K. C. *Hazard. Waste Hazard. Mater.* **1988,** *5*(1), 1.
23. Hanson, D. *Chem. Eng. News* September 16, 1991, p 21.
24. Geoservices, Inc. *Background Document on Bottom Liner Performance in Double-Lined Landfills and Surface Impoundments;* National Technical Information Service: Springfield, VA, 1987.
25. Johnson, R. L.; Cherry, J. A.; Pankow, J. F. *Environ. Sci. Technol.* **1989,** *23*(3), 340.
26. Hileman, B. *Chem. Eng. News* February 8, 1988, p 22.
27. Rose, J. *Environ. Science Technol.* **1994,** *28*(12), 512A.
28. Hanson, D. *Chem. Eng. News* September 12, 1994, p 7.
29. *The Economics of Recycling Municipal Waste: Background Analysis and Policy Approaches for the State and Local Governments;* staff report to New York State Legislative Commission on Solid Waste Management: Albany, NY, 1986; p 24.
30. Rathje, W. L.; Ho, E. E. *J. Am. Diet. Assoc.* **1987,** *87*(10), 1357.
31. Suflita, J. M.; Gerba, C. P.; Ham, R. K.; Palmisano, A. C.; Rathje, W. L.; Robinson, F. A. *Environ. Sci. Technol.* **1992,** *26*(8), 1486.
32. Thayer, A. M. *Chem. Eng. News* June 25, 1990, p 7.

33. Denison, R. E. *EDF Letter (Washington, D.C.)* **1993,** *XXIV*(6), 4.
34. Thayer, A. *Chem. Eng. News* January 30, 1989, p 7.
35. Layman, P. L. *Chem. Eng. News* October 4, 1993, p 11.
36. Layman, P. *Chem. Eng. News* March 28, 1994, p 19.
37. Young, J. E. *World Watch* **1989,** *2*(4), 8.
38. Office of Public Health. *Love Canal: Public Health Time Bomb;* Nailor, M. G.; Tarlton, F.; Cassidy, J. J., Eds.; New York State Department of Health: Albany, NY, 1978; special report to the governor and legislature.
39. Asante-Duah, D. K.; Saccomanno, F. F.; Shortreed, J. H. *Environ. Sci. Technol.* **1992,** *26*(9), 1684.
40. O'Sullivan, D. A. *Chem. Eng. News* April 3, 1989, p 21.
41. Walters, J. V.; Moffett, T. B.; Sellers, J. D.; Lovell, W. A. *Environ. Prog.* **1988,** *7*(4), 224.

12

Radioactive Pollution

Ionizing Radiation

Radiation that, on passage through matter, produces ions by knocking electrons out of their orbits is called *ionizing radiation.* This radiation is produced through decomposition of unstable, naturally occurring or synthetic elements referred to as radionuclides.

Types of Radiation

The four types of radiation are α-particles, β-particles, γ-rays, and neutrons. The α-particles have a mass of two protons and two neutrons and a charge of +2; β-particles are electrons with a mass of 0.00055 atomic mass unit (amu) and a charge of −1; γ-rays and X-rays are high-frequency electromagnetic waves with no mass and no charge. The difference between γ-rays and X-rays is that γ-rays occur naturally, whereas X-rays are generated. In addition, γ-rays are of higher frequency than X-rays.

Release of an α-particle leads to the formation of a daughter element with an atomic number 2 units lower and an atomic weight 4 units lower than that of the parent nuclide. Similarly, release of a β-particle from the nucleus causes conversion of a neutron to a proton, producing a daughter element with the same atomic weight as the parent nuclide but with its atomic number increased by 1 unit.

Neutron radiation does not occur naturally and is released only from synthetic radionuclides. Neutrons, which have no charge, are formed from protons. This conversion is accompanied by the release of an orbital electron from the atom. Neutrons produce ions indirectly, by collisions with hydrogen atoms. The impact knocks out protons, which in turn produce ions on passage through matter. Capture of a neutron forms an isotope of the parent nuclide with its atomic weight increased by 1 unit.

Mode of Action and Penetration

The mode of action of particles (α and β) varies from that of photons (γ- and X-rays). When α- or β-particles travel through matter, their electric charges (positive or negative) cause ionization of the atoms in the matter. This is called a direct effect. Whereas the track of α-particles is short and straight, β-particles scatter, frequently producing a wavy track. Gamma- and X-rays act indirectly.

There are three ways by which photons can cause ionization: the photoelectric effect, the Compton effect, and pair production. The *photoelectric effect* occurs when the photon striking an electron in the innermost shell (K shell) has energy equal to or slightly higher than that of the electron. The electron is then released from the atom; its energy is equal to that of the photon diminished by the K-shell binding energy. The *Compton effect* occurs when a photon strikes an electron in the L-shell (the next to the innermost shell) with energy much in excess of that of the electron. The electron is then knocked out, but only part of the photon energy is transferred to the electron. The remainder is reradiated as a photon of lower energy. *Pair production* occurs when a photon having energy greater than 1.02 MeV strikes the nucleus and disintegrates into an electron and a positron (positively charged electron). The positron loses energy by ionizing atoms of the matter. Eventually it collides with an electron and annihilates itself, producing two photons, each having an energy of 0.511 MeV and traveling in opposite directions (*1*).

The penetration of ionizing radiation through tissue depends on the type of radiation (i.e., its mass and charge) and also on its energy. The amount of damage to the tissue is related to the linear energy transfer. When a particle or a ray travels through matter, it gradually loses energy by transferring it to the matter.

The initial energy of the incoming radiation (E_{max}) divided by the thickness of the matter required to dissipate all the energy is referred to as the average linear energy transfer (LET). For equal doses of radiation, the damage to the irradiated tissue increases with an increasing LET value.

Both γ-rays and X-rays have no fixed penetration range; they attenuate exponentially with depth of penetration. Therefore, in this case the LET is expressed as $E_{max}/(2 \times \text{HVL})$; HVL is the half-value layer, the thickness of matter necessary to attenuate the intensity of radiation by half (*2*). Table 12.1 presents a comparison of different types of radiation, their LET values for radiation of 100 keV, and their penetrability of tissue.

In practical terms it means that α-radiation, although very damaging to tissue, does not penetrate a sheet of paper or the stratum corneum of human skin. β-radiation can easily go through one or two centimeters of living tissue. In contrast, γ-radiation and X-rays can be stopped only by a thick slab of lead or concrete.

Table 12.1. Characteristics of Varying Types of Ionizing Radiation

Type of Radiation	*Mass*	*Charge*	*LET (keV/μm)*	*Tissue Penetration (μm)*
β	1*e*	−1	0.42	180
α	2p+2n	+2	260	1
Proton	1p	+1	90	3
γ	0	0	1.2[a]	40,500 (HVL in H_2O)

[a] $E_{max}/(2 \times \text{HVL})$.
SOURCE: Based on data from reference 2.

Measurement of Radioactivity

The radiodecay of nuclides is a zero-order reaction. The rate of decomposition is independent of the concentration of the radionuclide, according to equation 12.1.

$$N = N_0 e^{-kt} \quad (12.1)$$

where N and N_0 are the concentration of the radionuclide at times 0 and t, respectively, and k is the decay constant, a characteristic value for each radionuclide. Accordingly, the half-life, $t_{1/2}$, equals $(\ln 2)/k$.

Two types of units are used to measure emitted and absorbed radioactivity. The traditional units are still used in the United States, although they are gradually being phased out. International units (SI) are in use elsewhere.

The traditional unit of emitted radioactivity is the curie (Ci). Originally, the curie was the amount of radioactivity emitted by 1 g of radium. This was later standardized to 2.2×10^{12} dpm (disintegrations per minute). The related SI unit, the becquerel (Bq), corresponds to 1 disintegration per second (dps).

The traditional unit of absorbed radioactivity is the rad, which is equal to 100 erg g^{-1} (2.38×10^{-6} cal g^{-1}). The SI unit, the gray (Gy), corresponds to 1 J kg^{-1}. The dose-equivalent unit, the rem, is the absorbed dose weighted for the destructive potential of a given type of radiation. This potential is related, for each type of radiation, to its LET value. By definition, 1 rem has the same biological effect as 1 rad of "hard" X-rays. However, it must be multiplied by the quality factor of 20 for α-radiation.[1]

---

[1]Energy units:
erg = dyn × cm = g cm^2 s^{-2}
joule (J) = 10^7 erg
calorie (cal) = 4.19 joule
electron volt (eV) = 1.6×10^{-12} erg

Prefixes:

milli- (m) = 10^{-3}	kilo- (k) = 10^3
micro- (μ) = 10^{-6}	mega- (M) = 10^6
nano- (n) = 10^{-9}	giga- (G) = 10^9
pico- (p) = 10^{-12}	tera- (T) = 10^{12}

The SI unit replacing the rem is the sievert (Sv). One sievert corresponds to 1 J kg^{-1} multiplied by a quality factor. An earlier unit of exposure, roentgen (R), is based on the amount of ionization produced in the air by γ-rays or X-rays. One roentgen is approximately equal to 1 rad. This unit has been replaced in SI by coulombs per kilogram (C kg^{-1}). The relationship between traditional and SI units is shown as follows:

Traditional Unit	SI Unit
1 Ci	37×10^9 Bq
27×10^{-12} Ci	1 Bq
1 rad	1×10^{-2} Gy
100 rad	1 Gy
1 R	285×10^{-6} C kg^{-1}
3876 R	1 C kg^{-1}

Sources of Radiation

The sources of radiation can be divided into natural and anthropogenic. Natural sources involve cosmic radiation and radioactive elements produced by three disintegration series originating from ^{238}U, ^{232}Th, and, to a lesser extent, ^{235}U (actinouranium, also known as actinium) (Figure 12.1).

Uranium is encountered in certain rocks, soil, and phosphate deposits. Radon, the gaseous decay product of ^{238}U and ^{232}Th, is of great concern. The two isotopes of radon (^{222}Rn and ^{220}Rn) are responsible for 54% of the earth's background radiation (*3*).

Radon is not equally distributed around the globe. The great majority of people live in areas where the outdoor radon exposure rate varies from 0.3 to 0.6 mSv per year. However, in certain areas of Brazil, India, and Iran the exposure is between 8 and 400 mSv per year (*4*).

The high occurrence of radon in those areas results from a soil rich in thorium. Elevated levels of radon have also been found in some areas of Florida because of the high ^{238}U content of phosphate deposits. Radon occurs in concentrations sufficiently high to create a health hazard in uranium mines and mine tailings. In the basements of some residential dwellings and office buildings, it can accumulate to concentrations greatly exceeding those of the outdoor background.

Radon, an α-emitter, is a noble gas. As such, it is very unreactive, and when inhaled it does not persist in the lungs long enough to cause any damage. However, it decomposes to its daughter elements, polonium isotopes 218 and 216, which originate from ^{238}U and ^{232}Th, respectively. These isotopes are solid α-emitters, with half-lives of 3 min and 0.16 s, respectively. They and their disintegration products may be trapped in the lungs and cause damage to the tissue.

Other natural sources of radioactivity are ^{40}K and ^{67}Rb. ^{40}K is a β- and γ-

URANIUM 238		URANIUM 235		THORIUM 232	
NUCLIDE	HALF-LIFE	NUCLIDE	HALF-LIFE	NUCLIDE	HALF-LIFE
$^{238}_{92}$U	4.47 x 10^9 y	$^{235}_{92}$U	7.1 x 10^8 y	$^{232}_{90}$Th	1.39 x 10^{10} y
α		α		α	
$^{234}_{90}$Th	24.1 d	$^{231}_{90}$Th	25.5 h	$^{228}_{88}$Ra	6.7 y
β		β		β	
$^{238}_{91}$Pa	1.17 min.	$^{231}_{91}$Pa	3.25 x 10^4 y	$^{228}_{89}$Ac	6.13 h
β		α		β	
$^{234}_{92}$U	245000 y	$^{227}_{89}$Ac	21.6 y	$^{228}_{90}$Th	1.9 y
α		β		α	
$^{230}_{90}$Th	8000 y	$^{227}_{90}$Th	18.2 d	$^{224}_{88}$Ra	3.64 d
α		α		α	
$^{226}_{88}$Ra	1600 y	$^{223}_{88}$Ra	11.43 d	$^{220}_{86}$Rn	54.5 sec.
α		α		α	
$^{222}_{86}$Rn	3.823 d	$^{219}_{86}$Rn	4.0 sec.	$^{216}_{84}$Po	0.16 sec.
α		α		α	
$^{218}_{84}$Po	3.05 min.	$^{215}_{84}$Po	1.78 x 10^{-3} sec.	$^{212}_{82}$Pb	10.6 h
α		α		β	
$^{214}_{82}$Pb	26.8 min.	$^{211}_{82}$Pb	36.1 min.	$^{212}_{83}$Ac	60.5 min.
β		β		β α	
$^{214}_{83}$Bi	19.7 min.	$^{211}_{83}$Bi	2.15 min.	$^{212}_{84}$Po $^{208}_{81}$Tl	3 x 10^{-7} sec., 3.1 min.
β				α β	
$^{214}_{84}$Po	0.000164 sec.	$^{207}_{81}$Tl	4.79 min.	$^{208}_{82}$Pb	stable
α		β			
$^{210}_{82}$Pb	22.3 y	$^{207}_{82}$Pb	stable		
β					
$^{210}_{83}$Bi	5.01 d				
β					
$^{210}_{84}$Po	138.4 d				
α					
$^{206}_{82}$Pb	stable				

NOTE: The type of particle released is shown below the nuclide.

Figure 12.1. Disintegration series of ^{238}U, ^{235}U (actinouranium), and thorium.

emitter with a half-life of 1.3×10^9 years. It occurs in rocks and soil, as well as in the muscles of animals, where it represents about 0.01% of the total potassium. ^{87}Rb, a β-emitter with a half-life of 4.89×10^{10} years, occurs in certain minerals, seawater, and waters of many mineral springs and salt lakes.

Anthropogenic sources of radioactivity are related to the nuclear power industry (mining, processing, reactors, and nuclear waste); nuclear warfare and testing; nuclear accidents; the use of radionuclides in science and medicine; and medical X-rays.

Health and Biological Effects of Radiation

Ionizing radiation is lethal, even though the amount of energy involved in killing an organism is negligible. Studies of the effects of the atomic bomb explosions in Hiroshima and Nagasaki indicate that individuals exposed to 450 rad (0.00107 cal/g) died within 2 weeks of exposure (*5*). However, at equal total dose, fractionated doses are less toxic than a single large dose.

Free Radicals

The biochemical effect of radiation is believed to result from the formation of free ·OH and ·H radicals arising from collisions of ionizing particles or induced ions with water molecules.

The free radicals react with cellular macromolecules, or with each other, to form H_2O_2, a strong oxidizing agent. Another type of free radical, $\cdot HO_2$, is formed by interaction of the ·H radical with cellular oxygen. This may then be reduced to H_2O_2.

The interaction of these free radicals and H_2O_2 with cellular macromolecules such as nucleic acids, proteins, lipids, and carbohydrates leads to a variety of damage: DNA strand breaks, point mutations, chromosomal aberrations, and ultimately to cell death. Some organs are more susceptible to radiation damage than others. In general, rapidly dividing cells are the most radiosensitive. Thus, when the whole body is exposed to radiation, the risk-weighting factors for individual organs have to be considered. This evaluation is referred to as the effective dose equivalent. The risk-weighting factors for different tissues (*4*) are as follows:

Tissue	*Weighting Factor*
Total body	1
Bone marrow	0.12
Bone surfaces	0.03
Thyroid	0.03
Breast	0.15
Lungs	0.12
Ovaries and testes	0.25
Remainder	0.30

Radiosensitivity

Species Variation

Radiosensitivity varies widely among species. For instance, the LD_{50} values for a 30-day exposure to X-rays in rats, rabbits, goats, and dogs are 796, 751, 237, and 244 rad, respectively (*6*). Whereas in mammals sublethal irradiation leads to a decline in longevity, in adult insects it induces an increase in life span. Because insects have less of a requirement for cell renewal than mammals do, this difference suggests that radiation is detrimental to proliferating cells only, whereas it may be beneficial to nonproliferating cells (*7*). Similarly, developing organisms are more radiosensitive than adult ones. For instance, fish embryos have an LD_{50} of 50 R, but adult fish may tolerate as much as 800–900 R (*8*).

In the human population, children and fetuses are particularly sensitive to radiation. Relatively small doses may cause mental retardation, stunted growth, deformities, and cancer.

Clinical Symptoms

The clinical symptoms of radiation sickness have been studied extensively in the survivors of the Hiroshima and Nagasaki explosions (*5*). Early manifestations of radiation illness are nausea and vomiting. The time of the onset of the symptoms is related to the exposure dose. For instance, at doses of 100–300 R the first symptom (epilation [loss of hair]) appears only 3 weeks after exposure, whereas at an exposure of 400–700 R nausea and vomiting occur after 1 week and other symptoms after 2 weeks (*9*).

Nausea is followed by epilation (loss of hair) and purpura (redness of the skin). Both onset of epilation and intensity of purpura may be correlated with the intensity of the exposure. Other manifestations, such as diarrhea and hemorrhages of the mouth, rectum, and urinary tract, are typical symptoms of damage to the hematopoietic and gastrointestinal systems. At a very heavy exposure, death may occur shortly after exposure. At some lower exposure, the early symptoms are followed by a latent period and a secondary phase of illness during which death may occur.

Chronic Exposure

We have a wealth of information on the health effects of high doses of radiation. However, very little is known about the effect of chronic exposure to small doses such as may occur at the workplace or to which the general public may be exposed.

Most of the information in these areas originates from studies of clinical exposure to X-rays, occupational exposure, and animal experiments by extrapolating from high to low doses, as was described for chemical carcinogens in Chapter 6. The extrapolations are usually based on the assumptions that there is no threshold dose below which there is no risk and that the risk is proportional to the dose. However, in the absence of reliable human data, estimates of the health effects of low doses of radiation have to be considered hypothetical at best.

The long-term effect of external exposure to radiation is an increase in the incidence of certain types of cancer, such as leukemia and thyroid, breast, and lung cancers. The frequency of incidence of each of these malignancies (*10*) is as follows:

- leukemia, 1.6,
- thyroid cancer, 1.2,
- breast cancer, 2.1, and
- lung cancer, 2.0.

These numbers are the excess of cancer cases per million exposed people, per rad, per year, compared with an unexposed population. The data were obtained from a 30-year study of Hiroshima and Nagasaki survivors.

At equal doses of exposure, the latency period is shortest for leukemia, with the highest frequency occurring about 5–7 years after exposure, and decreasing thereafter. In contrast, the other types of cancer begin to appear only about 10 years after exposure (*4*). According to some sources, the latency period may be inversely related to the dose and length of exposure (*11*).

Radioisotopes may also be incorporated into the body and produce continuous damage to the tissues. In most cases these isotopes are produced by nuclear fission. Strontium-90, a β-emitter with a $t_{1/2}$ of 28.9 years, is incorporated into bones in place of calcium and thus may induce osteosarcoma. Also incorporated into the bones is ^{226}Ra, a member of the ^{238}U disintegration series, an α- and β-emitter with a $t_{1/2}$ of 1590 years; it occurs naturally in soil and rocks. Cesium-137, a β-emitter with a $t_{1/2}$ of 30.2 years, is incorporated into muscles in place of potassium, and iodine-131, a β- and γ-emitter with a $t_{1/2}$ of 8.1 days, is incorporated into the thyroid gland.

Phosphate fertilizers are another source of internal exposure to radiation. Because most of the world's phosphate deposits contain high concentrations of uranium, crops grown on soil treated with phosphate fertilizers become contaminated with radioactive materials. Runoff from fields so fertilized may carry radioactivity into the watershed.

Plants

The sensitivity of plants to radiation damage varies within a 1000-fold range. The most resistant plants are "prostrate" and "recumbent" (herbaceous plants growing near the ground). In field experiments (*12*), certain plants in this category survived exposure to more than 3000 R per day. On the other hand, the higher plants, such as trees and bushes in the forest, did not survive exposure exceeding 350 R per day. The pattern of radiation damage to a forest exposed to γ-rays for 6 months is shown as follows:

Exposure (R per 20 h)	*Effect*
<2	No effect
2–20	Pines damaged
20–70	Pines destroyed
70–160	Oaks destroyed
160–350	Evergreen shrubs (heath) destroyed
>350	Sedge destroyed (all species dead)

In general, a negative correlation has been found among plant species between the size of chromosomes and radiosensitivity: the larger the chromosomes, the greater the damage to a species.

Nuclear Energy

The theoretical basis of a nuclear reactor is a chain reaction that originates when a slow neutron interacts with the uranium isotope ^{235}U. Each collision produces a fission of the uranium atom, which disintegrates into a number of products having smaller atomic weights. In addition, α-, β-, and γ-radiation and one or more high-energy neutrons are released. The neutrons, after slowing down, interact with other ^{235}U atoms to produce a chain reaction. The amount of energy released in each collision is 200 MeV, or 3.2×10^{-4} erg.

Because ^{235}U represents only 0.7% of crude uranium and is enriched to between 2 and 4% in nuclear fuel, most of the uranium (^{238}U) remains unused. For this reason (as well as for the production of fissionable material for nuclear weapons), breeder reactors were developed. Breeder reactors use the prevalent isotope of uranium. ^{238}U per se is not a fissionable material because it cannot sustain a chain reaction. However, it is a "fertile substance" that can be converted to nuclear fuel. This conversion proceeds as depicted in equation 12.2.

$$n + {}^{238}_{92}U \rightarrow {}^{239}_{92}U \xrightarrow{\beta} {}^{239}_{93}Np \xrightarrow{\beta} {}^{239}_{94}Pu \quad (12.2)$$

Both conversions of uranium into neptunium and neptunium into plutonium are fast reactions, with a $t_{1/2}$ of 23 min. Plutonium is a fissionable material. Thus, breeder reactors not only provide fission energy but also supply their own fuel. The ratio of fuel production to fission is higher than 1.

Nuclear Fuel

Mining

The sources of fuel for nuclear reactors are two uranium ores: uranium dioxide (UO_2, called pitchblende) and potassium uranovanadate ($K_2O \cdot 2U_2O_3 \cdot V_2O_5 \cdot 3H_2O$, called carnotite). In the United States, about half of the supply of these ores is obtained from underground mines, and the other half is obtained by strip mining.

Underground mining presents a health problem for the miners in the form of exposure to radon gas. As mentioned earlier, the cause for concern is not radon itself but rather its daughter element, ^{218}Po. A high incidence of lung cancer and other respiratory diseases among uranium miners has been observed both in Europe and in the United States (*6*). In strip mining, radon is of less concern because it is distributed in the atmosphere. However, both miners and the environment may be exposed to windblown radioactive dust.

Another environmental concern is leaching of large quantities of radioactive materials with mine drainage. This leaching creates a hazard to the watershed and groundwater contamination.

Processing

Processing of the ore involves milling, followed by chemical separation of uranium from the accompanying radium. Uranium is converted into ammonium diuranate [$(NH_4)_2U_2O_7$, referred to as yellow cake], whereas radium remains with the ore and is deposited in tailing ponds for storage.

Tailing ponds present an environmental problem because of the continuous radon emission. Moreover, the dry radioactive residue remaining after the water has evaporated may be windblown and thus contaminate large areas. Tailing ponds are considered to be the main contributors to radioactive pollution in the whole process of nuclear fuel production.

The next step is enrichment of ^{235}U to a level suitable for reactor fuel. The conventional method involves conversion of uranium into gaseous uranium hexafluoride (UF_6). The separation of ^{235}U from ^{238}U is based on different diffusion rates of the fluorides through a porous membrane. A new, more economical separation method called atomic vapor laser separation (AVLS) is based on selective absorption of a specific color of laser beam by the ^{235}U isotope. The ^{235}U then becomes ionized and can be separated from ^{238}U in a magnetic field.

The enriched uranium hexafluoride is converted to uranium dioxide and made into pellets that are loaded into zircaloy tubes. (Zircaloy is an alloy of zirconium made especially as casing for nuclear fuel.) The finished products are fuel rods. Very little radioactivity is released during the separation process and the fabrication of the fuel rods.

Nuclear Reactors

The heart of a nuclear reactor is the reactor core, which is an arrangement of several thousand fuel rods immersed in circulating water. Between the fuel rods are boron rods, which may be moved up and down. The core is set in a stainless steel pressure vessel through which cooling water is circulated.

A primary source of neutrons is needed to initiate the chain reaction. The neutrons produced in the fission of ^{235}U are highly energetic. To increase the chance of collision with ^{235}U atoms and thus make a sustained chain reaction possible, part of the neutrons' energy has to be dissipated before they strike the next fuel rod. This dissipation of energy is referred to as moderation; it is achieved by the interaction of neutrons with water molecules.

The movable boron rods absorb neutrons. Their purpose is to regulate the energy output and to allow the shutdown of the reactor when needed.

The heat produced in the fission is exchanged with water under high pressure and circulating at high velocity through the pressure vessel of the reactor. The water temperature reaches slightly over 300 °C. Steam, to drive a turbine, is produced either directly (in boiling water reactors) or through heat exchange (in pressurized water reactors). Breeder reactors, which do not require slowing down (moderation) of neutrons, use liquid sodium rather than water as the heat-exchange fluid.

The escape of some radioactivity from the reactors is unavoidable. Some radioactive fission products leak into the cooling water through pinholes in the fuel rod casings. Collisions of neutrons and protons with oxygen in circulating water, and of neutrons with the corrosion products of the system, produce additional radioisotopes that either escape or are purposely released into the environment. Table 12.2 shows an estimate, prepared by the United Nations Scientific Committee on Effects of Atomic Radiation (UNSCEAR), of short-term human exposure to radioactivity emitted during various phases of the fuel cycle. This estimate does not include radioactivity emitted from the tailing ponds.

Table 12.2. Short-Term Human Exposure to Radioactivity

Operation	*Workers*	*Public*[a]
Mining	0.9	0.5
Milling	0.1	0.04
Fuel fabrication	1.0	0.0002
Reactors	10.0	4.0

Note: All values are given as dose equivalents in men-sieverts per gigawatt of electricity produced per year.
[a]Almost all of the exposure is received by the population within a few thousand kilometers from the plant.
Source: Adapted from data in reference 4.

Nuclear Waste

The major problem of the nuclear energy industry is the disposal of spent fuel. About one-third of the nuclear fuel in use is replaced every year. For a 1000-MW reactor, this amounts to about 33 metric tons of highly radioactive material (about 5×10^9 Ci) that will be an environmental and health hazard for as long as 10,000 years. Figure 12.2 shows accumulation of the high-level nuclear waste (spent fuel rods) worldwide and in the United States.

For the first 150 days, the spent fuel remains in storage at the reactor site. In this "cooling-off" period, the initial radioactivity is allowed to decay somewhat. Although the initial decay may be significant, the amount of radioactivity remaining is still formidable (about 1.4×10^8 Ci for a 1000-MW reactor).

The crowding of reactor-site storage pools recently became such a problem that the Nuclear Regulatory Commission relaxed safety regulations concerning the storage procedures for spent fuel rods. It is now permissible to store them 12 inches apart, instead of 20 inches as required previously. After the cooling period, the radioactive material is transported to government interim storage facilities.

Storage

No permanent storage facilities for spent commercial reactor fuel are available anywhere in the world. This is probably the greatest dilemma of the nuclear

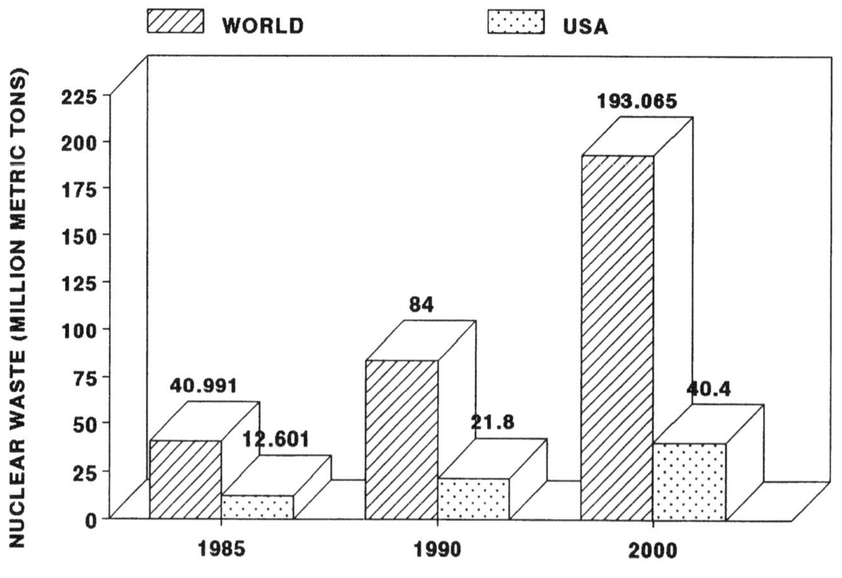

Figure 12.2. Accumulation of high-level nuclear waste, worldwide and in the United States. (Adapted from data presented in reference 13.)

power industry. The U.S. government is exploring storage possibilities at various sites, but this effort is frequently hampered by state and local opposition. The major environmental problems associated with above-ground nuclear waste storage facilities are the escape of gaseous fission products such as tritium and krypton-85, migration of waste by leaching or earthquake, and spontaneous heating of the radioactive materials.

Serious problems are arising as many early nuclear power plants begin aging. After 30 to 40 years of operation, the stainless steel reaction vessel, the pipe system, and the concrete shell surrounding the nuclear core become brittle because of the continuous exposure to nuclear radiation. Because such old plants cannot meet the required safety standards, they have to cease operations and be decommissioned. The decommissioning of a highly radioactive assembly presents a major problem, especially if there is no place to deposit the dismantled plant for permanent rest (*14*).

Reprocessing

In the early stages of nuclear energy development, plans were made to reprocess the spent fuel. According to these plans, the radioactive materials in the spent fuel rods would be chemically separated. Uranium would be enriched to ^{235}U and reused as fuel; the plutonium would be used in breeder reactors. The remaining byproducts would be permanently encased in glass or concrete and buried.

In the United States the plan was never instituted, mainly for economic reasons. Uranium proved to be in ample supply and the feasibility of breeder reactor technology was questioned. At present only three fuel-reprocessing plants are in operation worldwide, one in the United Kingdom and two in France.

Waste from Weapons Facilities

Many nuclear weapons facilities were designed and constructed in the 1940s and 1950s when there was little understanding and concern for environmental problems. Nuclear waste was then disposed of in a way that does not conform to contemporary environmental standards. The U.S. Department of Energy estimates that it may cost as much as $70 billion to bring air and water pollution under control and to clean up contaminated soil at nuclear weapons facilities (*15*).

A permanent storage facility for highly radioactive waste from nuclear arms production has been constructed. This facility, the Waste Isolation Pilot Plant (WIPP), is located in the salt flats of the Chihuahuan Desert, near Carlsbad, New Mexico. The storage facility was excavated 2150 ft below the desert floor in the rock salt.

According to Department of Energy expectations, the slowly moving salt formation will eventually surround and cover the waste-containing drums and seal them permanently. However, this has never been done before, so what will really happen is anybody's guess. The WIPP facility is completed but not yet opened for use. The EPA approved limited testing of the facility for 5 years, but the tests are presently on hold because there appears to be a problem with the salt brine seeping into the waste storage compartments.

Whereas WIPP is destined as a repository for waste from nuclear weapons production, the Department of Energy is focusing on Nevada's Yucca Mountain as a repository for high-level waste from nuclear power plants. It is envisioned that for safety reasons the storage facilities will be located more than 300 m above the water table. However, there is a concern that this area is earthquake prone, and a (currently dormant) volcano is located 20 km from the proposed site. An earthquake or volcanic eruption could raise the water table, bringing water in contact with the hot radioactive waste and producing steam explosions that would blow open the repository and spread its radioactive contents (*13*).

Low-Level Radioactive Waste (LLRW)

The term "low level" does not necessarily indicate that the amount of radioactivity in the waste is insignificant; it is used to distinguish the waste from the "high level" waste that refers to spent nuclear fuel. About 70 to 80% of the LLRW is the waste material generated in nuclear plants, and the rest is the radioactive waste from medical and academic laboratories and pharmaceutical plants.

In the United States, federal law requires that by January 1, 1993, each state provide for disposal of LLRW generated within the state; alternately, several states may enter an agreement to form "compacts" for a common disposal site. The LLRW burial sites involve a variety of designs, from shallow ditches to more sophisticated lined disposal units, or concrete vaults fitted with groundwater-monitoring devices. The main concern about LLRW is the danger of groundwater and soil contamination. Since some isotopes in the LLRW have a very long half-life, the burial sites would have to provide leak-proof confinement for hundreds of years; the fear is that this may not be possible, even with the most sophisticated design presently available. Experience shows that out of six official radioactive disposal sites operated over the last 50 years, three are now closed because they have radioactivity leaking off-site (*16*).

No data are currently available linking leakages from LLRW disposal sites to radiation exposure and any health effects. Estimates of the average individual exposure to radiation from all sources in the United States and worldwide are presented in Table 12.3.

Nuclear Accidents

Although the fission process appears extremely simple on paper, nuclear reactors are complicated machines. A simple malfunction of a pump or a leaking valve may have disastrous consequences. Therefore, elaborate and redundant systems are required. Despite this caution, accidents may happen because of complacen-

Table 12.3. Average Individual Exposure to Radiation

	Exposure	
Source	*United States*[a]	*Worldwide*[b]
Natural background	1	2
Medical radiation	0.9	0.4
Mining, buildings, etc.	0.05	Unknown
Consumer products	0.003	Unknown
Nuclear weapons fallout	0.05–0.08	0.02
Nuclear power	0.0028	0.01

NOTE: All values are for exposure in millisieverts per year.
[a]1987 data from reference 2.
[b]1985 data from reference 4.

cy, human error, negligence, system failure, sabotage, forces of nature, or any combination of these factors.

The most serious malfunction is loss of cooling water, even for 1 min. Even if the chain reaction were stopped immediately, the decaying radioactive materials would produce enough heat to melt the reactor core, the pressure vessel, and the concrete base. Fire and violent explosions of steam and hydrogen would eject tons of radioactive debris. The fallout would contaminate the environment, soil, crops, water, forests, livestock, wildlife, and people. Strontium-90 and cesium-137 deposited on grass would remain there for decades and would enter the food chain through grazing livestock.

The history of the nuclear energy industry includes several accidents. Some of the minor ones were covered up by the authorities so as not to spread antinuclear sentiment among the population. However, two major accidents received worldwide publicity: those at Three Mile Island near Harrisburg, Pennsylvania, in 1979 and Chernobyl in the Soviet Union in 1986.

Three Mile Island

At Three Mile Island a partial meltdown occurred, but without fire and explosion. According to industry disclosure, most of the radioactive contamination was confined to the reactor containment building; however, it appears that a considerable portion of the radioactivity also escaped to the environment. Although there were no immediate casualties, the long-term health effects of the exposed population have begun to surface. Unofficial surveys have indicated an elevated incidence of leukemia and other cancers within a radius of up to 20 miles from the plant. There were no accurate measurements of radioactivity during and immediately after the incident. However, on the basis of the damage to the vegetation, it may be estimated that many residents of the affected area were exposed to 200–300 rem. The cleanup took nearly 10 years and its cost exceeded $1 billion.

Chernobyl

The Chernobyl accident on April 26, 1986, was a major catastrophe (*17*). A complete meltdown of the reactor was accompanied by fire and explosions. The fact that Soviet reactors are moderated by graphite, rather than water, contributed to the fire.

The cost in human suffering and material loss was astronomical. There were 31 deaths and 1000 immediate injuries; 135,000 people had to be evacuated. The projected increase in cancer deaths is as high as 100,000. Direct financial losses are estimated at more than $3 billion.

According to Soviet estimates, the amount of debris released into the atmosphere amounted to 7000 kg containing 50–100 million curies. The fallout was not restricted to the Soviet Union, but it spread as far north as the Arctic Circle, as far south as Greece, and as far west as Great Britain. The area covered by the fallout and the fallout density depended on wind direction and the pattern of precipitation. Agricultural losses of the affected European countries were considerable.

Future of Nuclear Power

As of 1986, there were 366 nuclear power plants worldwide. They have a generating capacity of 255,670 MW and supply, on the average, 15% of the world's electricity (*17*). In the United States, the corresponding figures were 92 plants with a total capacity of 78,618 MW, supplying 16% of the nation's electricity.

Considering that nuclear power technology has been in existence for slightly more than 30 years, this productivity appears to be an impressive achievement. However, the future of the nuclear power industry is very uncertain. In 1972 the International Atomic Energy Agency (IAEA) projected that by the year 2000 the worldwide energy produced by nuclear fission would reach 3,500,000 MW. In 1986 these projections were scaled down to 500,000 MW (*17*).

Economics

The two reasons for this decline are economics and politics. Originally it was thought that the electricity produced by nuclear fission would be "too cheap to bother to meter it". In reality, it turned out to be the most expensive way of producing electricity.

According to figures of the nuclear energy industry, the average cost of electricity from nuclear plants is 12 cents per kilowatt-hour, as compared to 6 cents per kilowatt-hour from coal-powered plants. In addition, because of elaborate safety measures, the construction cost rose steadily from $200 per kilowatt in the 1970s to $3200 per kilowatt in 1986 (*17*). At the same time, the rate of growth of electricity consumption declined.

The National Energy Strategy of 1991, proposed by the Bush administration, contains provisions for stimulation of development of nuclear power. However, there are objections to this strategy, both in the legislature (House and Senate) and among environmental groups.

Safety

Concern about safety has also contributed to the decline of the nuclear power industry. Before the Chernobyl disaster, opposition to nuclear power in most countries was limited to grass-roots environmental organizations. After Chernobyl the situation changed. The grass-roots opposition increased, and the governments of many countries began to reassess the wisdom of further development of fission energy.

Chernobyl demonstrated the transboundary characteristics of nuclear accidents and the fact that no country has any contingency for dealing with such disasters. In addition, attempts by governmental or corporate officials to conceal the true extent of nuclear accidents (as was the case at the Windscale nuclear plant disaster in the United Kingdom in 1957, at Three Mile Island in 1979, and at Chernobyl in 1986) have undermined society's confidence in the truthfulness and competence of its leaders. This distrust has hardened antinuclear opposition.

Proponents of nuclear energy argue that, as compared to coal mining, relatively few lives have been lost in nuclear accidents. Although this is undoubtedly true, the difference is that coal mine accidents are limited to local areas. Nuclear accidents endanger the lives and property of the general public throughout vast areas, frequently beyond national borders.

Another argument in favor of nuclear energy is that, at present, it is the only practical large-scale source of energy that does not contribute to the greenhouse effect.

Inherently Safe Reactors

Even before Chernobyl, the nuclear power industry, some academic institutions, and governmental bodies had begun to analyze the causes of the nuclear power debacle. Between September 1983 and summer 1984, four organizations (the Massachusetts Institute of Technology, the Congressional Office of Technology Assessment, the Institute for Energy Analysis, and the Atomic Industrial Forum) published reports on the status of nuclear power and recommendations for a possible revival of the industry. Three of these groups recommended, among other things, radically changing the design of reactors (*18*).

Process-Inherent Ultimate-Safety Reactor

Two of these "inherently safe reactors" were singled out as possible alternatives to the conventional LWRs (light-water reactors) that are presently in use. The water-moderated, water-cooled PIUS (process-inherent ultimate-safety) reactor, which was developed in Sweden, contains several innovative safety features.

The stainless steel pressure vessel is embedded in a reinforced concrete structure. In an emergency the core is automatically flooded with borated water, which instantly stops the fission. The heat generated by the decay of the radioactive fission products is dissipated by convection currents and evaporation of a large pool of water. This reactor is designed so that the core would be covered by water for about a week, giving enough time for remedial action before any meltdown could occur. The power-generating capacity of a single PIUS reactor is limited to 400 MW.

High-Temperature Gas-Cooled Reactor

A graphite-moderated, helium-cooled HTGR (high-temperature gas-cooled reactor) was developed in the United States by GA Technologies. This small reactor is limited in its power-generating capacity to less than 100 MW per unit. The fuel consists of uranium oxide particles embedded in chunks of graphite and scattered among graphite blocks. Helium gas is used as both a coolant and a heat-transfer medium. Because the fuel is widely scattered, the reactor has a high heat capacity. Thus, it will heat up slowly in the case of coolant loss. Its operating temperature is around 1000 °C, well below the 2000 °C that the graphite can withstand.

The German version of HTGR is smaller. Its fuel is in the form of pebbles coated with graphite and contained in graphite balls. New fuel is loaded from the top, and the spent fuel emerges from the bottom. This arrangement allows refueling without shutting the reactor down. A few HTGR reactors are in operation in Germany, the United Kingdom, and the United States.

Power Reactor Inherently Safe Module (PRISM)

This reactor, developed by General Electric, is fundamentally different from those previously described because it uses liquid metal rather than water or gas as the coolant. The fuel rods are submerged in liquid sodium. Sodium boils at 900 °C and thus in the case of overheating, the coolant pool can absorb the excess heat. In addition, the rise in temperature causes the fuel and the coolant to expand, which slows the fission. Each module has a power-producing capacity of only 155 MW (*19*).

Whether this second generation of nuclear reactors will be more acceptable to the public is not certain. As stated in the report of the Office of Technology Assessment, nuclear energy has no future without public support (*18*). In any case, no matter how safe the new reactors become, the problem of radioactive waste disposal will remain.

References

1. Noz, M. E.; Maguire, G. O., Jr. *Radiation Protection in the Radiologic and Health Sciences;* Lea and Febiger: Philadelphia, PA, 1979.

2. *Low-Level Radiation Effects: A Fact Book;* Brill, A. B.; Adelstein, S. J.; Saenger, E. L.; Webster, E. W., Eds.; Society of Nuclear Medicine: New York, 1982.
3. Hanson, D. J. *Chem. Eng. News* February 8, 1988, p 7.
4. *Radiation: Doses, Effects, Risks;* United Nations Environment Programme: Nairobi, Kenya, 1985.
5. Okita, T. *J. Radiat. Res. Suppl.*, review of 30 years of study of Hiroshima and Nagasaki atomic bomb survivors; Okada, S.; Hamilton, H. B.; Egami, V.; Okajima, S.; Russell, W. J.; Takeshita, K., Eds.; Japan Radiation Society: Chiba, Japan, 1975; Chapter II A, p 49.
6. Hobbs, C. H.; McClellan, R. O. In *Cassarett and Doull's Toxicology;* Klaassen, C. D.; Amdur, M. O.; Doull, J., Eds.; MacMillan: New York, 1986; Chapter 21, p 669.
7. Ducoff, H. S. In *Biological Environmental Effects of Low-Level Radiation;* proceedings of a symposium organized by the International Atomic Energy Agency and the World Health Organization; International Atomic Energy Agency: Vienna, Austria, 1976; Vol. I, p 103.
8. Grosh, D. S. In *Introduction to Environmental Toxicology;* Guthrie, F. E.; Perry, J. J., Eds.; Elsevier Science: New York, 1980; Chapter 4, p 44.
9. Grosh, D. S.; Hopwood, L. E. *Biological Effects of Radiation;* Academic: New York, 1979; p 253.
10. Beebe, G. W.; Kato, H. *J. Radiat. Res. Suppl.*, review of 30 years of study of Hiroshima and Nagasaki atomic bomb survivors; Okada, S.; Hamilton, H. B.; Egami, V.; Okajima, S.; Russell, W. J.; Takeshita, K., Eds.; Japan Radiation Society: Chiba, Japan, 1975; Chapter II E, p 97.
11. Schilling, C. W. *Atomic Energy Encyclopedia in the Life Sciences;* W. B. Saunders: Philadelphia, PA, 1964; pp 235 and 239.
12. Woodwell, G. M. *Science (Washington, D.C.)* **1967**, *156*(3774), 461.
13. Lenssen, N. In *State of the World 1992;* Brown, L. R., Ed.; W. W. Norton: New York, 1992; Chapter 4, p 46.
14. Pollock, C. *Worldwatch Paper 69;* Worldwatch Institute: Washington, DC, 1986.
15. Long, J. *Chem. Eng. News* July 11, 1988, p 6.
16. *Rachel's Hazardous Waste News* September 9, 1992, no. 302.
17. Flavin, C. In *State of the World 1987;* Brown, L. R., Ed.; W. W. Norton: New York, 1987; Chapter 5, p 81.
18. Manning, R. *Environment* **1985,** *27,* 12–17.
19. World Resources Institute in collaboration with the United Nations Environment Programme and the United Nations Development Program. *World Resources 1990–91;* Oxford University: New York, 1990; Chapter 9, p 141.

13 Population, Environment, and Women's Issues

Present Trends in Population Growth

Ultimately, the necessity to supply food, energy, habitat, infrastructure, and consumer goods for the ever-growing population is responsible for the demise of the environment. Remedial actions for pollution abatement, and further technological progress toward energy efficiency, development of new crops, and improvements in manufacturing processes may help to mitigate the severity of environmental deterioration. However, we can hardly hope for restoration of a clean environment, improvement in human health, and an end to poverty without arresting the continuous growth of the world population.

According to the United Nations' *State of the World Population* report of June 1991, the world population reached about 5.4 billion by mid-1991 (*1*). At an average annual growth rate of 1.73% (*2*), doubling of the present number is expected to occur in 40 years (*see* footnote 2 in Chapter 1).

The rate of population growth and the fertility rates by continent, as well as in the United States and Canada, are presented in Table 13.1. It can be seen that the fastest population growth occurs in the poorest countries of the world. Despite the worldwide decrease in fertility rates between 1965 and 1990, the rate of population growth did not change very much because, as a result of the high fertility rates in the previous decades, the number of women of childbearing age had increased.

Table 13.1. Population Growth and Fertility Rate by Continent and in the United States and Canada

Continent	Annual Growth (%)	Doubling Time[a] (years)	Fertility Rate 1965	Fertility Rate 1990
Africa	3.00	23.4	6.7	6.2
Asia	1.85	37.7	5.7	3.5
South America	2.07	33.7	5.2	3.6
Oceania	1.44	48.3	3.5	2.6
North America (including Central America)[b]	1.28	54.3	5.2	3.5
Europe	0.23	300.0	2.5	1.7
United States[c]	0.82	84.5	2.6	1.8
Canada[c]	0.88	78.8	2.5	1.7

NOTE: The fertility rate is the number of children per woman per lifetime.
[a]Assuming that the present trend continues.
[b]Includes the United States and Canada.
[c]Includes immigration. The portion of population growth in the United States due to immigration is estimated at 28% for the period 1980–1985.
SOURCE: Adapted from reference 2.

Historically, the preference for large families in the developing nations was in part a result of either cultural or religious traditions. In some cases there were practical motivations, as children provided helping hands with farm chores and a security in old age.

At present the situation is changing. A great majority of governments of the developing countries have recognized that no improvement of the living standard of their citizens will ever be possible without slowing the explosive population growth (*3*). By 1985, a total of 70 developing nations had either established national family planning programs, or provided support for such programs conducted by nongovernmental agencies; now only 4 of the world's 170 countries limit access to family planning services (*1*). As result, 95% of the developing world population lives in the countries supporting family planning (*4*). Consequently, the percentage of married couples using contraceptives increased from less than 10% in 1960 to 51% in 1991.

It is estimated that to maintain the United Nations' medium population projection (8.504 billion by the year 2025), it is necessary to extend modern family planning services to 59% (567 million) of all married women of reproductive age by the year 2000. This should decrease the fertility rate in developing countries from 3.8 to 3.3 children (*2*). The ultimate goal, of course, would be to provide family planning for all couples of the world.

Status of Women and Population Growth

Despite the favorable trends in fertility rates across the world, the problem of rapid population growth is far from being solved. Even if it were possible to decrease the fertility rate everywhere in the world to the mere replacement rate, i.e., two children per woman, the population would still increase for 60 years or so because of the increasing number of women entering childbearing age.

The high fertility rate in many developing countries is linked to the low social standing of women, their poor education, and general poverty. According to the report prepared for the Population Institute, "when women feel socially and personally insignificant, they frequently become pregnant to feel that they are more than merely existing" (*5*). It has been shown that the higher the educational and socioeconomic status of women, the fewer children they produce.

In many countries discrimination against women is institutionalized. The laws make women ineligible for credit and land possession. Women perform a magnitude of chores (firewood gathering, cooking, tending farms and gardens, and caring for children) that are not counted in the gross national product (GNP). Although women's unpaid labor is estimated to be worth $4 trillion worldwide [about a third of the world annual economic product (*6*)], women are entirely dependent on their husbands. They are not allowed to obtain jobs or use contraceptives without their husbands' permission. In many developing countries pregnancy constitutes a high risk of death. It is estimated that in 1991, the death rate due to complications of pregnancy and childbirth was 1 in 21 in Africa, 1 in 38 in South Asia, and 1 in 73 in South America. Corresponding rates are one in 7,000 and 1 in 10,000 in North America and Northern Europe, respectively (*7*).

The relationship between education of women and their fertility rates has been established. It has been found that women with 7 years of schooling tend to marry on the average 4 years later and have on the average 2.2 fewer children than women without any schooling (*6*). Yet education is frequently denied to women. The average illiteracy rate in developing countries is 49% for women and 28% for men (*6*).

Another problem is poverty. The present trend in the developing countries, to shift family planning from the public to the private sector, has made contraceptives hardly affordable for many couples because of the cost. According to the report published by the Population Crisis Committee, in some African and Asian countries, the cost of condoms varies from 3.5 to 48% of the per capita GNP, pills from 4.8 to 37%, intrauterine devices (IUDs) from 2.8 to 71.3%, and sterilization from 7.1 to 261% (*8*). In contrast, in Western industrialized countries the corresponding expenditures are less than 1% of the per capita GNP (*8*).

In 1979 the United Nations drafted a global treaty for women's rights. This treaty requires that ratifying nations incorporate into their legal systems provisions for equal rights for women in education, employment, health care, and politics, and equal legal status. As of mid-1990, 101 nations ratified this treaty. Although some did so only to appease the women's movement and are hesitant to alter their discriminatory way of life (*9*), eventually they will have to implement the treaty's provisions. It is disturbing that the United States, the champion of human rights and civil liberties, has not yet ratified the treaty.

Population Growth and the Global Food Supply

Grain Supply

Since the mid-sixties, there has been a dramatic increase in world food production, especially in the developing countries. This boon in world nutrition, referred to as the green revolution, was possible thanks to the development of high-yield grains, heavy application of fertilizers and pesticides, irrigation, increased use of machinery, and augmentation of land area under cultivation. Figure 13.1 presents the growth of the world population and of cereal production from 1960, as projected to the year 2000. Because about 50% of the calories in the human diet is supplied by cereals, production of cereals is the best indicator of the nutritional status of the world population.

The benefits of the green revolution were not equally distributed throughout all countries of the world. The greatest success was achieved in the Asian centrally planned economies, where grain production increased by 114% between 1965 and 1988. The smallest gains were in Africa, with only a 40% increase during the same period. These small gains, coupled with a 77% increase in population, resulted in a decrease of per capita grain production from 118 to 108 kilograms.

Although the green revolution provided food for millions, it had some negative impact on the environment. Salinization, alkalinization, waterlogging, and

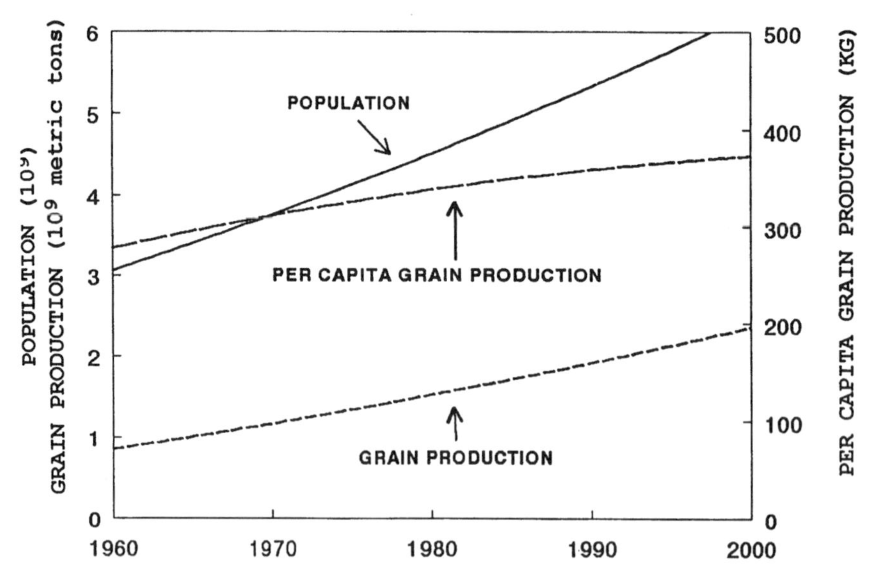

Figure 13.1. Growth of the world population and of cereal production between 1960 and 2000. The estimates of population were obtained by extrapolating from the population data of 1950, 1957, and 1990 presented in reference 11. (Data are from reference 10. Reproduced with permission from reference 28. Copyright 1994 S. F. Zakrzewski.)

depletion of groundwater were results of improper or excessive irrigation (*12*). Land erosion and runoff of fertilizers and pesticides caused water pollution, and in some cases cropland expansion may have contributed to deforestation. In addition, the green revolution had socioeconomic repercussions; the need for fertilizers and pesticides increased the cost of farming, forcing small farmers out of business.

Despite the success of the green revolution, many of the world's people remained undernourished (Figure 13.2). Although the percentage of undernourished people remained about constant throughout the period of 1969–1985, the absolute number was increasing. (The data shown in Figure 13.2 are rather conservative; some sources estimate the number of hungry people at 1 billion.) As may have been expected, the problem of hunger was worst in Africa, with 29 to 32% of its population undernourished.

The cause of world hunger is not a food shortage but rather unequal food distribution, poverty, and in some cases, political unrest. Poverty is certainly the most pervasive cause, and it can hardly be remedied as long as population growth is out of control. Present world food production could provide nutrition, albeit a mostly vegetable diet, to more than six billion people, much in excess of

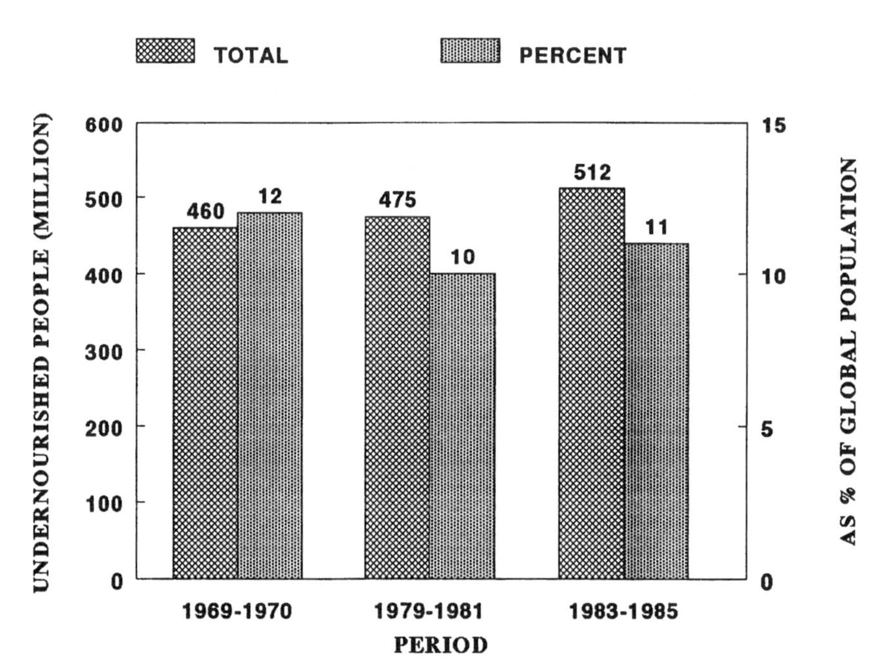

Figure 13.2. Estimated number of undernourished people in the world. The U.N. Food and Agriculture Organization considers people "undernourished" when their total daily caloric intake is below 1.4 times the basal metabolic rate; the basal metabolic rate is defined as the energy requirement while fasting and at complete rest. (Data are from reference 10.)

the present population of 5.4 billion (*13*). However, about one-third of world cereal production, the staple food of large masses, is used to feed livestock to produce high-caloric foods (eggs, dairy products, and meat), which are beyond the reach of the poor of the world (*14*).

Although the global production of cereals is still increasing, albeit at a slower rate than during the sixties, seventies, and early eighties, the per capita increase shows a distinct leveling trend (*see* Figure 13.1).

It appears that the peak gains of the green revolution may have occurred in the past. Since the strains of grains presently cultivated reached the limits of their responsiveness to fertilizers, not much can be gained by increasing the application rate of fertilizers (*15*). By 2025 the per capita cropland area is expected to decrease by about 40% from the present (*11*), unless new agricultural acreage is added. Asia has little potential for additional cropland because 82% of the available land is already under cultivation. In some other areas, such as sub-Saharan Africa and Latin America, large land reserves are available but the soil is of marginal quality (*10*). Dennis Avery of the Hudson Institute in Indianapolis argues that there is enough idle cropland in the United States and Argentina alone to provide food for an extra 1.4 billion people (*16*). Obviously this will not solve the world's needs for very long because at the present rate of population growth, 1.4 billion people will be added to the earth's population by the year 2009. Besides, Avery's assertion does not consider the long-term effects of land and water degradation caused by agriculture. According to some sources, desertification, salinization, waterlogging, and erosion may render as much agricultural land useless each year as is added (*10*). Leaving no soil reserves to allow reclamation of degraded land is a nearsighted policy. Nor does Avery take into consideration the fact that there is a growing water shortage. An ample supply of freshwater is necessary for successful agriculture, yet in many parts of the world, including the United States, some aquifers are beginning to be depleted owing to excessive use of groundwater for irrigation (*17*).

Another problem is loss of biodiversity. Modern agriculture is based on planting high-yield, monoculture crops. The genetic similarity within each type of grain makes the crops highly sensitive to a pest invasion, requiring increasing use of pesticides. This not only increases the cost and the energy requirement, but also has a detrimental effect on the environment. At the same time, encroachment of human settlements and agriculture on fallow land causes disappearance of native grasses, which represent genetic material for development of new varieties of grains. Although international seed banks have been created to preserve as much biodiversity as possible, it is doubtful that these seed collections will completely prevent the disappearance of species (*16*).

Meat Supply

Trends similar to those observed in grain production were also noticed in meat production. After 1950 world meat production was increasing. The peak was reached in 1987, when the total was 161 tons or 32 kilograms per capita. However, there was no further increase thereafter, and in 1992 per capita production declined. Overgrazing of pastures and the inability of remaining grasslands to support more cattle and sheep seem to be at the root of this slowdown (*18*).

Fish Supply

Fish is an important source of protein, especially for the population of the developing countries. Whereas in North America and Western Europe fish consumption contributes 6.6 and 9.7%, respectively, of the animal protein intake, it contributes 21.1% in Africa and 21.7% in Asian centrally planned economies (*19*). Between 1950 and 1989 the fish catch kept expanding, reaching 100 million tons, which translated into 19 kilograms per capita. Since 1989, despite sophisticated fishing technologies and a large number of fishing vessels prowling the seas, the catch has declined (*18*). It is believed that the oceans have reached their limits. There are two reasons for the declining catch. Pollution of coastal waters, the breeding areas of many species, affected fish reproduction. At the same time, overfishing depleted the fish stocks faster than they could be replaced. The depletion of the fish stocks not only put many fishermen out of work, but also raised prices, making fish less accessible to poor people. This is specially detrimental to the population of developing countries.

With continuous growth of population, the outlook for the future food supply is grim or at least very uncertain. Additional factors that threaten future food supply are urbanization, which takes land away from agriculture; damage to crops by excessive ultraviolet radiation (a consequence of stratospheric ozone depletion); and, possibly, a change of the climate caused by emissions of greenhouse gases.

Effect of Overpopulation on the Environment

The term "overpopulation" is not necessarily related to population density, but rather to the area carrying capacity. An area is considered overpopulated if it cannot sustain its population without permanently destroying natural resources (*20*).

Of course, the earth is resilient, and depleted resources may renew themselves given sufficient time and lack of interference from the human population. However, if preservation of our society is the goal, "permanent" has to be considered within the frame of a human life span. For instance, destruction of a tropical forest has to be considered permanent, even if it may regrow itself after several hundred years. So is desertification of land or depletion of groundwater.

The biologist Paul Ehrlich devised a formula that describes the impact of a society on the environment: "$I = PAT$", where I stands for impact, P for population, A for affluence (i.e., consumption), and T for technology (*21*).

In developing countries, where the majority are poor and technology is not well developed, the production and consumption of goods are low. The environmental deterioration is mainly due to the large number of poor people and their quest for lumber, firewood, cropland, and grazing land. Deforestation, especially by slash and burn, contributes to an increase of greenhouse gas emissions; it also affects the hydrological cycle and increases soil erosion. According to a publication by the United Nations Population Fund (*12*): "Between 1971 and 1986, world arable land expanded by 59 million hectares, while forest shrank by 125 million hectares. Over the same period, land going to settlements, roads, industries, office buildings and so on, may have expanded by more than 50 million hectares to cope with the needs of expanding urban centers. . . . Growing population may be responsible for as much as 80% of the loss of forest cover."

Loss of biodiversity is a direct consequence of deforestation and of human encroachment on the wildlife habitat. Although disappearance of species is a natural evolutionary phenomenon, the present rate of species extinction is estimated to be 400 times the natural rate. Such rapid extinction disrupts the ecological balance and may greatly affect the future global economy. Whereas in the past the reasons for extinction were competition between species and overexploitation, presently the destruction of habitat is the predominant factor. A relationship between population growth and species loss is shown in Figure 13.3.

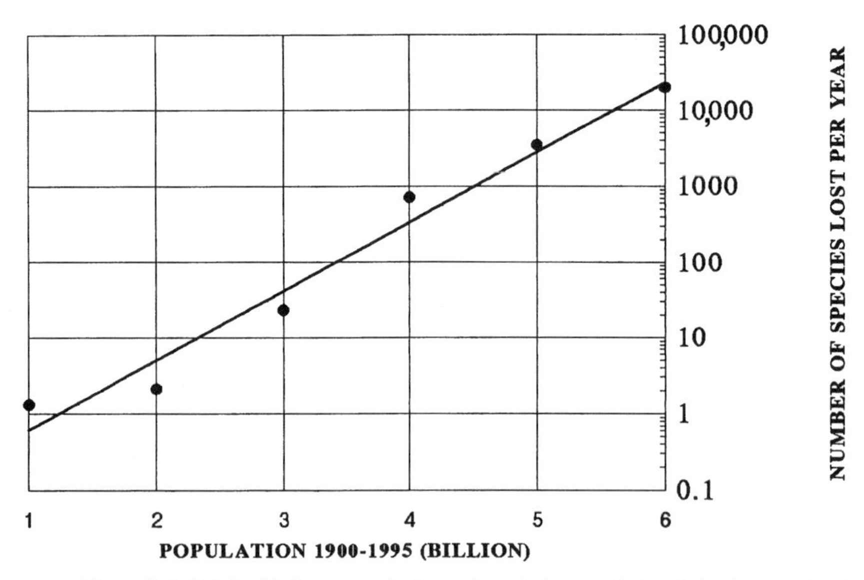

Figure 13.3. Relationship between species loss and population growth. Population data were based on reference 29, and estimates of species loss were taken from reference 30.

Changes in precipitation patterns due to deforestation, cultivation of marginal land, and overgrazing lead to land erosion and desertification. The problem of salinization, alkalinization, and waterlogging has been discussed in the preceding section. The situation is frequently aggravated by pervasive poverty, unequal land distribution between a few wealthy families and the poverty-stricken masses, and inefficient agricultural technologies.

Other consequences of overpopulation are loss of water resources and deterioration of water quality. Because freshwater resources are finite, the per capita availability of water is related to the number of people competing for the same water source. Since 1850, the global freshwater resources declined from 33,000 cubic meters per capita per year to the present 8500 cubic meters (*12*). It has been determined that a society is affected by water shortage when the amount of available freshwater declines to 500 cubic meters per person per year. In 88 developing countries, comprising 40% of the world population, water resources are presently dwindling to a level that imposes constraints on development (*12*). In addition, in many areas water quality is deteriorating because of industrial development or because of raw sewage discharges into lakes and rivers.

In the industrialized world, it is not so much population pressure as the volume of manufacturing and consumption of goods that has a detrimental impact on the environment. The demand for energy, mostly fossil fuels, necessary to drive our sophisticated technology puts additional stress on resources. Industries, power generation, and transportation pollute the air, land, and water. Acid rain and a demand for lumber destroy forests. Moreover, the high consumption of goods produces large amounts of municipal and industrial waste. The waste pollutes groundwater when buried, or air and surface water when incinerated.

Suburban developments are another source of environmental deterioration. Especially in the United States, where city centers are deteriorating, progressively more development occurs in the suburbs. Agricultural land is taken for commercial and residential construction. Large areas are paved over for shopping centers, parking lots, and highways. This development alters the natural hydrologic cycle; the rainwater runoff from streets and parking lots contributes to water pollution and in some instances augments flooding potential. Lacking viable public transportation, the sprawling suburbs increase our dependence on the private automobile for commuting. This adds further to air and water pollution and carbon dioxide emissions, and further enhances the demand for fuel.

In an industrialized society, because of the high demand for resources and energy, even a modest rate of population growth is undesirable. The economies of the industrialized world, with their dependence on consumption for prosperity, and geared for continuous growth, face a dilemma. A high rate of production and consumption of goods creates a prosperous economy, but it stresses the environment and natural resources. On the other hand, when consumption slows down, the economy goes into recession, resulting in unemployment and human suffering. Conventional economic theories do not consider finiteness of natural resources and do not consider depletion of these resources as depreciation in the

GNP. It would be a challenge for modern economists to devise a prosperous "no growth" economic system based on recycling rather than depletion of resources. A treatise on this subject has been published recently by the World Bank (*22*). The reader is also referred to the book *Beyond the Limits* (*23*).

Overpopulation, Urban Sprawl, and Public Health

In the second half of the twentieth century, growth of cities in the developing countries assumed catastrophic proportions. Table 13.2 shows population growth in selected cities since 1950, and Table 13.3 gives the change of the urban population in selected countries between 1960 and 1990. According to the predictions of the United Nations Population Fund, by the year 2000, 10 out of 12 of the world's largest cities will be in the developing countries. It is estimated that the population of each of these megacities will range between 13 and 26 million (*3*). This rapid growth of cities is attributed mainly to a high rate of birth among the city dwellers, but migration from the rural areas also contributes significantly. Uneven land distribution, land fragmentation, and decreased land fertility compel a landless, poverty-stricken rural population to migrate to the cities in search of employment. Because jobs are scarce, the people usually end in shanty towns or as homeless. Among the millions of homeless, many are children. In Latin America alone, the number of so-called "street children" is estimated at more than 20 million (*24*).

The infrastructure of the megacities of the developing world is completely overwhelmed by the number of people. Municipal authorities are unable to cope with the multitude of problems created by the bursting population. Urban

Table 13.2. Population Growth in the Fastest Growing Cities in Developing Countries

	Population (million)				
City	*1950*	*1951*	*1985*	*2000*[a]	*Increase, 1950–1985*[b] (%)
Sao Paulo	2.7	—[c]	15.9	24.0	489
Mexico City	3.05	—[c]	17.3	25.8	467
Delhi	—[c]	1.4	7.4	13.2	429
Manila	1.78	—[c]	7.9	11.1	273
Jakarta	1.45	—[c]	7.9	13.2	279
Bombay	—[c]	3.0	10.1	16.0	237
Cairo	2.5	—[c]	7.7	11.1	208

[a]Estimated values.
[b]From 1950 or 1951, respectively, to 1985.
[c]Data are not available.
SOURCE: Adapted from reference 3.

Table 13.3. Increase in Urban Population in Selected Countries Between 1960 and 1990

	Urban Population as a Percentage of the National Total		
Country	*1960*	*1975*	*1990*
Mexico	50.8	62.8	72.6
Brazil	44.9	61.8	79.9
India	18.0	21.5	28.0
Indonesia	14.6	19.4	28.8
Egypt	37.9	43.5	48.8
Philippines	30.3	35.6	42.4

SOURCE: Adapted from reference 25.

sprawl frequently occurs at the expense of agricultural land, reducing available cropland and further aggravating rural poverty. Inability of the municipalities to supply water and sanitary facilities for a large percentage of urban poor has significant public health repercussions. Data presented in Table 13.4 show the accessibility of clean drinking water and sanitary services to the urban population in developing countries. The percentage of people without these facilities has decreased in the last decade and is projected to decrease even more by the year 2000. However, because of the continuous growth of the urban population, the situation keeps deteriorating as far as the total number of people is concerned. The questionable purity of available water and the lack of hygienic facilities create the danger of waterborne diseases. The cholera epidemic that was spreading throughout Latin America in 1991 was undoubtedly the consequence of urban blight; this epidemic claimed 1500 lives by mid-1991 (*24*).

Industrialization and an increase in the number of motor vehicles frequently add to the plight of the urban population in the developing world. Antiquated technology and lack of antipollution devices create air pollution problems of

Table 13.4. Urban Population Without Access to Safe Drinking Water and Sanitation Services

	Number of people[a] (million)		
Population	*1980*	*1990*	*2000[b]*
Total urban population	972	1385	1972
Without water supply	368 (38)	447 (32)	500 (25)
Without sanitation	684 (71)	868 (63)	1026 (52)

[a]Figures in parentheses indicate percentage of the total urban population without water supply or sanitation; the percentages are rounded up to the nearest unit.
[b]Projected values.
SOURCE: The reported numbers were calculated from data in reference 26.

dangerous proportions. In Mexico City, for instance, the air is so polluted that women of the diplomatic corps are regularly advised not to have children during their stay in Mexico (*27*). The annual death toll due to air pollution in Latin America alone is estimated at 24,000 (*24*).

Rapid and uncontrolled population growth and the ever-growing gap in the distribution of wealth between the rich and the poor are the most critical problems facing humanity.

International Cooperation on Population Issues

Since 1964 international conferences on population have been organized by the United Nations every 10 years. At the 1984 conference in Mexico City, the United States abrogated its responsibility toward the world community. At that time the Reagan administration took the stand that the problem of population growth is a neutral issue—neither good nor bad. Under the pretext that the United States' contributions to international population programs are used to promote abortion in China, the United States withdrew its financial support for the United Nations Population Fund and International Planned Parenthood. This so-called "Mexico City" policy was reversed by the Clinton administration, and funding of the United Nations and Planned Parenthood programs was restored in 1993.

Between September 5 and 13, 1994, representatives of 180 nations met again, this time in Cairo, Egypt, for the United Nations International Conference on Population and Development (ICPD). Unfortunately, the media focused on the minor issue of a disagreement between the Vatican and the United States about abortion, whereas the real achievements of the conference were not publicized much. The Programme Action signed by 175 nations emphasized commitment by the signatories to promotion of reproductive health services and to the empowerment of women as the best means of stabilizing population growth. The document reiterated principles of equality that apply to women, children, and migrants and asserted women's right to control their own fertility. Further, the Programme Action called for better access to education for women and girls, elimination of violence against women, access to family planning services, and involvement of women in policy-making.

The annual cost for family planning services was estimated at $17 billion by the year 2000 and $21.7 billion by 2015. One-third of the estimated $17 billion is expected to come from industrialized countries and the rest from developing countries. Germany and Japan pledged $2 and $3 billion, respectively, to be spent over the next seven years. The United States pledged $595 million for fiscal year 1995, with subsequent increases in the following years.

Time will show whether this international effort will succeed in containing the world population to 7.25 billion by the year 2015, as predicted by the United

Nations. Thereafter, the population should begin to decrease. If we fail, a global disaster is looming for the future of humanity.

References

1. *Popline*; Vol. 13, May–June 1991; Population Institute: Washington, DC; p 4.
2. World Resources Institute, International Institute for Environment and Development in collaboration with U.N. Environment Programme. *World Resources 1990–1991;* Oxford University: New York, 1990; Chapter 16, p 253.
3. *The Global Ecology Handbook;* Corson, W. H., Ed.; Global Tomorrow Coalition: Washington, DC, 1990; Chapter 3, p 23.
4. "Family Planning Reduces World Population", *Popline;* Vol. 13, January–February 1991; Population Institute: Washington, DC; p 4.
5. "Women's Status a Factor in Global Warming", *Popline;* Vol. 13, January–February 1991; Population Institute: Washington, DC; p 4.
6. *Great Decisions*; Foreign Policy Association: New York, 1991; p 63.
7. Jacobson, J. L. *World Watch* Vol. 4, No. 1, January–February 1991; Worldwatch Institute: Washington, DC; p 9.
8. *1991 Report on World Progress Toward Population Stabilization: Access to Affordable Contraceptives;* Population Crisis Committee: Washington, DC, 1991.
9. McCoy-Thompson, M. *World Watch* Vol. 3, No. 3, May–June 1990; Worldwatch Institute: Washington, DC, 1990; p 7.
10. World Resources Institute, International Institute for Environment and Development in collaboration with U.N. Environment Programme. *World Resources 1990–1991;* Oxford University: New York, 1990; Chapter 6, p 83.
11. World Resources Institute, International Institute for Environment and Development in collaboration with U.N. Environment Programme. *World Resources 1990–1991;* Oxford University: New York, 1990; Chapter 4, p 19.
12. *Population and the Environment: The Challenge Ahead;* U.N. Population Fund: New York, 1991.
13. World Resources Institute, International Institute for Environment and Development in collaboration with U.N. Environment Programme. *World Resources 1988–1989;* Basic Books: New York, 1988; p 51.
14. Ehrlich, P. *The Population Explosion;* Simon and Schuster: New York, 1990; p 66.
15. Brown, L. R. In *State of the World 1991;* Brown, L. R., Ed.; W.W. Norton: New York, 1991; p 3.
16. Linden, E. *Time* August 19, 1991, p 48.
17. *The Global Ecology Handbook;* Corson, W. H., Ed.; Global Tomorrow Coalition: Washington, DC, 1990; p 155.
18. Brown, L. R. *World Watch* Vol. 6, No. 4, July–August 1993; Worldwatch Institute: Washington, DC; p 19.
19. Weber, P. In *State of the World 1991;* Brown, L. R., Ed.; W.W. Norton: New York, 1991; p 41.
20. Ehrlich, P. *The Population Explosion;* Simon and Schuster: New York, 1990; p 37.
21. Ehrlich, P. *The Population Explosion;* Simon and Schuster: New York, 1990; p 58.
22. *Environmentally Sustainable Economic Development Building on Bruntland;* Goodland, R.; Daly, H.; El Serafy, S., Eds.; World Bank, Environment Department: Washington, DC, 1991; Environment Working Paper No. 46.
23. Meadow, D.; Meadow, D.; Randers, J. *Beyond the Limits;* Chelsea Green: Post Mills, VT, 1992.

24. *Popline;* Vol. 13, July–August 1991; Population Institute: Washington, DC; p 3.
25. World Resources Institute, International Institute for Environment and Development in collaboration with U.N. Environment Programme. *World Resources 1990–1991;* Oxford University: New York, 1990; p 126.
26. World Resources Institute, International Institute for Environment and Development in collaboration with U.N. Environment Programme. *World Resources 1990–1991;* Oxford University: New York, 1990; p 65.
27. French, H. F. *World Watch* **1989,** *2*(4), p 39; Worldwatch Institute: Washington, DC.
28. Zakrzewski, S. F. *People, Health, and Environment;* SFZ Publishing: Amherst, NY, 1994.
29. Kussi, P. *This World of Man;* Pergamon: Oxford, England, 1985; p 191.
30. *Gaia, An Atlas of Planet Management;* Mayers, N., Ed.; Anchor Books: Garden City, NY, 1984.

14

Regulatory Policies and International Treaties

The National Environmental Policy Act

The purpose of the National Environmental Policy Act (NEPA) is to ensure that all federally administered or assisted programs are conducted so as to take the environmental impact of their activity into consideration. The scope of NEPA includes privately financed and conducted projects for which federal licensing is required. The law also establishes a presidential advisory group called the Council on Environmental Quality (CEQ).[1]

The crucial section of the act (*U.S. Code*, Title 102, Pt. 2c), which concerns the environmental impact statement (EIS), states, in part, that

> The Congress authorizes and directs that, to the fullest extent possible . . . all agencies of the Federal Government shall . . . include in every recommendation or report on proposal for legislation and other major Federal actions significantly affecting the quality of the human environment, a detailed statement by the responsible official on

[1]In February 1993, President Clinton proposed replacing the CEQ with the White House Office of Environmental Policy (OEP). At the same time he proposed elevating the Environmental Protection Agency to Cabinet status. The functions of the CEQ would then be split between the new EPA and OEP. The bill for this reorganization was approved by the Senate but has stalled in the Congress, leaving the proposed reorganization in limbo. In view of this turn of events, the proposed reorganization has not been implemented (*1*).

1. the environmental impact of the proposed action,
2. any adverse environmental effects which cannot be avoided should the proposal be implemented,
3. alternatives to the proposed action,
4. the relationship between local, short-term uses of man's environment and maintenance and enhancement of long-term productivity, and
5. any irreversible and irretrievable commitments of resources which would be involved in the proposed action should it be implemented.

Environment in this context refers not only to wilderness, water, air, and other natural resources. It has a broader meaning that includes health, aesthetics, and pleasing surroundings.

Although the law requires an EIS, it does not say anything about what conditions would be required in order to carry out the project. Moreover, NEPA does not give more weight to environmental considerations than it gives to other national goals. Thus the decision about implementation of a program is left to the courts.

In practice, few projects have ever been halted by a court decision under NEPA. However, some projects have been abandoned or modified, before being challenged in court, because of NEPA (*2*).

Environmental Regulatory Framework

Figure 14.1 shows the framework of the federal environmental regulatory structure. Four federal agencies cover the environmental aspects of the national policy.

The Environmental Protection Agency (EPA) is an independent unit not subject to the authority of any of the federal departments, but responsible directly to the U.S. Congress. The EPA administrator is nominated by the president.

The following acts are under the administration of EPA.

- Clean Water Act (CWA);
- Safe Drinking Water Act (SDWA);
- Clean Air Act (CAA);
- Federal Insecticide, Fungicide, and Rodenticide Act (FIFRA);
- Toxic Substances Control Act (TSCA);
- Resource Conservation and Recovery Act (RCRA);
- Comprehensive Environmental Response, Compensation, and Liability Act (CERCLA, also referred to as "Superfund"); and
- Food Quality Protection Act (FQPA).

These acts will each be discussed in this chapter.

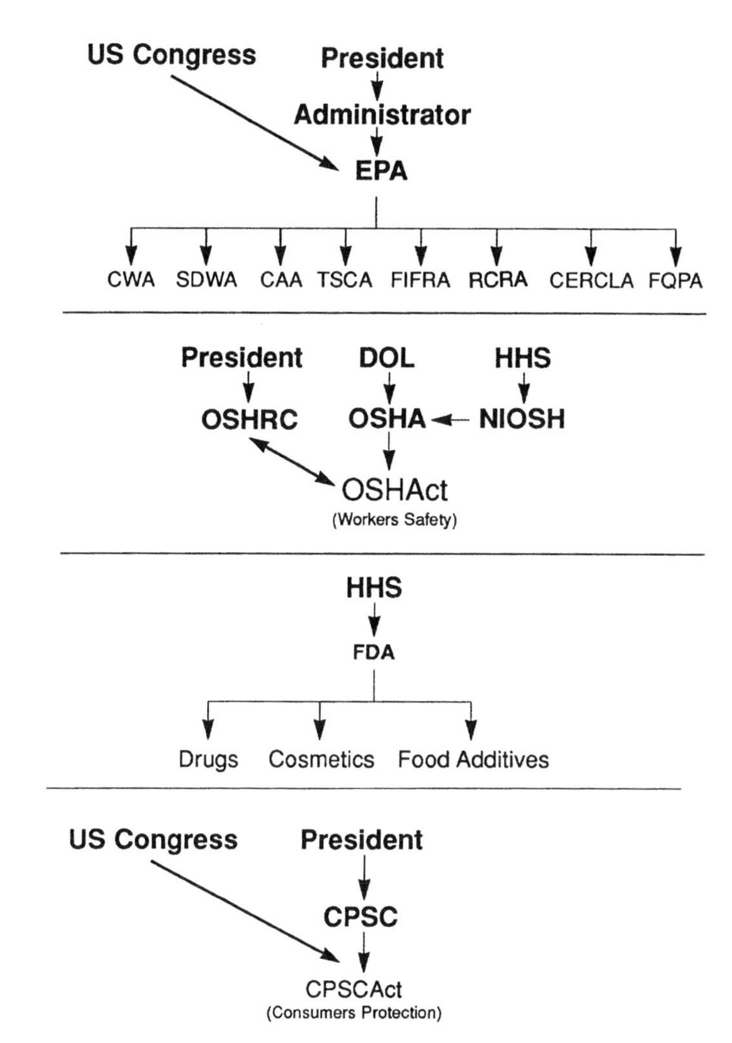

Figure 14.1. The framework of the federal environmental regulatory structure. Key: EPA, Environmental Protection Agency; CWA, Clean Water Act; SDWA, Safe Drinking Water Act; CAA, Clean Air Act; TSCA, Toxic Substances Control Act; FIFRA, Federal Insecticide, Fungicide, and Rodenticide Act; RCRA, Resource Conservation and Recovery Act; CERCLA, Comprehensive Environmental Response Compensation and Liability Act (Superfund); FQPA, Food Quality Protection Act; DOL, Department of Labor; HHS, Department of Health and Human Services; OSHA, Occupational Safety and Health Agency; OSHAct, Occupational Safety and Health Act; NIOSH, National Institute of Occupational Safety and Health; OSHRC, Occupational Safety and Health Review Commission; FDA, Food and Drug Administration; CPSC, Consumer Product Safety Commission; CPSCAct, Consumer Product Safety Commission Act.

The Occupational Safety and Health Agency (OSHA) is an agency within the Department of Labor. The Assistant Secretary of Labor serves as the agency's head. OSHA is responsible for administration of the Occupational Safety and Health Act, which is concerned with healthy and safe working conditions. A related organization is the National Institute of Occupational Safety and Health (NIOSH), an agency within the Department of Health and Human Services. NIOSH is a research unit responsible for the development and recommendation of occupational health and safety standards. The Occupational Safety and Health Review Commission (OSHRC) is a quasi-judicial review board consisting of three members nominated by the president for a period of 6 years. The duty of OSHRC is to mediate disputes and rule on challenges concerning OSHA enforcement actions.

To protect the health and safety of consumers, Congress created a Consumer Product Safety Commission (CPSC) in 1972. The role of this five-member commission is to ensure the safety of consumer products by mandating labeling, restricting use, or banning unsafe products. The commissioners are nominated by the president.

The Food and Drug Administration (FDA) is an agency of the Department of Health and Human Services. It serves as a controlling body concerned with the safety and licensing of drugs, cosmetics, and food additives.

EPA and Its Responsibilities

Before EPA

Several federal environmental laws designed to protect air and water were administered by a variety of agencies prior to 1970. The oldest federal legislation prohibiting disposal of refuse into navigable rivers and into New York Harbor is the Rivers and Harbor Act (R&HA) of 1899. The intent of this legislation is not to protect the environment, but rather to protect navigable waterways for purposes of national defense. Thus the administration of R&HA was entrusted to the Secretary of the Army.

This law assumed environmental significance only later when, in the absence of other easily enforceable laws, it was frequently invoked by the courts in environmental litigation. More comprehensive legislation, designed specifically to protect water from pollution, was enacted in 1948 and amended in 1965. This amendment requires the states to classify all waters within their territory by their intended use, to establish ambient water quality standards as appropriate for the designated use, and to present an implementation plan for federal approval. This legislation, known as the Federal Water Pollution Control Act (FWPCA), turned out to be useless because of enforcement difficulties. Frequently, several polluters discharged pollutants into the same river or lake. This shared responsibility made it difficult to indict any particular source.

The first federal law concerning air pollution was enacted in 1955, when the Congress passed legislation offering technical and financial assistance to states to aid in pollution abatement. In 1967 the Air Quality Act (AQA) was passed to supplement the 1955 law. This act authorizes federal agencies to interfere directly when interstate air pollution is involved and to supervise the enforcement of the state-imposed pollution-abatement measures. Despite these efforts, there was not much progress in air quality improvement in the 1960s.

EPA's Creation and Mandate

In 1970, a presidential order known as Reorganization Order No. 3 created the EPA. The EPA is an independent unit dedicated to the implementation and supervision of environmental laws and regulations and to the pursuit of environmentally oriented research.

Administration of the FWPCA, the AQA, and the newly enacted Clean Air Amendments of 1970 were entrusted to this agency. With increasing public awareness of environmental deterioration, more and more environmental protection legislation was enacted over the next two decades.

At times the agency has been criticized for its lack of effectiveness in enforcing the regulations. In fact, the effectiveness of the EPA depends to a great extent on the political climate. The law, by using such terms as "in his opinion" or "as he finds necessary", gives the EPA administrator considerable leeway in promulgating the standards and regulations. In addition, under certain provisions of the law, the administrator may grant or refuse exemptions to some regulations or postpone the deadlines of compliance.

Because the EPA administrator is nominated by the president, EPA attitudes toward the environment usually reflect those of the federal administration. The environmental neglect of the 1980s is the best example of political influence on the effectiveness of the EPA.

Clean Water Act

The present Clean Water Act (CWA) was enacted in 1972 as the FWPCA. It was amended in 1977, when it was renamed the CWA, and amended again in 1987.

The FWPCA of 1972 states as future goals the attainment of "fishable and swimmable waters" by 1983 and the complete elimination of discharges of pollutants into navigable waters by 1985. For this purpose, the 1972 act introduces the following measures:

- It retains the ambient water quality standards of the 1965 act.
- It superimposes on them nationally uniform, technology-based, effluent limitations for major point sources.
- It establishes deadlines for compliance.
- It introduces provisions for citizen suits that allow private citizens and or-

ganizations to initiate legal action against the polluting party, as well as against the EPA for not fulfilling its obligations by not enforcing the law.

- It outlines policies to deal with nonpoint pollution sources and groundwater protection.
- It establishes municipal waste-treatment grants.

Point Sources of Pollution

In dealing with discharges of pollutants by point sources, the act introduces three stages of economic and technological considerations (*3*): best practical technology (BPT), best conventional technology (BCT), and best available technology (BAT). BPT considers the total cost of existing technology versus effluent-reduction benefits. BCT "shall include consideration of the reasonableness of the relationship between the cost . . . and the effluent reduction benefits derived." BAT takes into consideration engineering aspects of control techniques, energy requirements, and nonwater environmental impact, but not the cost of application.

The law mandates that all industrial point sources must meet effluent limitations requiring application of the BPT by 1977. The standard becomes the BAT currently available by 1983. The act also introduces the National Pollutant Discharge Elimination System (NPDES), which involves issuance of permits to determine discharge limitations for each source. The administration of these permits may be delegated to the states. However, if the technology-based limitations are not sufficient to meet the water quality standards of 1965, the administrator can impose additional limitations.

The law requires that all municipal wastewater purification sources must achieve at least secondary-stage treatment by 1977 and BAT by 1983. New sources, defined as those for which construction began after implementation of this law, have to comply with BAT standards. Industries intending to discharge their effluents through municipal wastewater treatment facilities may do so, provided that the effluents are prepurified before being discharged into the municipal facility.

The administrator is empowered to establish special effluent limitations for toxic pollutants based upon toxicity, persistence, and degradability.

Nonpoint Sources of Pollution

The problem of nonpoint pollution is dealt with through federal cooperation with regional and local planning authorities. No citizen suits are permitted under this section.

Amendments

The most important amendments to the Clean Water Act of 1977 involve the following:

- Postponement for up to 6 years of the original 1977 deadline for achievement by municipalities of the secondary stage of wastewater treatment, if construction could not be completed because of the delay of federal funds.
- Modification of the 1977 deadline for industrial sources to achieve BPT by allowing an industry to apply for postponement if it acted in "good faith". Applications for postponement were to be considered on an individual basis, with the provision that compliance be achieved at the earliest "possible date", but in no case later than April 1979.
- Modification of 1983 BAT limitations by classifying industrial pollutants into three categories (conventional, toxic, and nonconventional) and by applying different limitations to each category. Biological oxygen demand (BOD), suspended solids, fecal coliform organisms, pH changes, and waste oil are classified as conventional pollutants. The discharge limitation for these pollutants was set as BCT, with a compliance date of no later than July 1, 1984. For pollutants classified as toxic (originally a list of 129 chemicals), BAT was mandated to be achieved by July 1, 1980; no exceptions were to be allowed. Nonconventional pollutants are any not classified as conventional or toxic; the limitations required BAT by July 1, 1984. The compliance date with these limitations could be modified by extending the deadline to July 1, 1987, provided that there was consent of the state and that water quality standards were not compromised.

The Water Quality Act of 1987 introduced the following amendments: (1) the deadline for compliance (achievement of BAT for nonconventional pollutants) was extended until 1989, and (2) the deadline for establishment of secondary treatment of wastewater by municipalities was postponed until 1988 in cases where construction could not be completed for reasons beyond the control of the owner.

The definition of secondary treatment is relaxed to include all biological treatment facilities such as oxidation ponds, lagoons, and ditches. Stricter effluent limitations are imposed, when necessary, to attain water quality standards. Also, procedures are provided for classification of waters as to their intended use and needed purity.

A new section is added concerning nonpoint water pollution. This section requires that states identify the waters where purity standards cannot be achieved because of nonpoint sources of pollution and that management programs be established to control nonpoint pollution.

Safe Drinking Water Act

The Safe Drinking Water Act (SDWA), enacted in 1974, directs the EPA to establish regulations for the protection of drinking water. Two types of standards are mandated: federally enforceable primary standards designed for health protection, and state-regulated secondary standards relating to the aesthetic appearance of drinking water. The primary standards prescribe either maximum contaminant levels or a treatment technique; the secondary standards are to be developed according to state guidelines.

SDWA also introduces regulations for the protection of groundwater by controlling underground injection of contaminants. In response to a 1984 report from the Office of Technology Assessment, which identified more than 200 chemical contaminants in groundwater, the act was amended in 1986 to set limits for contaminant levels in public water systems.

Primary responsibility for implementation and enforcement of this law could be delegated to the states if they request it and if they provide satisfactory monitoring and enforcement procedures.

The 1996 amendment to the SDWA requires that the EPA implement a screening and testing program for endocrine disrupters that may occur in drinking water (*4*).

Clean Air Act

The present Clean Air Act (CAA) consists of the Air Quality Act of 1967, the Clean Air Act of 1970, technical amendments to the Clean Air Act of 1973, and the Clean Air Act amendments of 1977 and 1982.

Purpose

The problem addressed by the CAA is stated in *U.S. Code*, Title 42, Pt. 1857 et seq., Section 101(a), as follows:

> that the growth in the amount and complexity of air pollution brought about by urbanization, industrial development and the increasing use of motor vehicles, has resulted in mounting danger to the public health and welfare, including injury to agricultural crops and livestock, damage to and deterioration of property and hazards to air and ground transportation.

Subsections 3 and 4 of Section 101(a) divide responsibilities between the states and the federal government by stating that "the prevention and control of air pollution at its source is the primary responsibility of state and local governments", whereas "Federal financial assistance and leadership is essential for the development of cooperative Federal–State, regional, and local programs to prevent and control air pollution."

To this effect, the administrator is charged to publish and from time to time revise a list of air pollutants and to establish national ambient air quality standards (NAAQS). Two types of standards (primary standards concerning human health and secondary standards concerning public welfare, such as structures, crops, and animals) are to be established for each of the seven pollutants (CO,

SO_2, NO_x, O_3, hydrocarbons, particulates, and Pb). In 1982 the EPA rescinded the standard for hydrocarbons, as it was considered unnecessary.

State Implementation Plan

Within 9 months after promulgation of the standards, each state is to submit a state implementation plan (SIP) for the administrator's approval. The air quality required by the primary standards is to be achieved no later than 3 years after the approval of the SIP, and that required by the secondary standards must be reached within "a reasonable time". In addition, each state is to ensure that "after June 30, 1979, no major source shall be constructed or modified in a nonattainment area."

Each SIP should also contain provisions for periodic inspection and testing of motor vehicles "to enforce compliance with applicable emission standards" (*see* the regulations concerning mobile sources, discussed later in this chapter).

Dispersion Techniques

The Clean Air Amendments of 1977 address the issue of dispersion techniques (i.e., the use of tall stacks as a means of compliance with NAAQS; *see* Chapter 8). The "tall stacks" provision states that the "degree of emission limitation required for control of any air pollutants shall not be affected in any manner by so much of the stack height of any source as exceeds good engineering practice (GEP)." GEP is interpreted as the height necessary to prevent excessive concentration of pollutants in the vicinity of the source due to atmospheric downwash (*5, 6*). This translates, in practical terms, to 2.5 times the height of the source.

New or Modified Sources

Different rules apply to new or modified sources and to existing ones. Existing sources have to comply with NAAQS. In addition, new sources are required to conform to the nationally uniform emission standards. The New Sources Performance Standards (NSPS) require that all new or modified sources use the best available technological system for continuous emission reduction. This standard allows consideration of the cost, any environmental effects unrelated to air quality, and energy requirements.

In connection with modification, the legality of the so-called "bubble effect" has been challenged in courts. The bubble effect refers to reduction of emissions from one part of a source while simultaneously increasing emissions from another part, so that the total emission from a source remains constant. The lower courts decided in two cases against, and in one case for, the legality of the bubble effect. Eventually the Supreme Court ruled that the legal system lacks the technical expertise to rule on this matter and that the decision should be left to the discretion of the EPA administrator (*6*).

In addition to the existing regulations, the 1977 amendments require that the emission of SO_2, NO_x, and particulates be reduced by a specific percentage of what would be emitted if no control devices were employed. Switching to low-sulfur coal is not considered satisfactory compliance with the law.

Prevention of Deterioration

Another innovation of the 1977 amendments is the principle of prevention of significant deterioration (PSD). According to PSD, the regions of the country affected by NAAQS are divided into three classes. Varying degrees of air quality are permitted: class I (national parks and wilderness areas), very little deterioration of air quality is allowed; class II (all other areas), moderate deterioration is allowed; and class III (areas destined for industrial development), considerable deterioration is allowed as long as NAAQS are not exceeded.

Air pollutants for which no NAAQS were set, but which may be harmful to human health, are classified as "hazardous" and are subject to the National Emission Standards for Hazardous Air Pollutants (NESHAP). The EPA administrator is authorized to establish a list and standards for these pollutants. Such standards are equally applicable to new and existing sources. Until very recently (*see* the new Clean Air Act, later in this chapter) there were seven substances, or classes of substances, on the list: beryllium, asbestos, mercury, vinyl chloride, benzene, arsenic, and radionuclides.

The act provides for noncompliance penalties that are tailored individually to each case. This flexibility is intended to take away any financial advantage of noncompliance.

Citizen suits are permitted against polluters, as well as against the EPA for lack of enforcement.

Mobile Sources

The mobile sources section (in Title II) of the CAA authorizes the EPA administrator to establish "standards applicable to the emission of any air pollutant from any class or classes of new motor vehicles or new motor vehicle engines, which in his judgment causes or contributes to . . . air pollution which endangers the public health or welfare. . . . Any regulation prescribed under this subsection shall take effect after such period as the Administrator finds necessary to permit the development of the requisite technology, giving appropriate consideration to the cost of compliance within such period."

According to this authorization, the following standards for light-duty vehicles and engines are established. In vehicles manufactured during and after 1975, emissions of CO and hydrocarbons are to be reduced by at least 90% of the emissions of 1970 models. In addition, in vehicles manufactured during or after 1976, the emission of NO_x is to be reduced by at least 90% of the emission of 1971 models.

The law provides that suspension of the standards may be granted if:

- such suspension is essential to the public interest,
- good-faith efforts to meet the standards have been made,
- the manufacturer establishes that the appropriate technology is not available, and
- the study and investigation conducted by the National Academy of Sciences establishes that the appropriate technology is not available.

Results of the CAA

The CAA succeeded in reducing urban air pollution as far as SO_2 and particulates were concerned. However, it failed in many cities to meet NAAQS with respect to CO and ozone. In addition, the act does not address the problem of acid deposition away from urban centers. A section in the act deals with the interstate transport of pollutants, and attempts have been made to use this law for control of acid rain. Nevertheless, the EPA has refused to act on this problem, and its position has been upheld by the courts.

The New CAA

In November 1990 a new CAA was signed into the law. The main provisions of this act are as follows:

- SO_2 emission from stationary sources has to be reduced by 50% of the 1990 emission, to 10,000 tons annually, by the year 2000.[2] Starting in 1992, NO_x emission has to be reduced by 33% of the present level, to 4 million tons annually.
- Emissions of NO_x and hydrocarbons from passenger cars have to be reduced by 60 and 40%, respectively, by the year 2003. Pollution-control devices on motor vehicles must have a useful life of no less than 10 years. Further, in the most polluted cities (Baltimore, Chicago, Hartford, Houston, Los Angeles, Milwaukee, New York, Philadelphia, and San Diego), cleaner-burning automotive fuel must be available by the year 2000. In California, 1 million vehicles must either use "cleaner" fuel or be provided with special emission-reducing equipment.
- A 90% reduction in emission of 198 toxic and carcinogenic chemicals is required by the year 2003.
- Production and use of chlorofluorocarbons and other ozone-depleting chemicals have to be eliminated completely by the year 2000 (*8*).

---

[2]A provision was introduced that permits the trading of SO_2 in the pollution allowances system. Thus, a utility that reduces its emissions below required limits may sell its allowance to another, less efficient company (*7*).

Federal Insecticide, Fungicide, and Rodenticide Act

Whereas the regulations discussed so far deal with air and water pollution problems, the four acts still to be discussed deal specifically with problems of production, handling, and disposal of toxic substances.

The federal law specifically directed toward regulation of toxic substances is the Federal Insecticide, Fungicide, and Rodenticide Act (FIFRA). The gist of the act is contained in Sections 135 and 136. Section 135, which deals with "economic poisons", requires registration with EPA and proper labeling of these poisons, if they are to be distributed in interstate commerce. The label must contain a warning as to the product's effect on health and the environment.

Section 136 deals specifically with pesticides. It authorizes the EPA to restrict or prohibit the use of a pesticide if it finds that the pesticide presents an unreasonable environmental risk. This determination requires consideration of harm versus benefit and requires reevaluation of the registration every 5 years. In addition, the EPA is authorized to issue emergency suspension of a registration, which takes effect immediately, if the toxicity or the environmental impact of a pesticide warrants such drastic action.

Toxic Substances Control Act

This comprehensive legislation, covering all toxic substances not covered by either the CAA, CWA, or FIFRA, was introduced in 1976.

Summary of the Law

The Toxic Substances Control Act (TSCA) is summarized in the following policy statement.

It is the policy of the United States that

1. adequate data should be developed with respect to the effect of chemical substances and mixtures on health and the environment and that the development of such data should be the responsibility of those who manufacture and those who process such chemical substances and mixtures.
2. adequate authority should exist to regulate chemical substances and mixtures which present an unreasonable risk of injury to health or the environment, and to take action with respect to chemical substances and mixtures which are imminent hazards.
3. authority over chemical substances and mixtures should be exercised in such manner as not to impede unduly or create unnecessary economic barriers to technological innovation while fulfilling the primary purpose of this Act, to assure that such innovation and commerce in such chem-

ical substances and mixtures do not present an unreasonable risk of injury to health or the environment.

Authority over Manufacturers

In essence, the Act gives the EPA administrator authority over manufacturers as specified in *U.S. Code*, Title 15, Sections 4–6. Section 4 requires manufacturers to test manufactured substances if:

- insufficient data are available,
- the substance may "present an unreasonable risk",
- the substance may "enter the environment in substantial quantities", or
- the substance presents the likelihood of "substantial human exposure".

Section 5 requires a manufacturer to notify the EPA 90 days prior to manufacturing or importing new substances. This premanufacturing notification (PMN) must contain information on chemical identity, proposed use, anticipated volume of production, expected byproducts, estimated workers' exposure during production, and methods of disposal. Toxicologic testing data are not required (if not available), unless the substance is covered under Section 4 or is on the EPA list of hazardous substances. In the latter case, the manufacturer must submit data showing that the substance does not present any unreasonable risk of injury. If the EPA is not convinced, it may request additional data.

Section 6, which applies to new and old substances alike, authorizes the EPA administrator to impose a number of restrictions (such as to ban manufacturing, prohibit certain uses, require labeling, or require a change of the manufacturing process). If the administrator determines that a substance to be produced presents unreasonable risk, the proposed Section 6 rule, which prohibits manufacturing until proper restrictions can be issued under Section 6, may be invoked.

The act provides for enforcement of the regulations with civil and criminal penalties and for citizen suits against violators and against the EPA for lack of enforcement.

Resource Conservation and Recovery Act

The Resource Conservation and Recovery Act (RCRA) deals with the generation, transport, and disposal of hazardous waste. It was enacted in 1976 in response to public concern over seepage of toxic substances from chemical waste dumps into groundwater and into basements of residential dwellings (*9*).

List of Hazardous Substances

The act directs the EPA to establish a list of hazardous substances "taking into account toxicity, persistence and degradability in nature, potential for accumulation in tissue", as well as corrosiveness and flammability. Further, the EPA is authorized to establish standards for generation, transport, and disposal of hazardous waste, and to require record-keeping at each stage. Disposal sites are required to obtain permits and to conform to certain engineering standards, such as double liners, leachate-collection systems, and groundwater-monitoring facilities.

Waste Disposal

RCRA 1984 amendments introduce requirements that an operator of a dump site has to provide either liability insurance or some sort of guarantee as an assurance of financial responsibility.

The amendments also mandate the EPA to promulgate rules for treatment of hazardous waste before such waste can be disposed of in landfills. After a 5-year effort to design such rules, the final hazardous-waste regulation came into effect on May 8, 1990. This regulation deals with almost 350 types of hazardous waste, including waste from industrial and academic research laboratories. Despite its wide coverage, this regulation is being criticized by environmental groups for being too lenient to industry by allowing disposal practices that may harm the environment (*10*).

Underground Storage Tanks

Another provision of these amendments requires inventory, inspection, and replacement of underground storage tanks. The responsibility for this inventory and inspection is delegated to the states. Owners or operators are to be held responsible for any damage to the public or environment caused by spills from leaky tanks.

Comprehensive Environmental Response, Compensation, and Liability Act

The Comprehensive Environmental Response, Compensation, and Liability Act (CERCLA, popularly known as "Superfund") was enacted in 1980 and amended by the Superfund Amendments and Reauthorization Act of 1986 (SARA). The purpose of the legislation was the cleanup of old, improperly constructed, hazardous-waste disposal sites.

The act includes four essential elements (*11*). First, it establishes an information-collecting system to enable the government to locate and characterize hazardous-waste disposal sites and to establish the national priority list (NPL) for cleanup. The owners and operators of such sites are required to notify EPA of the amount and type of hazardous substances deposited and of any release, or suspected release, of these substances into the environment.

Remedial Action

The second element of Superfund evolved from Section 311 of the Clean Water Act. This national contingency plan (NCP) concerns the cleaning up of toxic-waste sites. It authorizes the president to revise the NCP to include a new hazardous-substances response plan containing standards and procedures for either removal of the hazardous substances or appropriate remedial measures. The remedial actions should be cost-effective, and priorities should be based on the relative risk to health, welfare, and the environment. Federal remedial action is restricted to those cases in which no party responsible for disposal of the hazardous waste can be located or in which the responsible party takes no action.

The third element establishes the hazardous-substances trust fund to bear the cost of removal or confinement of the hazardous waste. The original appropriation of funds for the first 5 years was $1.6 billion. This was upgraded by SARA amendments in 1986 to $8.5 billion for the next 5 years.

Financial Liability

The fourth element discusses financial responsibilities and liabilities. In essence, the persons responsible for the release of hazardous waste are made responsible for the cleanup. This includes generators, transporters, and owners and operators of disposal facilities. The responsibility covers not only the cost of cleanup incurred by federal and state governments, but also any damages to people and natural resources that may have resulted from these activities. Except for acts of God or war, the liability law applies, even if no negligence or faulty performance can be demonstrated.

Superfund performance has been highly criticized not only by environmental groups, but also by Congress and the Office of Technology Assessment. Critics charge that the program wastes money and does not adequately protect the environment. Frequently, decision-making regarding the best remedial action for site cleanup (remedial investigation, feasibility study, or RIFS) is delegated by the EPA to the polluter. Not surprisingly, the polluter decides in favor of personal interest rather than the community and environment (*12*). The EPA is also under attack for lack of efficiency in recovering money from polluters for cleanup costs (*13*).

Food Quality Protection Act

The Food Quality Protection Act (FQPA) was signed into law on August 3, 1996, and took effect as of the date of signing, with no phase-in period. The act rescinded the 1958 amendment to the Food and Cosmetic Act of 1938, known as the Delaney Clause, which was until then under the jurisdiction of the U.S. Food the Drug Administration (*see* Chapter 4). The Delaney Clause was replaced with a health-based standard, allowing residues of pesticides on food, whether processed or not, if there is a reasonable certainty of no harm. As far as cancer is concerned, this standard means no more than 1 excess case of cancer in 1 million exposed people. However, the new law goes beyond the requirement of testing for carcinogenicity in adults. It requires special consideration for infants and children, and it requires the testing of pesticides for endocrine-system disrupting activity. It also provides for expanded consumer right to know (*14*).

OSHA and Its Responsibilities

In 1970 the 91st Congress passed legislation called the Occupational Safety and Health Act (OSHAct). The purpose of this legislation was "to assure safe and healthful working conditions for working men and women". This was the first comprehensive legislation covering all employers and employees in all industries, commerce, and agriculture in the United States and in any territory administered by the United States. The regulatory provisions protecting workers in certain industries, such as maritime, mining, and construction, which existed prior to 1970, were taken over by this new act. Federal and state employees covered by the Atomic Energy Act of 1954 are exempt from OSHAct.

The administration of the act is entrusted to the Department of Labor via the Occupational Safety and Health Agency (OSHA). The administrative structure of OSHA, discussed earlier in this chapter, is depicted schematically in Figure 14.1.

The duties of employers and employees are specified in Section 5 of the U.S. Code as follows:

a. Each employer
   1. shall furnish to each of his employees employment and a place of employment which are free from recognized hazards that are causing or are likely to cause death or serious physical harm to his employees;
   2. shall comply with occupational safety and health standards promulgated under this Act.
b. Each employee shall comply with occupational safety and health standards and all rules, regulations, and orders issued pursuant to this Act which are applicable to his own actions and conduct.

The act requires that the Secretary of Labor promulgate, "during the period beginning with the effective date of this Act and ending two years after such date", occupational health and safety standards. Three types of standards are established:

1. Interim standards promulgated for a period of 2 years; these are not subject to rule-making procedures.[3]
2. Permanent standards that could modify or revoke existing standards or promulgate new standards; rule-making procedures are required.
3. Emergency standards that are promulgated when the secretary establishes that the workers may be exposed to a grave danger from a newly determined hazard. Emergency standards may be issued without prior notification, but they apply for 6 months only. After this period they have to be either revoked or made permanent.

The employee's exposure to toxic and hazardous substances is regulated by Title 29 of the *Code of Federal Regulations*, Part 1910, Subpart z. This section provides a list of hazardous substances and specifies the permissible exposure limits for each compound, using TLV standards (*see* Chapter 7) as guidelines.

OSHA is authorized to enforce health and safety standards through inspections, citations, monetary penalties, and, in extreme cases, imprisonment. Workplace inspections may be conducted at any time without prior notification. In situations that present imminent danger, the inspector will issue a citation that specifies the nature of the violation and prescribe a reasonable time for correction of the hazardous situation. The citation must be posted by the employer at the site of violation.

The act gives the employer the right to contest citations, periods of abatement, or penalties by requesting a hearing before OSHRC and to contest OSHRC rulings by filing legal action with the U.S. Court of Appeals.

Miscellaneous Environmental Acts and Treaties

The Endangered Species Act

The Endangered Species Act was passed by the U.S. Congress in 1973 and applies in the United States only. The purpose of the act is the protection of biodiversity. The act mandates:

---

[3]Federal rule-making procedures require that any new regulation or any proposed change be published in the *Federal Register* as a notice of proposed rule making (NPRM). Interested parties have 30 days to respond. The agency proposing the rule must then schedule a hearing and notify the public of the time and the place of the meeting. Within 60 days, upon completion of the hearing, the proposed rule must be either withdrawn or promulgated.

- Listing of endangered species.
- Protection of the endangered species and an effort to revive them.

The act expired in 1985 and is now extended from year to year. The future fate of the act is uncertain because its renewal or permanency depends on the political sentiment of Congress, which may change from election to election.

International Treaties Protecting the Marine Environment

Law of the Sea Convention of 1982 (LOS)

The purpose of this convention was to regulate the use of ocean resources. The main provisions of the act were as follows:

- Establishment of Exclusive Economic Zones (EEZ), which comprise an area up to 200 miles from the coast. To enter EEZ for economic purposes, foreign ships must obtain permission of the country controlling the EEZ.
- Upholding the traditional notion of "freedom of the seas".
- Establishment of the principle that all nations should benefit from deep seabed resources, and that the resources should be mined under supervision of the International Seabed Authority.

Although 159 nations signed the treaty, several industrialized nations, including the United States, the United Kingdom, and Germany did not accept the last provision of the treaty, selfishly maintaining that resources should be available on a first-come, first-served basis.

Marpol Convention of 1973

The purpose of this convention was to establish international laws protecting the seas from pollution. The provisions of the convention were as follows:

- establishment of a minimum distance from the shore for dumping sewage, garbage, and toxic waste,
- prohibition of the disposal of plastics from ships,
- limitation on disposal of other garbage, and
- requirement for ports to provide facilities for trash from ships.

The U.S. Navy was exempt from the dumping provision until 1994.

London Dumping Convention of 1975

This convention supplemented the Marpol Convention by:

- banning dumping from ships and aircraft of "blacklisted" substances (heavy metals, petroleum products, and carcinogens) and
- requiring a permit for dumping of "graylisted" substances (lead, cyanide, and pesticides).

In 1983, the London Convention was amended by issuing a moratorium on the dumping of low-level radioactive waste.

References

1. Ember, L. *Chem. Eng. News* August 15, 1994, p 5.
2. Findley, R. W.; Farber, D. A. In *Environmental Law;* West Publishing: St. Paul, MN, 1988; Chapter 1, p 22.
3. Findley, R. W.; Farber, D. A. In *Environmental Law;* West Publishing: St. Paul, MN, 1988; Chapter 2 C, p 108.
4. Hileman, B. *Chem. Eng. News* September 2, 1996, p 21.
5. Raffle, B. I. *Environment Reporter Monograph 26* **1978,** *8*(47), 1; Bureau of National Affairs: Washington, DC.
6. Findley, R. W.; Farber, D. A. In *Environmental Law;* West Publishing: St. Paul, MN, 1988; Chapter 2 B, p 66.
7. Ember, L. *Chem. Eng. News* November 18, 1991; p 20.
8. Lemonick, M. D.; Blackman, A. *Chem. Eng. News* November 5, 1990, p 33.
9. Thayer, A. *Chem. Eng. News* January 30, 1989, p 7.
10. Hanson, D. *Chem. Eng. News* May 28, 1990, p 19.
11. Findley, R. W.; Farber, D. A. In *Environmental Law;* West Publishing: St. Paul, MN, 1988; Chapter 3, p 169.
12. Office of Technology Assessment. *Cleaning Up: Superfund's Problems Can Be Solved;* U.S. Government Printing Office: Washington, DC, 1989.
13. Ember, L. *Chem. Eng. News* October 2, 1989, p 17.
14. Hanson, D. *Chem. Eng. News* September 23, 1996, p 38.

Appendix
Subjects for Student Seminars

The following topics are important environmental problems facing today's world. It is suggested that students select topics from this list for independent research and presentation in class seminars:

- ocean pollution;
- overfishing of oceans and destruction of marine mammals;
- nuclear energy, pro and con;
- destruction of tropical forests;
- loss of biodiversity;
- crises in Antarctica;
- energy conservation;
- renewable energy sources;
- freshwater crisis; and
- world poverty and overconsumption by industrialized nations.

Index

Copy editing by Jay C. Cherniak
Production by Paula M. Bérard
Jacket design by Linda Mattingly

Typeset by Ampersand Graphics, Ltd., Reisterstown, MD
Printed and bound by Maple Press Company, York, PA